CONSTRAINT REASONING FOR DIFFERENTIAL MODELS

Frontiers in Artificial Intelligence and Applications

Volume 126

Published in the subseries

Dissertations in Artificial Intelligence

Under the Editorship of the ECCAI Dissertation Board
Proposing Board Member: Luís Moniz Pereira

Recently published in this series

ISSN 0922-6389

Constraint Reasoning for Differential Models

Jorge Cruz

Centro de Inteligência Artificial (CENTRIA),
Departamento Informática, Faculdade de Ciências e Tecnologia,
Universidade Nova de Lisboa, Caparica, Portugal

IOS
Press

Amsterdam • Berlin • Oxford • Tokyo • Washington, DC

ISBN 1-58603-532-0
Library of Congress Control Number: 2005929309

Publisher
IOS Press
Nieuwe Hemweg 6B
1013 BG Amsterdam
Netherlands
fax: +31 20 620 3419
e-mail: order@iospress.nl

Distributor in the UK and Ireland
IOS Press/Lavis Marketing
73 Lime Walk
Headington
Oxford OX3 7AD
England
fax: +44 1865 750079

Distributor in the USA and Canada
IOS Press, Inc.
4502 Rachael Manor Drive
Fairfax, VA 22032
USA
fax: +1 703 323 3668
e-mail: iosbooks@iospress.com

Constraint Reasoning for Differential Models
J. Cruz
IOS Press, 2005

v

Acknowledgements

I am specially indebted to Pedro Barahona who supported me, not only as the supervisor of this work, but also in all my academic formation. He is an excellent supervisor which I greatly recommend to any one intending to start a PhD in constraint programming. He has the rare ability of being open minded with respect to new ideas and contribute diligently to their success. Despite his overbooked schedule, he was always present whenever needed. He is a good friend and is a pleasure to work with him.

I am very grateful to Frédéric Benhamou and all his research team for their good hospitality and their crucial support in the early phases of this work. Thanks to Frédéric Goualard and Laurent Granvilliers for their friendship and all our valuable discussions.

Special thanks to Luís Moniz Pereira, for his careful reading and comments on the introduction and conclusions of this work, but above all, for his excellent work in promoting Artificial Intelligence which attracted me in the first place for taking a computer science degree.

My acknowledgements to the Computer Science Department for having conceded me three sabbatical years which were extremely important for the progress of this work.

Many thanks to my colleagues in the Computer Science Department and CENTRIA for their companionship, trust and concern. In particular, I am grateful to the constraint research group for creating a stimulating working environment. Thanks to my good friends Anabela and Cecília for their sincere care and help, promptly sharing with me their own resources whenever I needed.

Thanks to all my family for their love and precious help in handling so many practical problems which allowed me to fully concentrate on my work. I am particularly grateful to my parents for their continuous support and belief on my own choices.

Finally, I am deeply grateful to my wife Teresa, who, from the beginning of this work, has always been by my side, giving me confidence and encouragement in difficult moments and a good reason to keep on searching for success. I can never thank her enough for all the days, weekends and holidays that she has sacrificed for helping me.

Abstract

The basic motivation of this work was the integration of biophysical models within the interval constraints framework for decision support. Comparing the major features of biophysical models with the expressive power of the existing interval constraints framework, it was clear that the most important inadequacy was related with the representation of differential equations. System dynamics is often modelled through differential equations but there was no way of expressing a differential equation as a constraint and integrate it within the constraints framework.

Consequently, the goal of this work is focussed on the integration of ordinary differential equations within the interval constraints framework, which for this purpose is extended with the new formalism of Constraint Satisfaction Differential Problems. Such framework allows the specification of ordinary differential equations, together with related information, by means of constraints, and provides efficient propagation techniques for pruning the domains of their variables. This enabled the integration of all such information in a single constraint whose variables may subsequently be used in other constraints of the model. The specific method used for pruning its variable domains can then be combined with the pruning methods associated with the other constraints in an overall propagation algorithm for reducing the bounds of all model variables.

The application of the constraint propagation algorithm for pruning the variable domains, that is, the enforcement of local-consistency, turned out to be insufficient to support decision in practical problems that include differential equations. The domain pruning achieved is not, in general, sufficient to allow safe decisions and the main reason derives from the non-linearity of the differential equations. Consequently, a complementary goal of this work proposes a new strong consistency criterion, Global Hull-consistency, particularly suited to decision support with differential models, by presenting an adequate trade-of between domain pruning and computational effort. Several alternative algorithms are proposed for enforcing Global Hull-consistency and, due to their complexity, an effort was made to provide implementations able to supply any-time pruning results.

Since the consistency criterion is dependent on the existence of canonical solutions, it is proposed a local search approach that can be integrated with constraint propagation in continuous domains and, in particular, with the enforcing algorithms for anticipating the finding of canonical solutions.

The last goal of this work is the validation of the approach as an important contribution for the integration of biophysical models within decision support. Consequently, a prototype application that integrated all the proposed extensions to the interval constraints framework is developed and used for solving problems in different biophysical domains.

Symbols and Notation

LOGIC

$\neg$	The negation operator
$\wedge$	The conjunction operator
$\vee$	The disjunction operator
$\Rightarrow$	The implication operator
$\forall$	The universal quantifier
$\exists$	The existential quantifier

SET THEORY

$\in$	An element of
$\notin$	Not an element of
$\subseteq$	A subset of
$\subset$	A proper subset of
$\cap$	The intersection operator
$\cup$	The union operator
$\uplus$	The union hull operator
$\emptyset$	The empty set

REAL VALUES, INTERVALS AND BOXES

r, r_i, k	A real value
a, b, c, d	A real value representing the bound or the centre of an interval
$\lfloor r \rfloor$	The largest F-number not greater than r
$\lceil r \rceil$	The smallest F-number not lower than r
$-\infty, +\infty$	The minus and the plus infinity F-numbers
$I, I_i. K$	An interval, either a real interval or an F-interval
IR, IR_i	A real interval
IF, IF_i	An F-interval
$<a..b>$	An interval bounded by a and b (closed, half closed or open)
$[a..b]$	A closed interval bounded by a and b
$(a..b], (-\infty..b]$	A left open interval
$[a..b), [a..+\infty)$	A right open interval
$(a..b), (-\infty..+\infty)$	An open interval
$[a..a], [a], \{a\}$	A degenerate interval with the singe real value a
B, B_i	A box, either an R-box or an F-box
$<I_1,...,I_n>$	A box with n intervals
$<IR_1,...,IR_n>$	A real box with n real intervals
$<IF_1,...,IF_n>$	An F-box with n F-intervals

INTERVAL BASIC FUNCTIONS AND APPROXIMATIONS

$left([a..b])$	The function that returns the left bound of $[a..b]$
$right([a..b])$	The function that returns the right bound of $[a..b]$
$center([a..b])$	The function that returns the mid value of $[a..b]$
$width([a..b])$	The function that returns the width of $[a..b]$

$cleft([a..b])$	The function that returns the leftmost canonical bound of $[a..b]$
$cright([a..b])$	The function that returns the right canonical bound of $[a..b]$
$I_{apx}(IR)$	The *RF*-interval approximation of the real interval *IR*
$S_{apx}(SR)$	The *RF*-set approximation of the real set *SR*
$I_{hull}(SR)$	The *RF*-hull approximation of the real set *SR*

VARIABLES, EXPRESSIONS AND FUNCTIONS

x_i	A real valued variable
X_i	An interval valued variable
$+$	The real or interval arithmetic operator of addition
$-$	The real or interval arithmetic operator of subtraction
$\times$	The real or interval arithmetic operator of multiplication
$/$	The real or interval arithmetic operator of division
Φ	A basic real or interval arithmetic operator
Φ_{apx}	The interval operator Φ evaluated with outward evaluation rules
E, E_i, E_c	A real or interval expression
e, e_i, e_c	A real expression
f, g	A real function
f', g'	A real function which is the derivative of f or g, respectively
$f^*(D)$	The range of the real function f over the domain D
D_f	The domain of the real function f
f_E, f_{E_i}	A real expression representing the real function f
F, F', G	An interval function
F_E, F_{E_i}	An interval expression representing the interval function F and the interval function resulting from its interval evaluation
F_n	The Natural interval extension of a real function wrt a real expression
F_c	The Centered interval extension of a real function wrt an interval
F_m	The Mean Value interval extension of a real function wrt an interval
$F_{t(m)}$	The Taylor (m) interval extension of a real function wrt an interval
F_d	The Distributed interval extension of a real function
N	The interval Newton function with respect to a real function
NS	The Newton Step function with respect to a real function
NN	The Newton Narrowing function with respect to a real function

CONSTRAINT SATISFACTION PROBLEMS

(X,D,C)	A CSP with the variables X ranging over D and constrained to C
2^D	The power set of D, with respect to a CCSP defined as (X,D,C)
A, A', A_i	An element of 2^D (wrt a CCSP defined as (X,D,C))
$<x_1,...,x_n>$	A tuple of n variables of a CSP
D_i	The domains of variable x_i of a CSP
$D_1 \times ... \times D_n$	The Cartesian Product of the domains of n variables of a CSP
d_i	A value form the domain of variable x_i of a CSP
$<d_1,...,d_n>$	A tuple of n values from the domains of n variables of a CSP
d, d_l, d_r	A tuple of values from the domains of some variables of a CSP
$\Diamond$	A symbol from $\{\leq,=,\geq\}$
$c=(s,\rho)$	A constraint establishing a relation ρ between the variables within s

$c \equiv e_c \diamond 0$	A constraint represented as a relation between a real expression and 0
$c \equiv e_c \diamond e_0$, $c \equiv e_1 \diamond e_0$	A constraint represented as a relation between two real expressions
$d[s]$	The projection from the values in d to the values of variables in s
$B[s]$	The projection from the intervals in B to the intervals of variables of s

CONSTRAINT PROPAGATION

NF, NF', NF_i	A narrowing function associated with a constraint of a CCSP
Domain$_{NF}$	The domain of the narrowing function NF
Codomain$_{NF}$	The codomain of the narrowing function NF
Relevant$_{NF}$	The set of variables used for defining the narrowing function NF
Fixed-Points$_{NF}(A)$	The set of all fixed-points of the narrowing function NF within A
$\cup$Fixed-Points$_{NF}(A)$	The union of all fixed-points of the narrowing function NF within A
Q, S, S_i	A set of narrowing functions associated with constraints of a CCSP
Ψ_{e_i}	The inverse interval expression of a constraint wrt the expression e_i
Ψ_{x_i}	The inverse interval expression of a constraint wrt the variable e_i
$\pi_{x_i}^{\rho}$	A projection function wrt a constraint $c=(s,\rho)$ and a variable $x_i \in s$
$\mathrm{BNF}_{x_i}^{\rho}$	A box-narrowing function wrt a constraint $c=(s,\rho)$ and a variable $x_i \in s$
$\Pi_{x_i}^{\rho B}$	The interval projection of constraint $c=(s,\rho)$ wrt $x_i \in s$ and the F-box B
$prune(Q,A)$	The function that propagates a set Q of narrowing functions over an element A of 2^D and returns a smaller element $A' \subseteq A$
$narrowBounds(I)$	The function that narrows an F-interval I accordingly to the narrowing strategy of the constraint Newton method
$intervalProjCond(I)$	The function that verifies if the interval projection condition is satisfied at the canonical F-interval I
$searchLeft(I)$	The function that returns the leftmost zero of an interval projection within the F-interval I
$searchLeft(I)$	The function that returns the rightmost zero of an interval projection within the F-interval I

List of Figures

Part I   INTERVAL CONSTRAINTS

Chapter 12 BIOMEDICAL DECISION SUPPORT WITH ODEs

List of Tables

Part I   INTERVAL CONSTRAINTS

Part II   INTERVAL CONSTRAINTS FOR DIFFERENTIAL EQUATIONS

List of Definitions

Part I   INTERVAL CONSTRAINTS

Part II INTERVAL CONSTRAINTS FOR DIFFERENTIAL EQUATIONS

List of Theorems

Part I   INTERVAL CONSTRAINTS

Table of Contents

Constraint Reasoning for Differential Models
J. Cruz
IOS Press, 2005

Chapter 1

Introduction

The original motivation for this work derives from our past experience in the design of decision support systems for medical diagnosis. Knowledge representation has always been a major concern in the design of decision support systems, namely those applied to the medical domain. The early medical knowledge based systems were designed to accomplish some specific medical task (typically diagnosis) and the medical knowledge was mostly embedded in the procedures designed to accomplish that task. This led to problems of consistency (the medical knowledge reflected the views of the medical experts that advised the design of such systems, which by no means was consensual within the health care community) and also of reuse (the same knowledge could not be used in two different tasks – e.g. diagnosis and treatment).

This problem was soon recognised and several approaches were proposed to overcome it, namely to represent medical knowledge declaratively. As such the systems could represent medical knowledge proper (i.e. anatomical, physiological or pathological knowledge) separately from task related knowledge (i.e. what are the necessary steps to be executed when performing diagnosis, treatment or monitoring tasks). Moreover, task knowledge could be formalised and (re)used in several specific medical domains (cardiology, neurology, etc.).

Such a view was particularly useful for systems based on logic, in that useful relations between medical concepts are stored as facts (e.g. causal relations, associations, risk factors) that could be handled by the reasoning process. Nevertheless, the medical concepts and relations represented were usually relatively high level abstractions of the underlying processes and this led to the problem of handling uncertainty - the more abstract are the concepts and relations the more uncertain are the statements that can be made about them.

This problem can be alleviated if the knowledge handled by the systems would represent less abstract concepts whose underlying uncertainty is more controlled. This is the approach taken by systems based on "deep medical knowledge". In contrast with systems that represent causal but "shallow" relations, say, between a disease and a symptom, these systems would represent a more detailed set of pathophysiological states and processes that could explain the shallow relation between the disease and the symptom.

This was the kind of approach used for the development of our own decision support system for the diagnosis of neuromuscular diseases [32, 34, 35]. Such an approach was based on a causal-functional model for the representation of anatomical and physiological domain knowledge which supported a diagnostic reasoning strategy that mimics the medical reasoning usually performed over those dimensions of medical knowledge.

However, the approach still presented certain difficulties, both with the representation of the domain knowledge and with the soundness of the diagnostic reasoning. To avoid complexity, the quantitative knowledge about the elementary physiological processes was abstracted into simpler qualitative relations where the continuous domains were partitioned into symbolic values. Such simplifications prevented, for example, an adequate modelling of the evolution of processes over time. On the other hand, the reasoning strategy was only justified by clinical practice and not by the underlying knowledge model, thus lacking automatic mechanisms for guaranteeing its soundness.

The difficulties found in our practical approach seem to generalise for decision support systems in other biomedical domains where there is a clear gap between theoretical and practical approaches. Despite the existence of "deep" biophysical models generally accepted by the biomedical community (e.g. cardiovascular models, respiratory models, compartment models, etc), they are not explicitly incorporated into decision support systems due to their complexity. They are often highly non-linear models based on differential equations and these are difficult to reason about with simple "logical" procedures. In most existing decision support systems, specialised on practical tasks such as diagnosis or prognosis, this "deep" theoretical knowledge is still implicitly hidden in the heuristics and rules used to perform those tasks.

In this context, constraint technology seems to have the potential to bridge the gap between theory and practice. The declarative nature of constraints makes them an adequate tool for the explicit representation of any kind of domain knowledge, including "deep" biophysical modelling. The constraint propagation techniques provide sound methods, with respect to the underlying model, that can be used to support practical tasks (e.g. diagnosis/prognosis may be supported through propagation on data about the patient symptoms/diseases). In particular, the interval constraints framework seems to be the most adequate for representing the non-linear relations on continuous variables, often present in biophysical models. Additionally, the uncertainty of biophysical phenomena may be explicitly represented as intervals of possible values and handled through constraint propagation.

The basic motivation of this work was then the integration of biophysical (or more general physical) models within the interval constraints framework for decision support. On the one hand, it would be necessary that biophysical models and phenomena could be represented as interval constraints. On the other, to be of any practical use, the underlying constraint propagation techniques should be efficient enough to support the decision making process.

Comparing the major features of biophysical models with the expressive power of the existing interval constraints framework, it was clear that the most important inadequacy was related to the representation of differential equations. System dynamics is often modelled through differential equations but there was no way of expressing a differential equation as a constraint and integrate it within the constraints framework.

Consequently, the goal of this work is focussed on the integration of ordinary differential equations within the interval constraints framework. In particular, in the context of managing uncertainty in biophysical models for decision support, there is a special interest in representing uncertainty in the model parameters by ranging them over intervals of possible values.

In this work we extend the interval constraints framework with a new approach for handling differential equations by means of Constraint Satisfaction Differential Problems (CSDPs). Such framework allows the specification of ordinary differential equations, together with related additional information, by means of constraints, and provides efficient propagation techniques for pruning the domains of their variables. Such techniques are based on existing reliable enclosure methods developed for solving ordinary differential equations with initial value conditions.

The introduction of this new approach enabled the integration of all such information into a specific constraint (of a different kind), relating several variables (e.g. representing trajectory point values, parameter values, trajectory maximum value, etc). Such variables may subsequently be used in other constraints of the model. The specific method used for pruning its variable domains can be combined with the pruning methods associated with the other constraints within an overall propagation algorithm for reducing the bounds of all model variables.

The application of a constraint propagation algorithm for pruning the variable domains of a constraint system can be regarded as enforcing some form of local-consistency, since it depends on the pruning methods associated with each individual constraint. The quality of such local-consistency, that is, the pruning that may be achieved, is highly dependent on the ability of these pruning methods (narrowing functions) for discarding value combinations that are inconsistent with the respective constraint.

Enforcing local-consistency turned out to be insufficient to support decision in practical problems that include differential equations. If uncertainty is included in the differential model, the domain pruning achieved by such techniques would not, in general, be sufficient to allow safe decisions since a wide range of possibilities would still be possible after propagation.

The main reason for this poor performance derives from the non-linearity of the differential equations. In case of parametric differential equations, parameter uncertainty is quickly propagated and increased along the whole trajectory. Such behaviour is not a consequence of the enclosure method adopted but rather an effect of the non-linearity of the differential equation which in the extreme case of chaotic differential equations prevent any reasonable, long-term, trajectory calculation (even without any significant initial uncertainty).

Insufficiency of local-consistency was already recognised in many practical problems not involving differential equations. In continuous domains, stronger consistencies were proposed for dealing with such problems. These are higher order generalisations of local-consistency criteria (Hull or Box-consistency for continuous domains) enforced by algorithms that interleave constraint propagation with techniques for the partition of the variable domains.

However, for many practical differential problems, such stronger consistency criteria are still not adequate for decision support. Either the pruning is unsatisfactory or the respective enforcing algorithms are too costly (computationally). Consequently, a complementary goal of this work proposes a new strong consistency criterion particularly suited to decision support with differential models, by presenting an adequate trade-of between domain pruning and computational effort.

This new strong consistency criterion, Global Hull-consistency, is a generalisation of Hull-consistency to the whole set of constraints, which is regarded as a single global constraint. The criterion relies on the basic concept of a canonical solution, aiming at finding the smallest domains enclosure that includes all canonical solutions. Several alternative algorithms are proposed for enforcing Global Hull-consistency and an effort was made to provide implementations able to supply any-time pruning results. This is particularly useful in the context of decision support where the domain pruning is not the ultimate goal in itself (the computation may be interrupted whenever pruning is sufficient to make safe decisions).

All the proposed enforcing algorithms combine constraint propagation with domains partition, and terminate whenever they find canonical solutions bounding each edge of the current domains box. To anticipate the finding of canonical solutions, and eventually the termination of the algorithm, it seemed natural to extend such algorithms with local search capabilities.

In this work we thus propose a local search approach that can be easily integrated with constraint propagation and domains partition. It is based on a technique commonly adopted in multidimensional root finding over the reals, namely, line search minimisation along a vector obtained by the Newton-Raphson method. In the context of a constraint system, the points of the search space are complete real valued instantiations of all its variables and the search is directed towards the simultaneous satisfaction of all its constraints.

All the extensions to the interval constraints framework were proposed in the context of integrating biophysical models within decision support. It would be important to validate our approach in the sense that it provides an important contribution along such a direction. Consequently, the final goal of this work is the application of our proposals to practical problems of decision support based on biophysical models.

In this work we have developed a prototype application that integrates all the proposed extensions to the interval constraints framework, and uses it for solving problems in different biophysical domains. In particular, the problems addressed (the diagnosis of diabetes, the tuning of drug design and the study of epidemics) are representative of the kind of applications our approach is suited for.

1.1   Contributions

This work extends the interval constraints framework to handle differential equations. It provides a new approach to model differential equations (subsection 1.1.1), a new consistency criterion for pruning the variable domains (subsection 1.1.2), and a new approach for integrating local search (subsection 1.1.3). A prototype was implemented and applied to practical biophysical problems with good results (subsection 1.1.4).

1.1.1   Interval Constraints for Differential Equations

The interval constraints framework is extended with a new formalism to handle ordinary differential equations (ODEs): the Constraint Satisfaction Differential Problem (CSDP). In this formalism, ODEs are included as constraints, together with other restrictions further required on its solution functions. Such restrictions may incorporate in the constraint model all the information traditionally associated with differential problems, namely, initial and boundary conditions. Moreover, the expressive power of this framework is extended to represent several other conditions of interest that cannot be handled by classical approaches. These include maximum, minimum, time, area, first, and last restrictions.

The CSDP framework includes a solving procedure for pruning the domains of its variables. A constraint may be defined as a CSDP and integrated with other constraints, using its solving procedure as a safe narrowing function. This allows, for the first time, the full integration of ODEs and related information within a constraint model.

1.1.2   Global Hull-consistency – A Strong Consistency Criterion

A new strong consistency criterion, dubbed Global Hull-consistency, is introduced for pruning the domains of the constraint variables.

Several different approaches are proposed for enforcing Global Hull-consistency:
- A higher order consistency approach with algorithm $(n{+}1)$*B-consistency*;
- Backtrack search approaches that include algorithms BS_0, BS_1, BS_2, and BS_3;
- Ordered search approaches that include algorithms OS_1 and OS_3;
- A tree structured approach, based on the *TSA* algorithm.

The *TSA* algorithm, the most competitive of the above algorithms, maintains a binary tree representation of the search space and allows for dynamic focussing on specific relevant regions, losing no information previously obtained in the pruning process.

1.1.3   Local Search for Interval Constraint Reasoning

A local search approach is proposed for integration with constraint reasoning in continuous domains. Despite their success in solving optimisation problems, local search techniques have not been applied for constraint reasoning with continuous domains. The link with the constraint model is achieved through the specification of a multidimensional function, defined for each point of the search space, quantifying at each component the "distance" from satisfying a required constraint.

The originality of the approach is to confine the local search procedure to specific boxes of the search space, relying on the generic branch and bound strategy of the constraint reasoning algorithm to overcome problems traditionally found in the local optimisers (local optimum traps).

1.1.4   Prototype Implementation: Applications to Biophysical Modelling

A prototype application has been implemented integrating all the proposed extensions to the interval constraints framework. It is written in C++ and based on the interval constraint language OpAC [61] (for enforcing box-consistency) and the software packages FADBAD [6] and TADIFF [7] (for the automatic generation of Taylor coefficients).

Three applications to biophysical modelling illustrate the potential of the proposed framework:

- The diagnosis of diabetes based on a parametric differential model of the glucose/insulin regulatory system;
- The tuning of drug design supported on a two-compartment differential model of the oral ingestion/gastro-intestinal absorption process;
- An epidemic study based on a parametric differential model for the spread of an infectious disease within a population.

1.2   Guide to the Dissertation

The dissertation is organised into two parts. The first one addresses the interval constraints framework, its basic concepts and techniques, together with our proposals, which can be incorporated in the framework independently from the context of differential equations. The second part is concerned with the representation of differential equations and their integration in the interval constraints framework.

PART I: INTERVAL CONSTRAINTS

Chapters 2 to 5 overview the interval constraints framework. In Chapter 2 we describe the Constraint Satisfaction Problem paradigm in general and characterise the particular features associated with continuous domains. Chapter 3 addresses interval analysis, focussing on the methods and properties that are useful for the interval constraints framework. Chapter 4 explains the constraint propagation techniques and how they take advantage from interval methods. Chapter 5 overviews the consistency criteria usually enforced in continuous domains. Chapter 6 discusses maintaining Global Hull-consistency as an alternative consistency criteria. Chapter 7 describes the integration of a local search procedure within the interval constraint propagation. Chapter 8 presents the experimental results.

Chapter 2: Constraint Satisfaction Problems

The generic paradigm of a Constraint Satisfaction Problem (CSP) is introduced, its main concepts are defined, and a solving procedure is presented in terms of a search process over a domains lattice. The special case of CSPs with continuous domains is addressed, leading to the basic notions of interval domains and Continuous Constraint Satisfaction Problems (CCSPs). Intervals are identified as elementary objects for representing continuous domains, their basic operations are defined, and different notions of interval approximations are presented. The interval concept is extended to the multidimensional case, leading to the definition of boxes. The generic solving procedure for CSPs is refined to CCSPs by considering the specificity of the continuous domains.

Chapter 3: Interval Analysis

Interval arithmetic is presented as an extension of real arithmetic for real intervals. Its basic operators are defined together with their evaluation rules and algebraic properties. Interval functions are introduced as the interval counterparts of real functions which can be represented by means of interval expressions. The soundness of the evaluation of interval expressions is stressed. The key concept of an interval extension of a real function, widely used in the interval constraints framework, is defined and related to the sound evaluation of its range. Several important forms of interval extensions are addressed, and the general properties of their intersection and decomposition are discussed. The overestimation problem, known as the dependency problem, regarding the evaluation of an interval expression is identified, and its absence is noted when the expression does not contain multiple occurrences of the same variable. The main interval methods used in interval constraints are presented. The interval Newton method is described and its fundamental properties analysed.

Chapter 4: Constraint Propagation

Constraint propagation for pruning the variable domains is described together with enforcing algorithms based on narrowing functions associated with the constraint set. The attributes of such narrowing functions are identified and the main properties of the resulting constraint propagation algorithms are derived accordingly. The main methods used in the interval constraint framework for associating narrowing functions with constraints are presented. Their common strategy of considering each projection with respect to each constraint variable is stressed, and their extensive use of interval analysis techniques for guaranteeing the correctness of the resulting narrowing functions is emphasised. In particular, the constraint decomposition method and the Newton constraint method are fully described.

Chapter 5: Partial Consistencies

Local consistency is defined as a property that depends exclusively on the narrowing functions associated with the constraint set. The main local consistency criteria used in continuous domains, Interval-, Hull- and Box-consistency, are defined and identified as approximations of Arc-consistency used in finite domains. The methods for enforcing such criteria are discussed: being Hull- and Box-consistency associated with the constraint decomposition method, and the Newton constraint method, respectively. The insufficiency of enforcing local consistency on some problems is illustrated. Higher order consistency criteria are then defined as generalisations of the local consistency criteria, and a generic enforcing algorithm is presented.

Chapter 6: Global Hull-Consistency

Global Hull-consistency is proposed as an alternative consistency criterion in continuous domains. Several approaches are devised for enforcing Global Hull-consistency. A suggested approach ($(n+1)B$-*consistency*) is based on existing higher order consistency criteria and the corresponding generic enforcing algorithm. Four different alternative approaches (BS_0, BS_1, BS_2, and BS_3) are proposed based on backtrack search over the space of possibilities. Two additional approaches (OS_1 and OS_3) are derived from the modification of the backtrack search into an ordered search of the space of possibilities. A final approach (TSA) is proposed based on a binary tree representation of the search space. For each of the above approaches the respective enforcing algorithm is explained and its termination and correctness properties justified.

Chapter 7: Local Search

A local search procedure is proposed for integration with the interval constraints framework. It is based on a line search minimisation along a direction determined by the Newton-Raphson method. All the underlying algorithms used by the approach are fully explained, and the termination and convergence global properties are derived. Alternative local search approaches are suggested and discussed. The integration of the proposed local search procedure with Global Hull-consistency enforcement is presented for each of algorithms discussed in the previous chapter.

Chapter 8: Experimental Results

Preliminary results on the application of the Global Hull-consistency criterion are presented. The need for strong consistency requirements such as Global Hull-consistency is illustrated with simple examples where weaker alternatives are clearly insufficient. A more realistic problem based on data from the USA census is fully discussed. It aims at finding parameter ranges of a logistic model such that the difference between the predicted and the observed values does not exceed some predefined threshold. The pruning and time results obtained with the Global Hull-consistency approach (with TSA algorithm) are compared with those obtained by enforcing 2B-, 3B-, and 4B-consistency. Similar comparisons are presented on the different problem of finding the structure of a (very simple) protein from distance constraints among its atoms. The integration of local search within the best Global Hull enforcing algorithms is discussed on another instance of the protein structure problem.

PART II: INTERVAL CONSTRAINTS FOR DIFFERENTIAL EQUATIONS

Chapter 9 introduces Ordinary Differential Equations (ODEs) and reviews the existing approaches for solving problems with ODEs. In chapter 10 we present our proposal of Constraint Satisfaction Differential Problems (CSDPs) for integrating differential equations within the interval constraints framework. Chapter 11 describes the procedure that is proposed for solving CSDPs. Chapter 12 tests our proposal on several biomedical problems for decision support with ODEs. In chapter 13 conclusions are discussed and future work is suggested.

Chapter 9: Ordinary Differential Equations

Ordinary differential equations and initial value problems (IVPs) are presented. Solutions of ODEs and IVPs are defined. Classical numerical approaches for solving IVPs are reviewed. Taylor series methods are addressed in more detail. Different sources of errors and its consequences in numerically solving an IVP are discussed. Interval approaches for solving IVPs are reviewed. Interval Taylor Series (ITS) methods are fully described. The existing approaches that apply interval constraints for ODE solving are reviewed. The early constraint approaches that maintain an extensive constraint network along the ODE trajectory are described. The recent proposals for solving IVPs with constraint propagation techniques are surveyed.

Chapter 10: Constraint Satisfaction Differential Problems

The Constraint Satisfaction Differential Problem (CSDP) is presented as a special kind of CSP for handling differential equations with related additional information expressed as a set of restrictions. Its definition is given and its expressive power illustrated with the definition and explanation of each possible restriction type. The Continuous CSP framework is extended for the inclusion of a new kind of constraint defined as a CSDP. The concept of canonical solution is redefined for such extended context and its consequences on Global Hull-consistency are discussed. The integration of CSDP constraints with local search is explained. The modelling capabilities of the extended framework are illustrated for representing parametric ODEs, interval valued properties and properties depending on the combination of different components of an ODE system.

Chapter 11: Solving a CSDP

The procedure proposed for solving a CSDP is described as a constraint propagation algorithm, based on a set of narrowing functions associated with its constraints, that maintains a safe enclosure for the whole set of possible ODE solutions. The representation of such an enclosure is explained and the narrowing functions for enforcing each type of CSDP restriction are fully characterised. Additional narrowing functions, based on reliable Interval Taylor Series methods, are defined for further reducing the uncertainty of the ODE trajectory. The combination of all such different narrowing functions into the constraint propagation algorithm is explained and the termination and correctness properties of the solving procedure are derived.

Chapter 12: Biomedical Decision Support with ODEs

The extended interval constraints framework is applied to biomedical decision support problems based on differential models. In a first problem, a parametric differential model of the glucose/insulin regulatory system is used for supporting the diagnosis of diabetes during a glucose tolerance test (GTT). This example illustrates the use of value restrictions for modelling initial and boundary conditions (provided by the blood exams) and their integration for reducing the parameter ranges allowing for decisions based on some non-linear combination of these parameters. In a second problem, the tuning of a drug design is supported based on a two-compartment differential model of its oral ingestion/gastro-intestinal absorption process. This example illustrates the usefulness of the integration in the constraint model of other important restrictions such as maximum, minimum, area, and time restrictions. The last problem is based on a parametric differential model for the spread of an infectious disease within a population. An epidemic study is accomplished for predicting the effects of an infectious disease and determining the vaccination rate necessary to guarantee some desirable conditions. This example illustrates the expressive power of non-conventional restrictions, such as first and last restrictions, and its inclusion into a more complex differential model for which there are no analytical solution forms.

Chapter 13: Conclusions and Future Work

The contributions of this work are analysed, some open problems are identified, and directions for future work are set.

Part I

INTERVAL CONSTRAINTS

Chapter 2

Constraint Satisfaction Problems

A constraint is a way of specifying a relation that must hold between certain variables. By restricting the possible values that variables can take, it represents some partial information about these variables, and can be regarded as a restriction on the space of possibilities.

Mathematical constraints are precise specifiable relations among variables, each ranging over a given domain, and are a natural way for expressing regularities upon the underlying real-world systems and their mathematical abstraction.

Many problems of the real world can thus be modelled as constraint satisfaction problems (CSPs), in particular, problems involving inaccurate data or partially defined parameters. The CSP is a classic Artificial Intelligence paradigm whose theoretical framework was introduced in the seventies [136, 98, 97].

A CSP is defined by a set of variables each with an associated domain of possible values and a set of constraints on subsets of the variables. A constraint specifies which values from the domains of its variables are compatible. These can be done explicitly, by presenting the consistent or inconsistent value combinations, or implicitly, by means of mathematical expressions or computable procedures determining these combinations. A solution to the CSP is an assignment of values to all its variables, which satisfies all the constraints.

More formally we will use the following general definitions (similar to definitions used by other authors [4, 83]).

Definition 2-1 (Constraint). A constraint c is a pair (s, ρ), where s is a tuple[1] of m variables $<x_1, x_2, \ldots, x_m>$, the constraint scope, and ρ is a relation of arity m, the constraint relation. The relation ρ is a subset of the set of all m-tuples of elements from the Cartesian product $D_1 \times D_2 \times \ldots \times D_m$ where D_i is the domain of the variable x_i:

$$\rho \subseteq \{<d_1, d_2, \ldots, d_m> \mid d_1 \in D_1, d_2 \in D_2, \ldots, d_m \in D_m\} \qquad \square$$

The tuples in the constraint relation (ρ) indicate the allowed combinations of simultaneous values for the variables in the scope. The length of these tuples (m) is called the arity of the constraint.

Definition 2-2 (Constraint Satisfaction Problem). A CSP is a triple $P=(X,D,C)$ where X is a tuple of n variables $<x_1, x_2, \ldots, x_n>$, D is the Cartesian product of the respective domains $D_1 \times D_2 \times \ldots \times D_n$, i.e. each variable x_i ranges over the domain D_i, and C is a finite set of constraints where the elements of the scope of each constraint are all elements of X. $\qquad \square$

In order to give a formal definition for the satisfaction of a particular constraint, it is necessary to identify from a tuple of elements (associated with all the CSP variables) those that are associated with the variables of the scope of the constraint. This is achieved by tuple projection with respect to the variables of the scope.

---

[1] Tuples with m elements $t_1, t_2, \ldots, t_m$ will be written in the form $<t_1, t_2, \ldots, t_m>$.

> **Definition 2-3 (Tuple Projection).** Let $X=\langle x_1, x_2, \ldots, x_n \rangle$ be a tuple of n variables, and $d=\langle d_1, d_2, \ldots, d_n \rangle$ a tuple[2] where each element d_i is associated with the variable x_i. Let $s=\langle x_{i_1}, x_{i_2}, \ldots, x_{i_m} \rangle$ be a tuple of m variables where $1 \leq i_j \leq n$. The tuple projection of d wrt s, denoted $d[s]$, is the tuple:
> $$d[s] = \langle d_{i_1}, d_{i_2}, \ldots, d_{i_m} \rangle$$ ❏

A tuple satisfies a constraint if and only if its projection wrt the scope of the constraint is a member of the constraint relation.

> **Definition 2-4 (Constraint Satisfaction).** Let $P=(X,D,C)$ be a CSP. Let (s,ρ) be a constraint from C and d an element of D:
> $$d \text{ satisfies } (s,\rho) \text{ iff}[3]\ d[s] \in \rho$$ ❏

A tuple is a solution of the CSP if and only if it satisfies all the constraints.

> **Definition 2-5 (Solution).** A solution to the CSP $P=(X,D,C)$ is a tuple $d \in D$ that satisfies each constraint $c \in C$, that is:
> $$d \text{ is a solution of } P \text{ iff } \forall_{c \in C}\ d \text{ satisfies } c$$ ❏

Two important notions in constraint satisfaction problems are those of consistency and equivalence.

> **Definition 2-6 (Consistency).** A CSP $P=(X,D,C)$ is consistent iff it has at least one solution (otherwise it is inconsistent):
> $$P \text{ is consistent iff } \exists_{d \in D}\ d \text{ is a solution of } P$$ ❏

> **Definition 2-7 (Equivalence).** Two constraint satisfaction problems with the same tuple of variables $P=(X,D,C)$ and $P'=(X,D',C')$ are equivalent iff both have the same set of solutions:
> $$P \text{ and } P' \text{ are equivalent iff}$$
> $$\forall_{d \in D} (d \text{ is a solution of } P \Rightarrow d \text{ is a solution of } P') \land$$
> $$\forall_{d' \in D'} (d' \text{ is a solution of } P' \Rightarrow d' \text{ is a solution of } P)$$ ❏

The definition of equivalence between two CSPs assumes that both of them have the same tuple X of variables, however these could easily be extended to the case where they both have the same set of variables. Moreover, it could be extended to the case where the set of variables of one CSP is a subset of the set of variables of the other CSP. In this case they would be equivalent wrt this subset if it is possible to define a bijective function between the set of solutions of each CSP mapping solutions that share the values of all the common variables[4].

---

[2] For simplicity, the same notation is used either if the element of d_i represents a particular value or a set of values from the domain of variable x_i.

[3] iff stands for: if and only if

[4] This extension will be later used, in subsection 4.2.1, for defining equivalent CSPs obtained by constraint decomposition, which includes necessarily new variables.

2.1　Solving a Constraint Satisfaction Problem

A CSP can have one, several or no solutions. In many practical applications the modeling of a problem as a CSP is embedded in a larger decision process. Depending on this decision process it may be desirable to determine whether a solution exists (verify the consistency of the CSP), to find one solution, to compute the space of all solutions of the CSP, or to find an optimal solution relative to a given objective function.

Solving a CSP can be seen as a search process over the lattice of the variable domains. For a given CSP (X,D,C) let us consider the complete domain lattice L defined by the elements obtained from the power set of D partially ordered by set inclusion ($\subseteq$) and closed under arbitrary intersection ($\cap$) and union ($\cup$). Figure 2.1 shows an example of a CSP P (with finite domains) and figure 2.2 the corresponding domain lattice L.

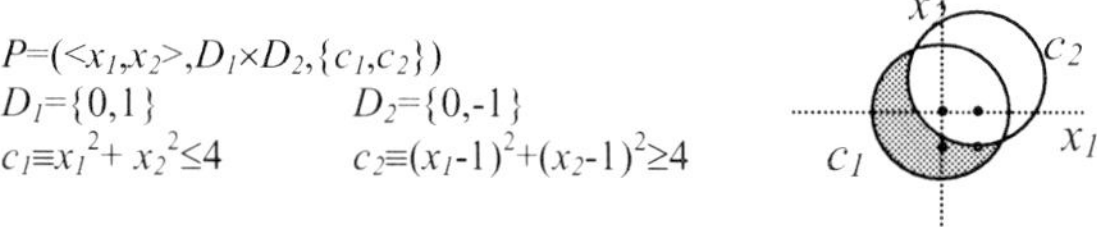

Figure 2.1 An example of a CSP with finite domains. The figure represents the two axes x_1 and x_2, the four points are the domain set, the circumferences are the two constraints (inside c_1 and outside c_2). The solutions are the two points ($\langle 0,-1\rangle$ and $\langle 1,-1\rangle$) inside the dashed area.

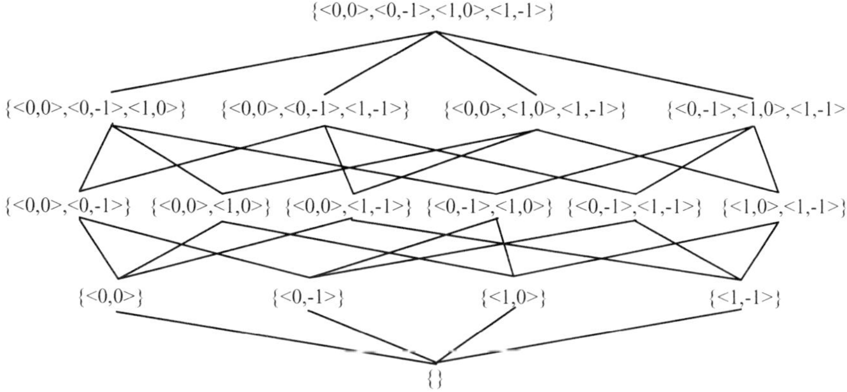

Figure 2.2 Domain lattice of the previous example partially ordered by set inclusion ($\subseteq$).

The search procedure starts at the top of the domain lattice (the original domain D) and navigating over the lattice elements it will eventually stop, returning one of them. If it returns the bottom element (the empty set $\{\}$) then the CSP has no solution.

In the example of figure 2.1 the returned element should be:

(i)　　$\{\langle 0,-1\rangle\}$ or $\{\langle 1,-1\rangle\}$ if the goal is to find at least one solution;

(ii)　　$\{\langle 0,-1\rangle,\langle 1,-1\rangle\}$ if the goal is to compute the space of all solutions;

(iii)　　$\{\langle 0,-1\rangle\}$ if the goal is to find the solutions that minimize $x_1^2+x_2^2$.

The navigation over the lattice elements usually alternates pruning with branching steps and ends whenever a stopping criterion is satisfied.

2.1.1　*Pruning*

The pruning consists on jumping from an element of the lattice to a smaller element (with respect to the set inclusion partial order) as a result of applying an appropriate filtering algorithm which eliminates some value combinations that are inconsistent with the constraints. All lattice elements containing value combinations that were eliminated in the pruning step will not be considered any further. The elements remaining form a sub-lattice with non-eliminated value combinations as a top element.

Figure 2.3 shows the result of applying a pruning step on the top element of the domain lattice of the above example. The combination of values $x_1=0$ and $x_2=0$, proved to be inconsistent with the constraints by some filtering algorithm, is absent in the resulting lattice.

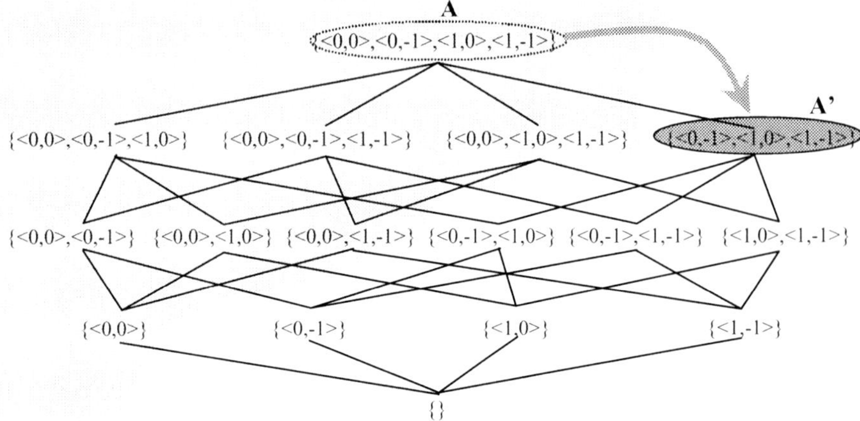

Figure 2.3 Pruning some value combinations. From the top element of a lattice (**A**), a new top element (**A'**) is obtained. In the example <0,0> was proved inconsistent by a filtering algorithm.

The filtering algorithm must guarantee that no possible solution is eliminated from the set of value combinations of the original lattice element. If the filtering algorithm were complete, all inconsistent combinations of values would be deleted and so, the new top element would contain all the solution space. However this is not generally the case, and several inconsistent combinations may still remain (in the example, combination <1,0> is inconsistent but it was not detected in the pruning step).

2.1.2 *Branching*

The branching step may be applied when the pruning step fails to further eliminate inconsistent combinations of values. The idea is to split a set of value combinations into smaller sets (two or more), for which the pruning step will hopefully result in better filtering. In the domains lattice the branching step corresponds to consider separately smaller elements whose union is the original element.

Figure 2.4 illustrates this for the above example, where the top element {<0,-1>,<1,0>, <1,-1>} was split into two smaller elements {<0,-1>} and {<1,0>,<1,-1>} for future consideration.

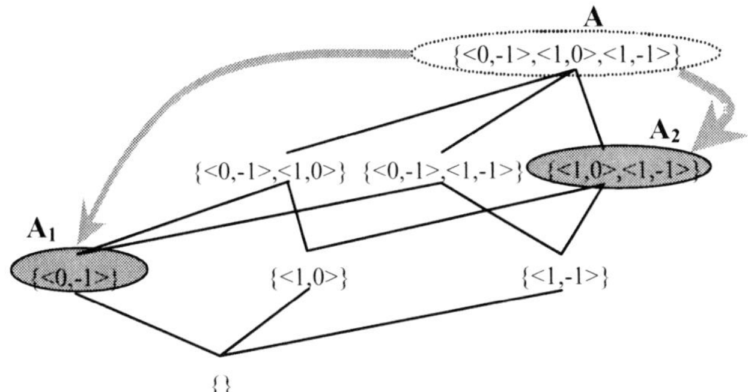

Figure 2.4 Branching a lattice element (**A**) into smaller elements (**A₁** and **A₂**). The original element is the union of all the new smaller elements (**A=A₁∪A₂**).

2.1.3   *Stopping*

The search over the different branches may be done concurrently or by some backtracking mechanism until a stopping criterion is attained. This stopping criterion may be the achievement of the intended goal (find one, all or the best solution) or the satisfaction of some specific properties imposed to avoid the complexity explosion of the search procedure. Figure 2.5 shows the final results after searching each branch of the previous example.

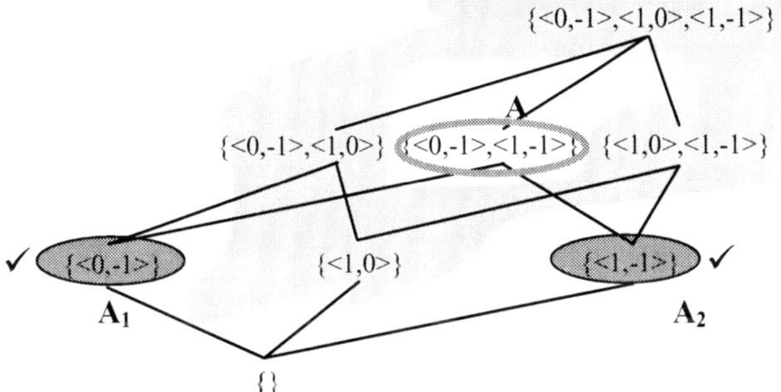

Figure 2.5 Stopping the search when the goal of finding all solutions (**A₁** and **A₂**) is achieved. The result is the top element (**A=A₁∪A₂**) of the remaining lattice.

In the example of figure 2.5, <0,-1> and <1,-1> were proved to be solutions of the CSP and <1,0> was proved to be inconsistent, thus, if the goal was to compute the solution space, the final result would be the top element {<0,-1>,<1,-1>} of the remaining lattice.

2.2   **Constraint Satisfaction Problems With Continuous Domains**

The notion of CSP was initially introduced to address combinatorial problems over finite domains. Thus in the original framework the domains of the variables were expected to be finite sets. However, the above definitions are general enough to represent constraint satisfaction problems with either finite or infinite domain sets.

Numeric CSPs (NCSPs), initially proposed by Davis in [44], are extensions of the earlier CSP framework to address variables with continuous domains. In our formalization NCSPs are a special kind of CSPs where the constraints cannot be given extensionally, they must be specified as numeric relations, and the domains are either integer domains or continuous domains.

Definition 2.2-1 (Numeric Constraint Satisfaction Problem). A NCSP is a CSP $P=(X,D,C)$ where

    i) $\forall_{D_i \in D}\ D_i \subseteq \mathbb{Z} \lor D_i \subseteq \mathbb{R}$

    ii) $\forall_{(s,\rho) \in C}\ \rho$ is defined as a numeric relation between the variables of s ❏

Further restrictions, either on the allowed variable domains or on the kind of numeric expressions used for specifying the numeric relations, may be imposed to address subclasses of problems and eventually to take advantage of their specific properties. For example, if only linear equations over the real numbers are allowed then the subclass of linear constraint satisfaction problems could be considered and some particular methods for solving systems of linear equations would be used. However, the expressive power of these restricted classes of CSPs would be decreased, possibly preventing the modelling of some important relationships among the problem variables. If only linear constraints are allowed, problems with non-linear relations between variables could not be easily represented.

Continuous CSPs (CCSPs), are an important subclass of NCSPs where all variable domains are continuous real intervals and all the numeric relations are equalities and inequalities. The following formal definition is based on [69]

Definition 2.2-2 (Continuous Constraint Satisfaction Problem). A CCSP is a CSP $P=(X,D,C)$ where each domain is an interval of $\mathbb{R}$ and each constraint relation is defined as a numerical equality or inequality:

    i) $D=<D_1,...,D_n>$ where D_i is a real interval $(1 \le i \le n)$

    ii) $\forall_{c \in C}\ c$ is defined as $e_c \Diamond 0$ where e_c is a real expression[5] and $\Diamond \in \{\le,=,\ge\}$ ❏

The CCSP framework is powerful enough to model a wide range of problems, in particular physical systems whose components may be described as sets of continuous valued variables, and whose relations among these variables may be defined by numerical equalities or inequalities, eventually with uncertain parameters.

2.2.1   *Intervals Representing Unidimensional Continuous Domains*

In CCSPs, the initial domains associated with the variables are infinite sets of real numbers called real intervals. The following is a general definition for any real interval, either open, half-open or closed interval.

Definition 2.2.1-1 (*R*-interval). A real interval is a connected set of reals. Let $a \le b$ be reals, the following notations for representing real intervals will be used:

$$[a..b] \equiv \{r \in \mathbb{R} \mid a \le r \le b\} \qquad (a..b) \equiv \{r \in \mathbb{R} \mid a < r < b\}$$

$$(a..b] \equiv \{r \in \mathbb{R} \mid a < r \le b\} \qquad [a..b) \equiv \{r \in \mathbb{R} \mid a \le r < b\}$$

$$[a..+\infty) \equiv \{r \in \mathbb{R} \mid a \le r\} \qquad (a..+\infty) \equiv \{r \in \mathbb{R} \mid a < r\}$$

$$(-\infty..b] \equiv \{r \in \mathbb{R} \mid r \le b\} \qquad (-\infty..b) \equiv \{r \in \mathbb{R} \mid r < b\}$$

$$(-\infty..+\infty) \equiv \mathbb{R} \qquad\qquad\qquad \varnothing \equiv \{\}$$

The notation $<a..b>$ will represent a nonempty real interval of any of the defined forms. ❏

---

[5] Real expressions will be defined later in section 3.2 (definition 3.2-1). Constraints expressed as $e_1 \Diamond e_2$ are also considered since they can be rewritten into the form $e_1 - e_2 \Diamond 0$.

In practice, computer systems are restricted to represent a finite subset of the real numbers, the floating-point numbers. Several authors [91, 14, 134, 74] have defined the set of machine numbers (*F*-numbers) as the set of floating-point numbers augmented with the two infinity symbols (-∞ and +∞). In the formal definition we will also include the real number 0 which is always a member of the set of floating-points.

Definition 2.2.1-2 (*F*-numbers). Let **F** be a subset of $\mathbb{R}$ containing the real number 0 as well as finitely many other reals, and two elements (not reals) denoted by -∞ and +∞:

$$\mathbf{F} = \{r0,\ldots,rn\} \cup \{-\infty,+\infty\} \qquad \text{with } 0 \in \{r0,\ldots,rn\} \subset \mathbb{R}$$

The elements of **F** are called *F*-numbers. ❑

F is totally ordered: any two real elements of **F** are ordered as in $\mathbb{R}$; for all real element r, $-\infty < r < +\infty$. If f is an *F*-number f^- and f^+ are the two *F*-numbers immediately below and immediately above f in the total order ($-\infty^- = -\infty$ and $+\infty^+ = +\infty$; $-\infty^+$ is the smallest real in **F** and $+\infty^-$ is the largest real in **F**).

Given the above definition, we can now define the subset of real intervals that can be represented by a particular machine as the set of real intervals bounded by *F*-numbers (*F*-intervals).

Definition 2.2.1-3 (*F*-interval). An *F*-interval is a real interval $\varnothing$ or $<a..b>$ where a and b are *F*-numbers. In particular, if $b=a$ or $b=a^+$ then $<a..b>$ is a ***canonical F*-interval**. ❑

Figure 2.6 illustrates the above concepts, showing a degenerate[6] and an half-open *R*-interval ($[r_1..r_1]$ and $(r_2..r_3]$ respectively), and two *F*-intervals ($[a..b]$ and $[c..d]$). $[a..b]$ is a canonical *F*-interval because a and b are consecutive *F*-numbers in the total order ($b=a^+$).

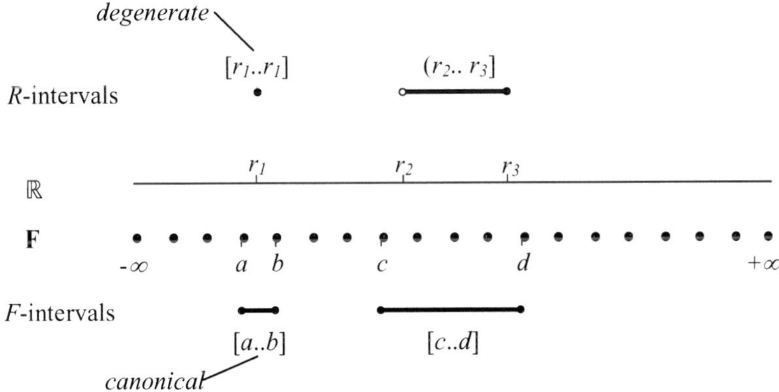

Figure 2.6 *R*-intervals and *F*-intervals.

In the rest of this work we further restrict the nonempty *F*-intervals to the forms with the closed bounds whenever the bound is a real value ($[a..b]$, $[a..+\infty)$, $(-\infty..b]$ and $(-\infty..+\infty)$) with a and b real values)[7]. However, the framework can be naturally extended to consider any of

---

[6] Degenerate intervals $[r..r]$ (either *R*-intervals or *F*-intervals) may also be denoted as $[r]$, $\{r\}$ or even r if it is clear in the context that it is an interval and not a real value.

[7] The notation $[a..b]$ will be generically used to represent any *F*-interval where $[-\infty..b]$, $[a..+\infty]$ and $[-\infty..+\infty]$ denote respectively $(-\infty..b]$, $[a..+\infty)$ and $(-\infty..+\infty)$.

the real interval forms defined in 2.2.1-1 (see [11] and [29] for a detailed discussion on this issue).

2.2.2   *Interval Operations and Basic Functions*

All the usual set operations, namely, intersection ($\cap$), union ($\cup$) and inclusion ($\subseteq$), may also be applied on intervals, either *R*-intervals or *F*-intervals, since intervals are connected sets of real values. A particularly useful operation between two intervals is the union hull ($\uplus$) where the result is the smallest interval containing all the elements of both interval arguments. The resulting interval is only different from the normal union operation when the intersection of the two arguments is the empty set, in which case all real values between the two interval arguments (not belonging to any of them) are also included.

Definition 2.2.2-1 (Union Hull). Let $I_1 = \langle_1 a_1..b_1 \rangle_1$ and $I_2 = \langle_2 a_2..b_2 \rangle_2$ be two intervals, either

R-intervals or *F*-intervals. The union hull operation ($\uplus$) is defined as:

$$I_1 \uplus I_2 = \begin{cases} I_1 \cup I_2 & \text{if} & I_1 \cap I_2 \neq \varnothing \\ \langle_1 a_1..b_2 \rangle_2 & \text{if} & \forall_{r_1 \in I_1} \forall_{r_2 \in I_2} \, r_1 < r_2 \\ \langle_2 a_2..b_1 \rangle_1 & \text{if} & \forall_{r_1 \in I_1} \forall_{r_2 \in I_2} \, r_2 < r_1 \end{cases}$$

$\square$

Several basic functions with an interval argument are usually defined for obtaining the extreme values (*left, cleft, right* and *cright*), the mid value (*center*) or the size (*width*) of an interval.

Definition 2.2.2-2 (Interval Basic Functions). Let $[a..b]$ be a closed interval, either an *R*-interval or an *F*-interval. The following basic functions return a real value and are defined as:

$$left([a..b]) = a \qquad\qquad right([a..b]) = b$$
$$center([a..b]) = (a+b)/2 \qquad\qquad width([a..b]) = b-a$$

Let $[a..b]$ be a closed *F*-interval. The following basic functions return a canonical *F*-interval and are defined as:

$$cleft([a..b]) = \begin{cases} [a] & \text{if } a=b \\ [a..a^+] & \text{if } a<b \end{cases} \qquad cright([a..b]) = \begin{cases} [b] & \text{if } a=b \\ [b^-..b] & \text{if } a<b \end{cases}$$

$\square$

The union hull operation together with the interval basic functions is exemplified in figure 2.7.

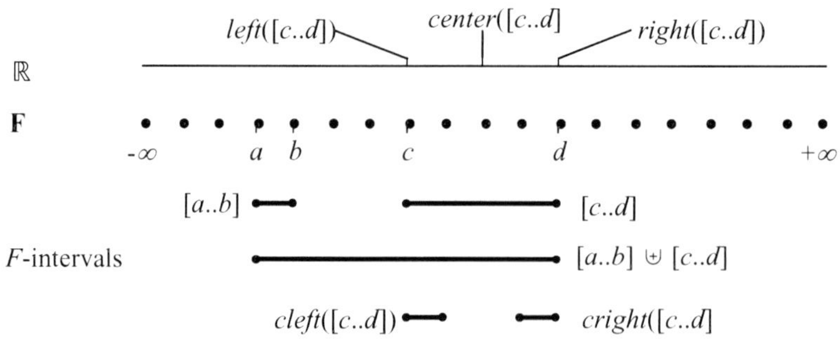

Figure 2.7 Interval operations and basic functions.

2.2.3   Interval Approximations

For any real number r we will denote by $\lfloor r \rfloor$ the largest F-number not greater than r and $\lceil r \rceil$ the smallest F-number not lower than r. Any real interval can be associated to an F-interval which is the closest approximation that can be represented by a particular machine.

Definition 2.2.3-1 (RF-interval approximation). Let $IR=\langle a..b\rangle$ be a real interval. The RF-interval approximation of IR, denoted $I_{apx}(IR)$, is the smallest F-interval including IR ($IR \subseteq I_{apx}(IR)$):

$I_{apx}(IR)=[\lfloor a \rfloor .. \lceil b \rceil]$[8].

In the special case where IR is a single real $\{r\}=[r..r]$ then $I_{apx}(IR)=[\lfloor r \rfloor .. \lceil r \rceil]$.   ❑

The definition of RF-set approximation extends the above definition to represent any set of real values, either connected or not.

Definition 2.2.3-2 (RF-set approximation). Let SR be a set of real values defined by the union of n real intervals ($SR=IR_1\cup...\cup IR_n$). The RF-set approximation of SR, denoted $S_{apx}(SR)$, is the set defined by the union of the n corresponding RF-interval approximations:

$S_{apx}(SR) = I_{apx}(IR_1) \cup...\cup I_{apx}(IR_n)$ ❑

If in the above definition the union operation is substituted by the union hull operation then, the result is the smallest F-interval containing all the elements of a set of reals.

Definition 2.2.3-3 (RF-hull approximation). Let SR be a set of real values defined by the union of n real intervals ($SR=IR_1\cup...\cup IR_n$). The RF-hull approximation of SR, denoted $I_{hull}(SR)$, is the F-interval defined by:

$I_{hull}(SR) = I_{apx}(IR_1) \uplus...\uplus I_{apx}(IR_n)$ ❑

Figure 2.8 summarizes the above interval approximation definitions.

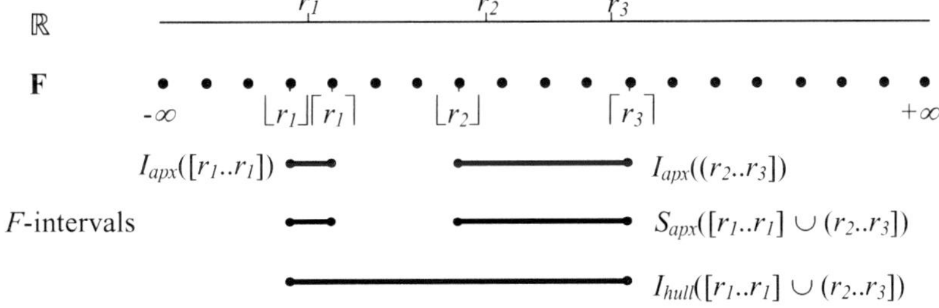

Figure 2.8 Interval approximation.

2.2.4   Boxes Representing Multidimensional Continuous Domains

Extending to several dimensions the concepts of an R-interval and F-interval, we will get respectively the notions of an R-box and F-box.

Definition 2.2.4-1 (R-box). An R-box BR with arity n is the Cartesian product of n R-intervals and is denoted by $\langle IR_1,...,IR_n\rangle$ where each IR_i is an R-interval:

$BR = \{\langle r_1, r_2, ..., r_m\rangle \mid r_1 \in IR_1, r_2 \in IR_2, ..., r_n \in IR_n\}$ ❑

---

[8] Extending the notation $\lfloor -\infty \rfloor$ and $\lceil +\infty \rceil$ to denote respectively $-\infty$ and $+\infty$.

> **Definition 2.2.4-2 (*F*-box).** An *F*-box *BF* with arity n is the Cartesian product of n *F*-intervals and is denoted by $<IF_1,...,IF_n>$ where each IF_i is an *F*-interval:
> $$BF = \{<r_1, r_2, ..., r_m> \mid r_1 \in IF_1, r_2 \in IF_2, ..., r_n \in IF_n\}$$
> In particular, if all the *F*-intervals IF_i are canonical then *BF* is a **canonical *F*-box**. ❏

During the solving process of a CCSP several value combinations between the variables may be discarded from the original domains box. Moreover, each variable domain may no longer be a connected set of real values but rather a disconnected set corresponding to the union of multiple connected sets.

Some approaches [77, 128] consider structures composed of several *F*-intervals to represent each variable domain. In [77] a sequence of disjunct *F*-intervals is organized in a structure called a division. In [128] a hierarchical arrangement of *F*-intervals constitutes a taxonomy. In these approaches the space of possibilities of several variables is represented by the set of the structures (divisions or taxonomies) associated with each variable and corresponds to the set of *F*-boxes obtained by all combinations of the possible *F*-intervals from the variable domains.

In [69] the feasible space between k variables is represented by 2^k-trees. A 2^k-tree is a hierarchical decomposition of the solution space into k-arity *F*-boxes which summarizes the subset of constraints between the k variables.

In most interval constraint approaches the basic structures are *F*-intervals and the solutions space is represented by enclosing *F*-boxes. In particular a single real value is represented by a canonical *F*-interval and the assignment of a single real value to each variable of a set of variables is represented by a canonical *F*-box. Consequently, canonical *F*-boxes are the closest representations of CCSP solutions. In practice, if the system is not able to prove the inconsistency of a canonical *F*-box then the box may contain a real solution that is not accessible due to precision limitations. We will call a *canonical solution* of a CCSP any canonical *F*-box that cannot be proved inconsistent (wrt to the CCSP) either because it contains solutions or due to approximation errors in the evaluation of the constraint set.

> **Definition 2.2.4-3 (Canonical Solution).** Let $P=(X,D,C)$ be a CCSP. Let *BF* be a canonical *F*-box included in D $(BF \subseteq D)$ and $E_c(BF)$ denote the *F*-interval obtained by the evaluation of the expression E_c with argument *BF* [9].
> $$BF \text{ is a canonical solution of } P \text{ iff } \forall_{c \in C} \; \exists_{r \in E_c(BF)} \; r \diamond 0$$ ❏

2.2.5 *Solving Continuous Constraint Satisfaction Problems*

In a CCSP $P=(X,D,C)$, since the initial variable domains are infinite sets (real intervals), the domains lattice obtained from the power set of D, is also infinite. Thus the search process for solving a CCSP is theoretically performed over an infinite space. In practice, due to the computer limitations for representing real values, only a finite subset of the domains lattice is representable, and so, the navigation process is limited to these elements. The search procedure starts at the top of the domain lattice[10] and navigates over the accessible elements of the lattice until eventually stopping, returning one of them. The accessible elements of the lattice are those representable by an *F*-box or by the union of several *F*-boxes.

---

[9] Each constraint c is defined as $E_c \diamond 0$ with $\diamond \in \{\leq,=,\geq\}$ (see definition 2.2-2). The evaluation of expressions with interval arguments will be addressed later in section 3.2.

[10] In practice, if $D=D_1 \times D_2 \times ... \times D_n$ then $D'=I_{apx}(D_1) \times I_{apx}(D_2) \times ... \times I_{apx}(D_n)$ will be the top element of the domain lattice defined as the power set of D'.

The pruning step consists on jumping from an element A of the lattice to a smaller element A', both representable by an F-box or by the union of several F-boxes. Despite dealing with F-boxes, the filtering algorithm must still guarantee that no possible real solution is lost, that is, any solution in A must also be in A'. Nevertheless, the filtering algorithm may be unable to prune some inconsistencies due to the limited representation power (i.e., if the solution set in A is not representable then the best pruning achievable is its tightest representable approximation which include several inconsistent real valued combinations). For example, if the only solution of a CCSP asserts the real value π to a variable then the best possible pruning is the canonical F-box including such π value for that variable, which also includes other nearby values that do not belong to any solution of the CCSP.

The branching step consists on splitting an element A of the lattice into n smaller elements $A_1,\ldots,A_n$, each representable by an F-box or by the union of several F-boxes. The union of all the smaller elements must be equal to the original element.

To simplify the domains representation, most solving strategies impose that the only lattice elements considered in the pruning and branching steps are representable by single F-boxes (as opposed to a union of F-boxes). Pruning corresponds to narrowing the original F-box into a smaller one where the lengths of some F-intervals are decreased by the filtering algorithm (eventually being zeroed, proving the original F-box to be inconsistent). The branching step usually consists on splitting the original F-box into two smaller F-boxes by splitting one of the original variable domains around an F-number (usually the F-number nearest to the mid value of the F-interval representing the domain).

With the above restriction on the search procedure, and noting that the top lattice element (the starting point of the search) is representable by a single F-box, all navigation is performed over the subset of the lattice elements representable by single F-boxes (the reachable sub-lattice). Nevertheless, the final result may be any representable lattice element, since it corresponds to the union of all elements remaining at the end of the search process.

Despite being a finite search space, the domains lattice of a CCSP usually contains a huge number of elements, and any strategy to navigate over it must be aware that the underlying real valued search space is infinite. To be effective, a solving strategy cannot rely exclusively on branching expecting the splitting process to stop eventually because the search space is finite. In fact, the splitting process is theoretically guaranteed to stop but the combinatorial number of necessary splits usually prevents such stopping from being achieved in a reasonable amount of time. One approach often adopted imposes conditions on the branching process, for instance, branching may only be performed on lattice elements with some variable domains larger[11] than a predefined threshold and this may only be done by splitting one of this domains.

2.3   Summary

In this chapter the paradigm of the Constraint Satisfaction Problem (CSP) was introduced and the particular case of Continuous CSPs was presented. Intervals were introduced as the elementary objects for representing continuous domains and boxes as their multidimensional counterpart. Several interval basic functions and operators were defined and different notions of interval approximation were given. The following chapter will address interval analysis, focussing in the methods and properties that are useful for the interval constraints framework.

---

[11] If the domain is represented by an F-interval $[a..b]$ its size may be given by its width $b-a$; if the domain is represented by the union of F-intervals its size may be given by the sum of their widths.

Chapter 3

Interval Analysis

Interval Analysis was introduced by Moore [99] with the purpose of providing upper and lower bounds for the effects of mathematical computation errors arising from different sources, rounding errors, approximating errors and uncertainty data errors. The main goal is to perform interval arithmetic operations to achieve sound mathematical computations over intervals (instead of reals). A major concern in Interval Analysis is to develop interval algorithms that make the interval bounds as narrow as possible.

The original goal of interval constraints [27] was to address the incorrectness of numerical computations due to the floating-point representation of real numbers, offering a sound computation model based on interval arithmetic. Additionally, the soundness of interval arithmetic computations provided the right tools for defining sound filtering algorithms to prune the variable domains when solving CCSPs. Hence, the pruning step is a proof that the real valued combinations removed from the original variable domains do not belong to a solution of the CCSP and the validation of this proof is guaranteed by the soundness of interval arithmetic. Moreover, efficient interval methods developed in Interval Analysis (e.g. the interval Newton method) are used in interval constraints to implement efficient filtering algorithms.

3.1   Interval Arithmetic

Interval arithmetic is an extension of real arithmetic for real intervals. The basic operations of real arithmetic, sum, difference, product and quotient, are redefined for real intervals. The intended meaning of these operations between pairs of intervals is the set obtained by applying them to all pairs of real numbers, one from each of the two intervals. The following formal definition is based on the original one given by Moore [99].

Definition 3.1-1 (Basic Interval Arithmetic Operators). Let I_1 and I_2 be two real intervals (bounded and closed)[1]. The basic arithmetic operations on intervals are defined by:

$$I_1 \, \Phi \, I_2 = \{ \, r_1 \, \Phi \, r_2 \mid r_1 \in I_1 \, \wedge \, r_2 \in I_2 \} \qquad \text{with} \qquad \Phi \in \{+,-,\times,/\}$$

except that I_1/I_2 is not defined if $0 \in I_2$.   ❑

Accordingly to the above definition a set of algebraic rules may be defined to evaluate the result of any basic arithmetic operation on intervals in terms of formulas for its bounds.

Definition 3.1-2 (Evaluation Rules for the Basic Operators). Let $[a..b]$ and $[c..d]$ be two real intervals (bounded and closed):

$$[a..b] + [c..d] = [a+c..b+d] \qquad\qquad [a..b] - [c..d] = [a-d..b-c]$$

$$[a..b] \times [c..d] = [\min(ac,ad,bc,bd)..\max(ac,ad,bc,bd)]$$

$$[a..b] \, / \, [c..d] = [a..b] \times [1/d..1/c] \qquad \text{if } 0 \notin [c..d]$$ ❑

---

[1] In definition 2.2.1-1 it would correspond to the form $[a..b]$.

Interval arithmetic is thus a generalization of real arithmetic. In the extreme case where both interval operands are degenerate (i.e. a single real number of the form $[r..r]$) interval arithmetic reduces to ordinary real arithmetic. Most algebraic properties of real arithmetic also hold for interval arithmetic. However, the distributive law is an important exception.

Definition 3.1-3 (Algebraic Properties of the Basic Operators). Let I_1, I_2, I_3 and I_4 be real intervals (bounded and closed). The following algebraic properties hold for the basic interval operations:

Commutativity:	$I_1+I_2=I_2+I_1$	(interval addition)
	$I_1{\times}I_2=I_2{\times}I_1$	(interval multiplication)
Associativity:	$(I_1+I_2)+I_3=I_1+(I_2+I_3)$	(interval addition)
	$(I_1{\times}I_2){\times}I_3=I_1{\times}(I_2{\times}I_3)$	(interval multiplication)
Neutral Element:	$I_1+[0..0]=I_1$	(interval addition)
	$I_1{\times}[1..1]=I_1$	(interval multiplication)
Subdistributivity:	$I_1{\times}(I_2+I_3)\subseteq I_1{\times}I_2+I_1{\times}I_3$	

Inclusion Monotonicity: $I_1{\subseteq}I_3 \wedge I_2{\subseteq}I_4 \Rightarrow I_1\Phi I_2 \subseteq I_3\Phi I_4$

(with: $\Phi\in\{+,-,\times,/\}$ and $I_3\Phi I_4$ defined)

Figure 3.1 shows an example of the subdistributivity property. The evaluation of both sides of the subdistributivity expression ($I_1{\times}(I_2+I_3)$ and $I_1{\times}I_2+I_1{\times}I_3$) is performed in parallel for a particular case ($I_1=[0..1]$, $I_2=[2..3]$ and $I_3=[-2..-1]$) by consecutively applying the evaluation rules defined in 3.1-2.

$$
\begin{array}{ccc}
I_1{\times}(I_2+I_3) & \subseteq & I_1{\times}I_2+I_1{\times}I_3
\end{array}
$$

$$
\begin{array}{l|cc}
I_1=[0..1] & [0..1]{\times}([2..3]+[-2..-1]) & [0..1]{\times}[2..3]+[0..1]{\times}[-2..-1] \\
I_2=[2..3] & [0..1] \quad \times \quad [0..2] & [0..3] \quad + \quad [-2..0] \\
I_3=[-2..-1] & [0..2] & \subseteq \qquad [-2..3]
\end{array}
$$

Figure 3.1 An example of subdistributivity.

The key idea of interval arithmetic is that despite the different sources of error in arithmetic computations the correct real values are always within the bounds of the resulting real interval. One of these error sources is the limitation of computer systems to the floating-point representation of real numbers. The solution of interval arithmetic is to represent real numbers as F-intervals (the F-interval approximation of a real number) and to evaluate the basic interval arithmetic rules by outward rounding. The outward rounding forces the result of any basic interval arithmetic operation to be an F-interval which is the F-interval approximation of the real interval that would be obtained by evaluating the corresponding rule with infinite precision. The following is the redefinition of the four basic interval arithmetic rules with outward rounding evaluation.

Definition 3.1-4 (Outward Rounding Evaluation Rules of the Basic Operators). Let $[a..b]$ and $[c..d]$ be two F-intervals (bounded and closed):

$$[a..b] + [c..d] = [\lfloor a+c \rfloor .. \lceil b+d \rceil]$$
$$[a..b] - [c..d] = [\lfloor a-d \rfloor .. \lceil b-c \rceil]$$
$$[a..b] \times [c..d] = [\min(\lfloor ac \rfloor, \lfloor ad \rfloor, \lfloor bc \rfloor, \lfloor bd \rfloor)..\max(\lceil ac \rceil, \lceil ad \rceil, \lceil bc \rceil, \lceil bd \rceil)]$$
$$[a..b] / [c..d] = [a..b] \times [\lfloor 1/d \rfloor .. \lceil 1/c \rceil] \qquad \text{if } 0\notin[c..d]$$

Outward rounding preserves the inclusion monotonicity property of interval arithmetic. In the following, if Φ is a basic interval arithmetic operator then Φ_{apx} denotes the corresponding outward evaluation rule. For an m-ary basic interval arithmetic operator Φ

and the real intervals $I_1,\ldots,I_m$: $\Phi_{apx}(I_1,\ldots,I_m)=I_{apx}(\Phi(I_1,\ldots,I_m))$. Moreover, when mentioning the interval arithmetic evaluation to be performed with infinite precision, this corresponds to the unrealistic extreme situation where all real numbers are also F-numbers and so: $\forall_{r\in\mathbb{R}}\ \lfloor r\rfloor=\lceil r\rceil=r$ making $\Phi_{apx}(I_1,\ldots,I_m)=\Phi(I_1,\ldots,I_m)$.

The correctness of the interval arithmetic computations is guaranteed by the inclusion monotonicity property because, if the correct real values are within the operand intervals then the correct real values resulting from any interval arithmetic operation must also be within the resulting interval. Moreover, the computation of any interval arithmetic expression (a successive composition of arithmetic operations over real intervals) will preserve the correct real values within the final resulting interval.

The inclusion monotonicity property of the interval arithmetic expressions (for the basic arithmetic operators +, -, × and /) and its consequences on the correctness of the interval arithmetic computations were firstly addressed by Moore in [99].

3.1.1 *Extended Interval Arithmetic*

Later generalizations of Moore's work included: extensions on the definition of the division operator; extensions on the forms of real intervals allowed as operator arguments; and extensions on the set of basic interval operators allowed in the arithmetic expressions.

The redefinition of the division operator, allowing the division by an interval containing zero, implies the consideration of unbounded intervals and the redefinition of other operators with possible unbounded arguments. The resulting arithmetic, called extended interval arithmetic, was first suggested in [68] and in [84] and later modified in [110] and [122]. Such approaches restricted the division operator to bounded interval arguments.

If unbounded intervals are allowed as arguments of the division operation then the resulting interval is no longer guaranteed to be a closed interval. For example, if the numerator is the degenerate interval [1..1] and the denominator is the unbounded interval [1..+∞), the resulting real interval is half open [1..1]/[1..+∞)=(0..1]. The generalization of the extended interval arithmetic to F-intervals, not necessarily closed, was proposed in [75].

In the interval arithmetic framework it is assumed that the result of any interval operator is a single real interval. This was true for the original definitions (3.1-1 and 3.1-2) of the basic interval operators but in extended interval arithmetic the result may be the union of two disjunct real intervals. For example [4..8]/[-2..1]=(-∞..-2]∪[4..+∞). In many approaches the pair of disjoint real intervals is replaced by the smallest single interval containing their union (their union hull as defined in 2.2.2-1). In the previous example the result would be (-∞..+∞). Other solution proposed in [111] is based on the observation that the pair of disjoint real intervals is normally only required for intersection with another single interval (see interval Newton method). Thus an additional three-argument operation could be considered representing the common operation $(I_1/I_2)\cap I_3$, and the result would be the smallest real interval containing it. Internally the implementation of the operator (the rules defining the bounds of the resulting interval) would consider the possible disjoint result of I_1/I_2. However the final result would be a single interval. If in the previous example the result of the division were to be intersected with [0..10] then the final real interval would be [4..10] (as opposed to [0..10] obtained by the other approaches).

Similarly, several other elementary functions (*exp, ln, power, sin, cos*) may be included as basic interval arithmetic operators. The definition of rules for these new operators may benefit from the study of the monotonic properties of these functions within their domains of application. For example, the exponential unary-operator (*exp*) may be defined by rule $exp([a..b])=[exp(a)..exp(b)]$ since the exponential function is increasing monotonic in $\mathbb{R}$.

3.2   Interval Functions

The sound evaluation of interval functions is perhaps the major contribution of Interval Analysis within the Interval Constraints framework. In order to understand this contribution it is necessary to make explicit the difference between a function (either a real or interval function) and the expression used to represent it in terms of the arithmetic operators, variables and constants. Although a function may be represented by several equivalent expressions, the interval arithmetic evaluation of these expressions may yield different interval results due to its approximate nature. Nevertheless, the soundness of interval arithmetic guarantees that all these intervals contain the intended correct function results.

The following are formal definitions of real/interval expressions, and how they may be used to represent real/interval functions.

Definition 3.2-1 (Real and Interval Expressions). An expression E is an inductive structure defined in the following way:

    (i)   a constant is an expression;

    (ii)   a variable is an expression;

    (iii)   if $E_1,\ldots,E_m$ are expressions and Φ is a m-ary basic operator then $\Phi(E_1,\ldots,E_m)$ is an expression;

A real expression is an expression with real constants, real valued variables and real operators. An interval expression is an expression with real interval constants, real interval valued variables and interval operators. ❏

If x_1, x_2 and x_3 are real valued variables then $(x_1+x_2)\times(x_3-\pi)$ is a real expression with three binary real operators (+, × and -) and a real constant (π). If X_1 and X_2 are real interval valued variables then $(X_1+cos([0..\pi]\times X_2))$ is an interval expression with two binary interval operators (+ and ×), a unary interval operator (cos) and an interval constant ($[0..\pi]$). Note that the above definition does not restrict a constant to be representable by an F-number or an F-interval (in both previous examples π and $[0..\pi]$ are non-representable constants). To improve readability, a degenerate interval constant $[a..a]$ will be represented within the interval expressions as a real constant a.

Definition 3.2-2 (Real and Interval Functions). A function is a mapping from elements of a set (the domain) to another set (the codomain). The subset of the codomain consisting on those elements that are mapped by the function is called the range of the function. In an n-ary real function, f, the elements of the domain are n-ary tuples of real values and the elements of the codomain are real values. The range of a real function f over a domain D is denoted $f^*(D)$. In an n-ary interval function, F, the elements of the domain are n-ary tuples of real intervals (R-boxes) and the elements of the codomain are real intervals. An n-ary function may be represented by an expression, where each occurrence of the ith function argument is designated by a variable. The real expression representing the real function f is denoted by f_E and its variables by x_i. The interval expression representing the interval function F is denoted by F_E and its variables by X_i. ❏

Several different real expressions may represent the same real function. This is a direct consequence of the properties of the real operators used in the expression. For example, the real expressions $f_{E_1}\equiv(x_1+x_2)\times(x_3-\pi)$ and $f_{E_2}\equiv((x_1\times x_3)+(x_2\times x_3))-((\pi\times x_1)+(\pi\times x_2))$ both represent the same real function f due to the distributive and commutative properties of the addition (+) and multiplication (×) real operators.

However, the floating-point evaluation of a real expression for a particular tuple of real values is not guaranteed to return the correct real value (the element of the codomain associated with the tuple in the real function represented by the expression) but an F-number approximation of it. Moreover, different expressions representing the same function may yield different approximations of the correct real value, and without further numerical analysis it is not possible to assess the accuracy of the approximations.

When an interval expression F_E is used to represent an interval function F then this function maps an R-box into the smallest real interval containing all the real values that would be obtained for each real valued combination within the interval domains of the R-box.

Definition 3.2-3 (Semantic of an Interval Expression[2]). Let F be the n-ary interval function represented by the interval expression F_E, and B an n-ary R-box. The interval, denoted by $F(B)$, obtained by applying the interval function F to B, is the smallest real interval containing the range $f^*(B)$ of the real function f over B:

$$f^*(B)=\{f(<r_1,\ldots,r_n>) \mid <r_1,\ldots,r_n> \in B\} \subseteq F(B) \wedge \forall_{<a..b>} f^*(B) \subseteq <a..b> \Rightarrow F(B) \subseteq <a..b>$$

The real function f is expressed by the real expression f_E which is obtained by replacing in F_E each interval variable X_i by the real variable x_i and each interval operator by the corresponding real operator. ❑

Interval arithmetic provides a method for evaluating an interval expression by substituting each variable by its interval domain and applying recursively the basic operator evaluation rules. The evaluation process may be seen as an interval function as well, once it maps R-boxes into R-intervals (in particular into F-intervals).

Definition 3.2-4 (Interval Arithmetic Evaluation of an Interval Expression). Let F be the n-ary interval function represented by the interval expression F_E, and B an n-ary R-box. The interval arithmetic evaluation of F_E wrt B is an interval function recursively defined as:

$$F_E(B) = \begin{cases} I_{apx}(I) & \text{if } F_E \equiv I & (I \text{ is an interval constant}) \\ I_{apx}(B[X_i]) & \text{if } F_E \equiv X_i & (X_i \text{ is an interval variable}) \\ \Phi_{apx}(E_1(B),\ldots, E_m(B)) & \text{if } F_E \equiv \Phi(E_1,\ldots,E_m) & (\Phi \text{ is an interval operator}) \end{cases}$$ ❑

We will denote an interval expression with a subscripted capital letter (F_E) which is also the name of the interval function defined by its interval arithmetic evaluation (as in definition 3.2-4). The capital letter (F) refers to the intended interval function that is represented by the interval expression (as in definition 3.2-3).

The interval arithmetic evaluation of an interval expression provides a sound computation of the interval function represented by the expression.

Theorem 3.2-1 (Soundness of the Interval Expression Evaluation)[3]. Let F_E be an interval expression representing the n-ary interval function F, and B an n-ary R-box. The interval arithmetic evaluation of F_E with respect to B is sound:

$$F(B) \subseteq F_E(B)$$ ❑

Figure 3.2 illustrates the intended interval function F that is represented by the expression $F_E \equiv X_1 \times ([0.5..1.5]-X_1)$. Any interval obtained with a degenerate interval argument $F([r])$ corresponds in the figure to a vertical bar within the two solid lines. The intervals $F([0.5])$,

---

[2] This definition assumes that the interval expression does not contain any interval constants. Otherwise, if there are m interval constants, then f should have m more arguments with the domains defined by the interval constants.

[3] The demonstrations of theorem 3.2-1 and the other theorems of Interval Arithmetic are presented in Appendix A.

$F([1.0])$, $F([1.5])$ and $F([2.0])$ are explicitly represented in the figure. For non degenerate interval arguments $I=[a..b]$ the result $F(I)$ may be obtained by projecting on the vertical axis all the possible intervals obtained by any degenerate arguments within I ($F([r])$ $\forall r \in [a..b]$). This process is exemplified for the argument $I=[0.5..2.0]$ (the arrowhead dashed lines represent the intervals projection). The interval obtained by the interval arithmetic evaluation of the expression F_E with the same argument $I=[0.5..2.0]$ is also shown in figure. As expected, and accordingly to theorem 3.2-1, the evaluated interval encloses the intended interval: $F([0.5..2.0])=[-3.0..0.5625] \subseteq F_E([0.5..2.0])=[-3.0..2.0]$.

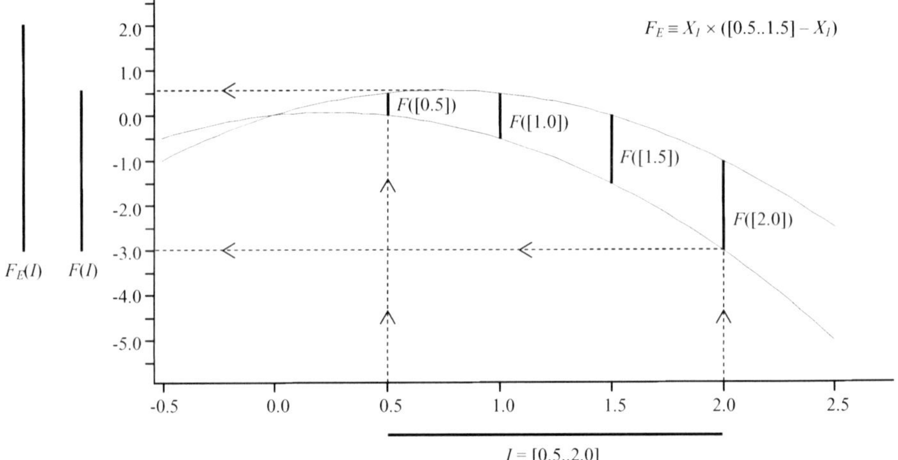

Figure 3.2 The intended interval function represented by an interval expression.

3.2.1   *Interval Extensions*

An important concept relating interval with real functions is that of interval extension.

Definition 3.2.1-1 (Interval Extension of a Real Function). Let f be an n-ary real function with domain D_f, and F an n-ary interval function. The interval function F is an interval extension of the real function f iff:

$$\forall_{<r_1,\ldots,r_n> \in D_f} \; f(<r_1,\ldots,r_n>) \in F(<[r_1..r_1],\ldots,[r_n..r_n]>) \qquad \square$$

Thus, if F is an interval extension of f then each real value mapped by f must lie within the interval mapped by F when the argument is the corresponding R-box of degenerate intervals. Consequently, F provides a sound evaluation of f in the sense that the correct real value is not lost. Moreover, the interval arithmetic evaluation of any expression representing an interval extension of a real function provides a sound evaluation for its range and is itself an interval extension of the real function.

Theorem 3.2.1-1 (Soundness of the Evaluation of an Interval Extension). Let F be an interval extension of an n-ary real function f, F_E an interval expression representing F, and B be n-ary R-box. Then, both $F(B)$ and $F_E(B)$, enclose the range of f over B:

$$f^*(B) \subseteq F(B) \subseteq F_E(B) \qquad \square$$

Figure 3.3 shows an example of a function f represented by the real expression $f_E \equiv x_I - x_I^2$, together with the interval function F presented in the previous example. Clearly F (and also

F_E) is an interval extension of f since $f(r) \in F([r])$ for any real value r (the thick solid line is always within the two thin solid lines). In addition, figure 3.3 illustrates the enclosing for the range of function f (with the argument x_1 between 0.5 and 2.0) obtained by both, the interval extension F and the interval arithmetic evaluation of its expression F_E: $f^*([0.5..2.0])=[-2.0..0.25] \subseteq F([0.5..2.0]) \subseteq F_E([0.5..2.0])$.

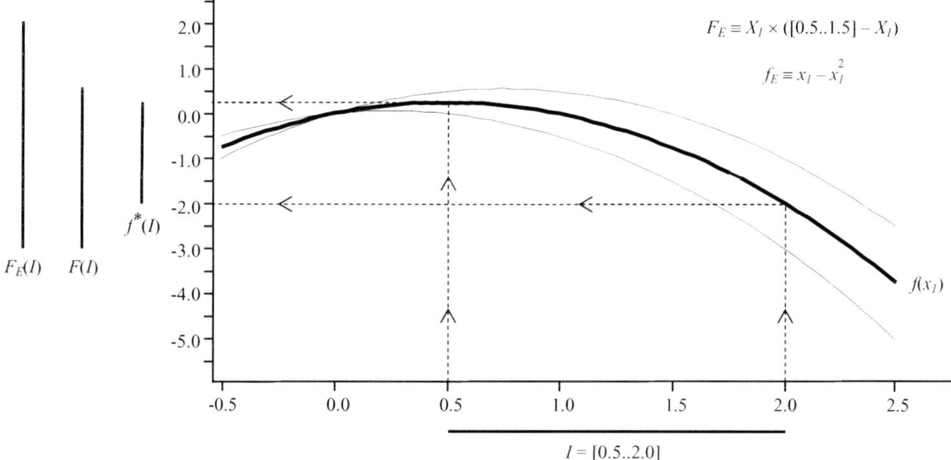

Figure 3.3 An interval extension of an interval function.

A particular interval extension of a real function, called Natural interval extension, is directly obtained from the real expression representing it.

Definition 3.2.1-2 (Natural Interval Expression). If f_E is a real expression representing the real function f, then its natural interval expression F_n is obtained by replacing in f_E: each real variable x_i by an interval variable X_i; each real constant k by the real interval $[k..k]$, and each real operator by the corresponding interval operator. ❏

Theorem 3.2.1-2 (Natural Interval Extension). Let f_E be a real expression representing the real function f, and F_n be the natural interval expression of f_E. The interval function F represented by F_n is the smallest interval enclosure for the range of f and the interval arithmetic evaluation of F_n is an interval extension of f denominated Natural interval extension with respect to f_E. ❏

As seen before, several equivalent real expressions may represent the same real function f. Consequently, the natural interval extensions wrt these equivalent real expressions are all interval extensions of f.

Figure 3.4 shows three equivalent real expressions (f_{E_1}, f_{E_2} and f_{E_3}) that represent the same real function f presented in the previous example and their natural interval expressions (F_{E_1}, F_{E_2} and F_{E_3}). The interval function F is the intended function represented by any of these interval expressions and, accordingly to theorem 3.2.1-2, is the smallest interval enclose for the range of f (in the figure this is illustrated for $I=[0.5..2.0]$ with $f^*(I)=F(I)=[-2.0..0.25]$). The interval arithmetic evaluation of different forms of interval extensions of f may lead to different accuracy for the resulting interval (the width of the interval may be different). Figure 3.4 shows the different enclosures obtained by the three different extensions for the range of function f (with the argument x between 0.5 and 2.0).

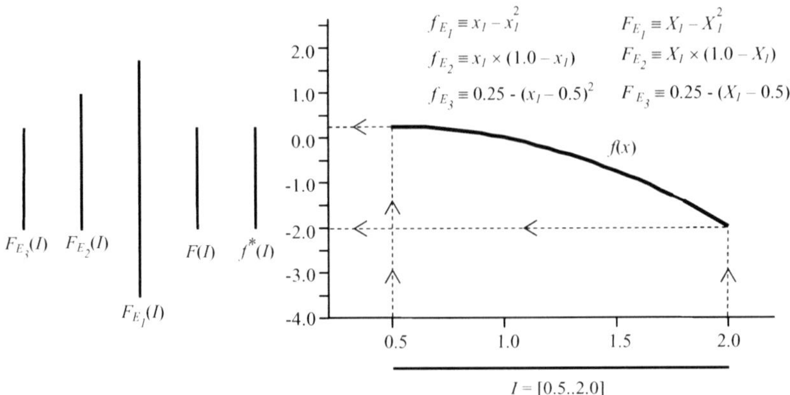

Figure 3.4 Natural interval extensions of a real function wrt three equivalent real expressions.

Because any interval extension evaluation returns an interval enclosure for the range of *f*, the intersection of different extension evaluations will possibly provide a better enclosure.

Theorem 3.2.1-3 (Intersection of Interval Extensions). Let F_1 and F_2 be two *n*-ary interval functions and *B* an *n*-ary *R*-box. Let *F* be an *n*-ary interval function defined by: $F(B)=F_1(B)\cap F_2(B)$. If F_1 and F_2 are interval extensions of the real function *f*, then *F* is also an interval extension of *f*.   ❏

Moreover, the overestimation of the enclosure may be reduced if, instead of evaluating an expression over an *R*-box, the box is firstly split into sub-boxes (whose union is the original *R*-box) and the union of all the intervals obtained by the separate evaluation of each sub-box is considered.

Theorem 3.2.1-4 (Decomposed Evaluation of an Interval Extension). Let *F* be an interval extension of the *n*-ary real function *f*, and F_E an interval expression representing *F*. Let *B*, B_1 and B_2 be *n*-ary *R*-boxes. If $B=B_1\cup B_2$ then:
$$F(B) \subseteq F_E(B_1)\cup F_E(B_2) \subseteq F_E(B)$$
❏

Figure 3.5 exemplifies the decomposed evaluation of the interval expression F_{E_1}.

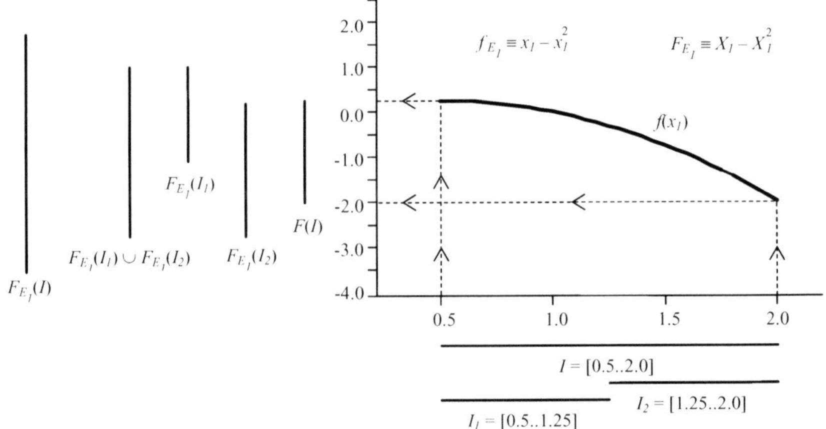

Figure 3.5 The decomposed evaluation of an interval expression.

It is shown that by dividing the argument I=[0.5..2.0] into I_1=[0.5..1.25] and I_2=[1.25..2.0] the overestimation of the $F(I)$ enclosure is reduced: $F(I) \subseteq F_{E_1}(I_1) \cup F_{E_1}(I_2) \subseteq F_{E_1}(I)$.

The quality of the enclosure for the ranges of a real function depends on the way that its interval extension is expressed. When a variable occurs repeatedly in an interval expression, its interval arithmetic evaluation usually produces an overestimated result due to the problem known as the dependency problem.

> **Definition 3.2.1-3 (Dependency Problem).** In the interval arithmetic evaluation of an interval expression, each occurrence of the same variable is treated as a different variable. The dependency between the different occurrences of a variable in an expression is lost. ❏

Informally, if each basic interval operator is able to compute the exact ranges for its interval arguments then the overestimation of the interval arithmetic evaluation must be a consequence of considering real valued combinations where different occurrences of the same variable are assigned with different real values. No overestimation would be obtained if an interval expression, where no variable occurs more than once, is evaluated with infinite precision. This is expressed in an important theorem of Interval Analysis, firstly proved by Moore [99].

> **Theorem 3.2.1-5 (No Overestimation Without Multiple Variable Occurrences).** Let F_E be an interval expression representing the *n*-ary interval function F, and B an *n*-ary *R*-box. If F_E is an interval expression in which each variable occurs only once then:
> $F(B) = F_E(B)$ (w/ exact interval operators and infinite precision arithmetic evaluation)❏

In the example of figure 3.4 we can see that no overestimation is obtained with the interval arithmetic evaluation of expression F_{E_3} because in this expression variable X_1 occurs only once.

For a particular real function the best interval extensions, in terms of minimal overestimation, are those that minimize the effects of the dependency problem. Although it is often impossible to get interval extensions without repeated variable occurrences, interval extensions with fewer multiple occurrences are usually preferable.

Several typical forms of interval extensions studied in interval analysis were used in interval constraints to take advantage of their specific properties. In the following we will consider different forms of interval extensions for univariate real functions, however these forms can be easily extended for the multivariate case (see [66] for further details on this issue).

Centered forms, introduced by Moore in [99], may be used to get tighter enclosures when the interval domains of the variables are sufficiently small. The key idea is that for any real function f there is a real function g such that:

$f(x_1) = f(c) + g(x_1 - c)$ where c is a real value at which f is defined.

For an original expression f_E representing f and a particular interval I, a generic procedure is described in [99] to obtain an expression f_c representing $f(c) + g(x_1 - c)$, with c defined as the center of I (c=*center*(I)). The Centered interval extension of f wrt this interval I is the Natural interval extension F_c of the expression obtained by algebraically simplifying the expression f_c. F_c is an interval extension of f since it is an interval extension of f_c which is equivalent to f_E.

Consider function f represented by the real expression $f_E \equiv x_1 - x_1^2$ (see figures 3.3 through 3.5). An equivalent representation of f for a given interval I=[$a..b$] is $f_c \equiv c - c^2 + (x_1 - c) \times ((1 - (a+b)) - (x_1 - c))$. For the particular interval [0.5..2], the center c is 1.25 and the Centered interval extension F_c is expressed by -0.3125+(X_1-1.25)×(-X_1-0.25) whose evaluation results in the interval [−2.375..1.75] (which is a better enclosure of $f^*([0.5..2])$ than the obtained by the natural interval extension of $x_1-x_1^2$). For the particular interval [0..1] the center c is 0.5 and the Centered interval extension F_c is expressed by $0.25-(X_1-0.5)^2$ whose evaluation results in the interval [0..0.25] (which is the exact range $f^*([0..1])$). Note that the Centered interval extensions are not guaranteed to be inclusion monotonic since the interval expression of F_c is different for different interval arguments.

It was proved in [64] that a Centered interval extension approximates the range of a function quadratically as the width of the interval argument tends to zero. This property makes them particularly useful when the width of the interval is small. With large intervals the enclosure obtained by the Centered interval extension may be even worst than the obtained by the Natural interval extension of the original form.

Another class of centered forms is derived from the mean value theorem [3, 120].

Consider a univariate function f differentiable over the interval I. By the mean value theorem, for any real value $r \in I$ and a fixed $c \in I$ there is a ξ between r and c such that:

$$f(r)= f(c)+f'(\xi)\times(r-c) \qquad \text{where } f' \text{ is the derivative of } f$$

Let F and F' be interval extensions of f and f' respectively. By definition of an interval extension $f(c)\in F([c])$ and $f'(\xi)\in F'([\xi])\subseteq F'(I)$ thus:

$$\forall_{r\in I}\; f(r) \in F([c])+F'(I)\times(r-c)$$

The interval function F_m defined by $F([c])+F'(I)\times(I-c)$ where $c=center(I)$ is called the Mean Value interval extension of f wrt I. The above property guarantees that F_m is a interval extension of f for interval I (see definition 3.2.1-1). The Mean Value interval extension F_m depends on the evaluation of the interval extension F' chosen for the derivative of f. Different expressions for F' lead to different values for the mean value form.

In the case of a real function f represented by the real expression $x_1-x_1^2$, consider the interval extensions F and F' represented respectively by $X_1-X_1^2$ and $1-2X_1$. The Mean Value interval extension of f for a given interval I is expressed by $c-c^2+(1-2X_1)\times(X_1-c)$ where $c=center(I)$. For the particular interval [0.5..2], F_m is expressed by $-0.3125+(1-2X_1)\times(X_1-1.25)$ whose evaluation results in the interval [−2.5625..1.9375]. For the particular interval [0..1], F_m is expressed by $0.25+(1-2X_1)\times(X_1-0.5)$ whose evaluation results in the interval [−0.25..0.75].

The Mean Value interval extension, which is proved to be inclusion monotonic, is easier to compute than the Centered interval extension (does not require any algebraic manipulations) and possesses the quadratic approximation property too [3, 120].

The Mean Value interval extension is just a special case of the Taylor interval extension, which can be derived from the Taylor series expansion of a function f about a point c. Assuming that f has continuous derivatives $f^{(i)}$ of any order $i\leq m+1$ within interval I then, for any real value $r\in I$ and a fixed $c\in I$ there is a $\xi\in I$ such that:

$$f(r) = f(c)+\sum_{i=1}^{m}\frac{(r-c)^i f^{(i)}(c)}{i!}+\frac{(r-c)^{m+1} f^{(m+1)}(\xi)}{(m+1)!}$$

Let F and $F^{(i)}$ with $1\leq i\leq m+1$ be interval extensions of f and $f^{(i)}$ respectively. The Taylor interval extension of order m of the real function f wrt I is the interval function, denoted $F_{t(m)}$, and defined by:

$$F([c])+\sum_{i=1}^{m}\frac{(X_1-c)^i F^{(i)}([c])}{i!}+\frac{(X_1-c)^{m+1} F^{(m+1)}(X_1)}{(m+1)!} \qquad \text{where } c=center(I)$$

The Taylor expansion guarantees that $F_{t(m)}$ is a interval extension of f for interval I, and in the particular case of $m=0$, the Taylor interval extension $F_{t(0)}$ is the Mean Value interval extension F_m.

In the previous example, $F^{(2)}$ is -2 and $F^{(i)}$ is 0 for $i>2$. Consequently, the Taylor interval extension of order $m\geq 1$ for a given interval I is expressed by $c-c^2+(1-2c)\times(X_1-c)-$

$(X_I-c)^2$ where $c=center(I)$. For the particular interval $[0.5..2]$, $F_{t(m)}$ (with $m\geq1$) is expressed by $-0.3125-1.5\times(X_I-1.25)-(X_I-1.25)^2$ whose evaluation results in the interval $[-2..0.8125]$. For the particular interval $[0..1]$, $F_{t(m)}$ (with $m\geq1$) is expressed by $0.25-(X_I-0.5)^2$ whose evaluation results in the interval $[0..0.25]$.

Non centered forms, as the distributed form, are also used in interval constraints [134]. An univariate real function f is expressed in the distributed form f_d if:

$$f_d \equiv \sum_{i=1}^{k} r_i x_1^{n_i} \quad \text{where } r_i \text{ and } n_i \text{ are real and natural numbers, respectively, and } n_i{\neq}n_j \text{ if } i{\neq}j$$

The Distributed interval extension F_d of a real function f is the Natural interval extension of its distributed form. Not all real functions can be expressed in the distributed form and so not all real functions have a Distributed interval extension. However, distributed forms can be used for the canonical representation of polynomials. The enclosures obtained by the Distributed interval extension are often worst than those obtained by the Natural interval extension of the original expression since it imposes several (k) occurrences of the same variable, possibly increasing the dependency problem. The use of the Distributed interval extension in interval constraints is justified, not by the precision of the obtained enclosures, but rather by its efficiency w.r.t. the constraint techniques that will be discussed in the next chapter, in particular, the constraint Newton method (subsection 4.2.2).

In the case of the example that has been presented, the real expression $f_E{\equiv}x_I-x_I^2$ is already in the distributed form, and so the Distributed interval extension F_d of the real function f is its Natural interval extension expressed by $X_I-X_I^2$. As can be seen in figure 3.4 this extension (F_{E_1} in the figure) obtains less accurate enclosures of the range of f for the interval $[0.5..2.0]$ than the other non distributed interval extensions (F_{E_2} and F_{E_3}).

3.3   Interval Methods

Interval methods for finding roots of equations with one variable are frequently used in interval constraints due to its efficiency and reliability. In particular, the interval Newton method, a generalization of the Newton's method for finding zeros of real functions, is the most commonly used since it presents several remarkable properties quite useful for filtering algorithms.

Extensions of the univariate root finding methods to the multivariate case are rarely used in interval constraints as they are more complex to implement. Instead interval propagation techniques are used, generally with better performance. However, in some situations, they can be effectively applied. In particular, in cases where the search space is a small box enclosing a single root, multivariate interval methods may be quite useful either for isolating the root or for proving its existence and uniqueness.

3.3.1   Univariate Interval Newton Method

The interval Newton method for searching zeros of univariate functions, introduced by Moore in [99], is based on the mean value theorem which can be formulated as:

$$\exists_{\xi\in[a..b]} \quad f(r_1)-f(r_2)=(r_1-r_2)\times f'(\xi) \quad\quad (a=min(r_1,r_2) \text{ and } b=max(r_1,r_2))$$

where f is a real function, continuous in $[a..b]$ and differentiable in $(a..b)$.

Let I be a closed real interval containing both r_1 and r_2. If the real function f is continuous and differentiable in I then from the mean value theorem:

$$\forall_{r_1,r_2\in I} \exists_{\xi\in I} \quad f(r_1)-f(r_2)=(r_1-r_2)\times f'(\xi)$$

If there is a zero of f in I ($\exists_{r_0 \in I}\ f(r_0)=0$), let us represent it by r_0 and instanciate the universal quantifier for r_2 with it in the above first order formula:

$\forall_{r_1 \in I}\ \exists_{\xi \in I}\ f(r_1) = (r_1 - r_0) \times f'(\xi)$

Instanciating r_1 with any real value c within I, for example its mid value ($c=center(I)$):

$\exists_{\xi \in I} f(c)=(c-r_0) \times f'(\xi)$

Solving the equation for r_0 and considering $g(x) = c - \dfrac{f(c)}{f'(x)}$ we obtain:

$\exists_{\xi \in I}\ [(r_0 = g(\xi) \wedge f'(\xi) \neq 0) \vee (f(c)=0 \wedge f'(\xi)=0)]$

Which is equivalent to:

$(f(c) \neq 0 \vee \forall_{r \in I}\ f'(r) \neq 0) \Rightarrow \exists_{\xi \in I}\ (r_0 = g(\xi) \wedge f'(\xi) \neq 0)$

The universal quantified condition $\forall_{r \in I}\ f'(r) \neq 0$ expresses that the real value zero does not belong to the range of f over I. Similarly, the right side of the implication $\exists_{\xi \in I}\ (r_0 = g(\xi) \wedge f'(\xi) \neq 0)$ implies that r_0 must be within the range of g over I. Consequently we will finally get the following formula:

$(f(c) \neq 0 \vee 0 \notin f'^*(I)) \Rightarrow r_0 \in g^*(I)$ **(1)**

From the above implication we can conclude that, if there is a zero r_0 of f within I and if the left side of the implication is true, then r_0 must also be within the range $g^*(I)$ of the real function g over the interval I. Thus, in this case, and according to theorem 3.2.1-1, any zero of f within I must also be within any extension G of the real function g. Moreover, if $I \subseteq G(I)$ when the left side of the implication is false then, in all cases, it follows that whenever a zero of f is within I then it must be also within $G(I)$: $\forall_{r_0 \in I}\ f(r_0)=0 \Rightarrow r_0 \in G(I)$.

The interval Newton method consists on applying function G over a real interval I and intersecting the result with the original interval to obtain a set S with the guarantee that if a zero is within I then it must also be within S. ($S=I \cap G(I)$).

The extension G of the real function g originally proposed in [99], is the Newton function:

Definition 3.3.1-1 (Newton Function). Let f be a real function, continuous and differentiable in the closed real interval I, and f' its derivative. Let F and F' be interval extensions of f and f', respectively. Let c be the mid value of the interval I ($c=center(I)$).

The interval Newton function N with respect to f is: $N(I) = [c] - \dfrac{F([c])}{F'(I)}$ ❏

In the original proposal the interval division operator was not defined if the denominator was an interval containing zero (see definition 3.1-1). Thus the Newton function was not defined if $0 \in F'(I)$ and so, it could only be applied when the left side of the implication in **(1)** was true because if $0 \notin F'(I)$ then necessarily $0 \notin f'^*(I)$ (according to theorem 3.2.1-1: $f'^*(I) \subseteq F'(I)$).

The Newton method would then consist on iterating the following Newton step only on intervals where the Newton function was defined.

Definition 3.3.1-2 (Newton Step). Let f be a real function, continuous and differentiable in the closed real interval I. Let N be the Newton function with respect to f. The Newton step function NS with respect to f is:

$NS(I) = I \cap N(I)$ ❏

If division by intervals containing zero is not allowed, then the result of the Newton function is a single interval. Consequently the result of the Newton step is also a single interval ($NS(I){\subseteq}I$) and the successive iteration of Newton steps until a fix point is obtained may be defined as an interval narrowing function NN (narrowing because $NN(I){\subseteq}I$).

Definition 3.3.1-3 (Newton Narrowing)[4]. Let f be a real function, continuous and differentiable in the closed real interval I. Let NS be the Newton step function with respect to f. The Newton narrowing function NN with respect to f is:

$$NN(I) = \begin{cases} \varnothing & \text{if} & NS(I){=}\varnothing \\ I & \text{if} & NS(I){=}I \\ NN(NS(I)) & \text{if} & NS(I){\subset}I \end{cases}$$ ❑

The original restriction on the domain of definition for the Newton function restrained the Newton narrowing to intervals where the Newton function is defined. Whenever an interval failed that restriction the narrowing could not be applied to the entire interval and further analysis would require its partition.

With extended interval arithmetic the division by an interval containing zero became possible and unrestricted approaches for the application of the Newton function were firstly proposed in [2] and [65]. Extended interval arithmetic not only guarantees that the Newton function N is an extension of the real function g if the left side of the implication in **(1)** is true but also, in the special case where it is false it guarantees that $I{\subseteq}N(I)$. This last property derives from the fact that if it is false then $f(c){=}0$ and $0{\in}f'^{*}(I)$, consequently both $F([c])$ and $F'(I)$ contain zero, and the evaluation of $F([c])/F'(I)$ would return the interval $(-\infty,+\infty)$ so, in this case, $I{\subseteq}N(I){=}(-\infty,+\infty)$.

Using extended interval arithmetic, the result of the Newton function is not guaranteed to be a single interval. As seen in section 3.1, the division by an interval containing zero may yield the union of two intervals. In this case the Newton function is not even an interval function in the form of definition 3.2-2 since it does not map an R-box into a single interval. Several different strategies were devised to deal with this problem.

One strategy is to work with the two obtained intervals and, in the Newton step, intersect their union with the original interval. If the result of the intersection is a single interval, the Newton narrowing can normally continue. Otherwise, the union hull ($\uplus$) of the obtained intervals could be considered[5] or else, the two obtained intervals would have to be considered separately. Other strategy is to redefine the extended interval arithmetic division rules to always obtain a single interval. If the result were the union of two intervals then the result is redefined to be the single interval obtained by their union hull.

Figure 3.6 shows an example of the application of the interval Newton method for isolating a zero of a real function within an interval. The real function f (solid line) is the same presented in figures 3.3 through 3.5, represented by the expression $f_E \equiv x_1{-}x_1^2$ and the initial interval that is narrowed without loosing any zero of f is [0.5..2.0]. The derivative of f is function f' (dashed line) represented by the expression $f'_E \equiv 1{-}2x_1$. The Natural interval extensions of f and f' are $F_E \equiv X_1{-}X_1^2$ and $F'_E \equiv 1{-}2X_1$, respectively. The Newton function N is defined according to the definition 3.3.1-1. The table in figure 3.6 summarizes the results obtained by successive application of the Newton step[6]. In each line i is shown the current interval (I_i), the intervals needed to compute the Newton function ($[c_i]$, $F_E([c_i])$ and $F'_E(I_i)$)

---

[4] This definition assumes that the result of the Newton step function is always an interval.

[5] As suggested in section 3.1, the whole Newton step function could be considered a interval arithmetic operator internally implemented with the extended interval arithmetic rules.

[6] It is assumed in all the interval arithmetic evaluations that the distance between any two consecutive F-numbers is the constant 0.001 (for any F-number f: $f^{+}{-}f = 0.001$)

and the interval obtained by applying the Newton function to the current interval ($N(I_i)$). The initial interval (I_0) is [0.5..2.0] and for each line $i>0$ the current interval is updated accordingly to the Newton step $I_i = I_{i-1} \cap N(I_{i-1})$. The process stops at line 4 since $I_5 = I_4 \cap N(I_4) = I_4$ (see definition 3.3.1-3 of the Newton narrowing function). As can be seen in the figure, the Newton narrowing with argument [0.5..2.0] quickly converges to the unique zero of the real function f within this interval: $NN([0.5..2.0]) = [1.000..1.000]$.

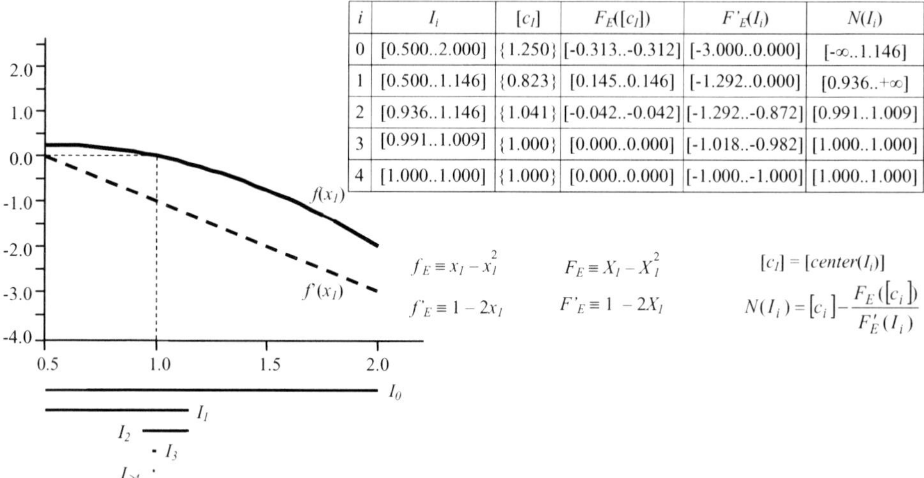

i	I_i	$[c_I]$	$F_E([c_I])$	$F'_E(I_i)$	$N(I_i)$
0	[0.500..2.000]	{1.250}	[-0.313..-0.312]	[-3.000..0.000]	[-∞..1.146]
1	[0.500..1.146]	{0.823}	[0.145..0.146]	[-1.292..0.000]	[0.936..+∞]
2	[0.936..1.146]	{1.041}	[-0.042..-0.042]	[-1.292..-0.872]	[0.991..1.009]
3	[0.991..1.009]	{1.000}	[0.000..0.000]	[-1.018..-0.982]	[1.000..1.000]
4	[1.000..1.000]	{1.000}	[0.000..0.000]	[-1.000..-1.000]	[1.000..1.000]

$$f_E \equiv x_I - x_I^2 \qquad F_E \equiv X_I - X_I^2 \qquad [c_I] = [center(I_i)]$$

$$f'_E \equiv 1 - 2x_I \qquad F'_E \equiv 1 - 2X_I \qquad N(I_i) = [c_i] - \frac{F_E([c_i])}{F'_E(I_i)}$$

Figure 3.6 An example of the application of the interval Newton method.

The remarkable properties of the interval Newton method, specially suited for filtering algorithms will be addressed in the following theorems (firstly formulated by Moore [99]).

Theorem 3.3.1-1 (Soundness of the Interval Newton Method with Roots). Let f be a real function, continuous and differentiable in the closed real interval I. If there exists a zero r_0 of f in I then r_0 is also in $N(I)$, $NS(I)$ and $NN(I)$, where N, NS and NN are respectively the Newton function, the Newton step function and the Newton narrowing function with respect to f:

$$\forall_{r_0 \in I} \; f(r_0)=0 \Rightarrow r_0 \in N(I) \wedge r_0 \in NS(I) \wedge r_0 \in NN(I) \qquad \square$$

The above theorem guarantees the soundness of the Newton narrowing function (and its interval arithmetic evaluation) for narrowing the search space of a possible zero of a real function. If a zero of a function is searched within I then it may be searched within $NN(I)$ which is possibly a narrower interval. This narrowing is a proof that no zero of the function was within the discarded values of I.

Theorem 3.3.1-2 (Soundness of the Interval Newton Method without Roots). Let f be a real function, continuous and differentiable in the closed real interval I. If $NS(I)=\varnothing$ or $NN(I)=\varnothing$ (where NS and NN are respectively the Newton step function and the Newton narrowing function with respect to f) then there is no zero of f in I:

$$NS(I)=\varnothing \vee NN(I)=\varnothing \Rightarrow \neg \exists_{r_0 \in I} \; f(r_0)=0 \qquad \square$$

The above theorem guarantees that whenever the result of the Newton narrowing function (or its interval arithmetic evaluation) is the empty set ($NN(I)=\varnothing$), I does not provedly contains any zero of the real function.

Despite its soundness, the method is not complete, that is, in case of non existence of a root within an *R*-box, the result of the Newton narrowing function is not necessarily the empty set. Therefore obtaining a non empty set does not guarantee the existence of a root. However, in some cases, the Newton method may guarantee the existence of a root. The next theorem is due to Hansen [64]:

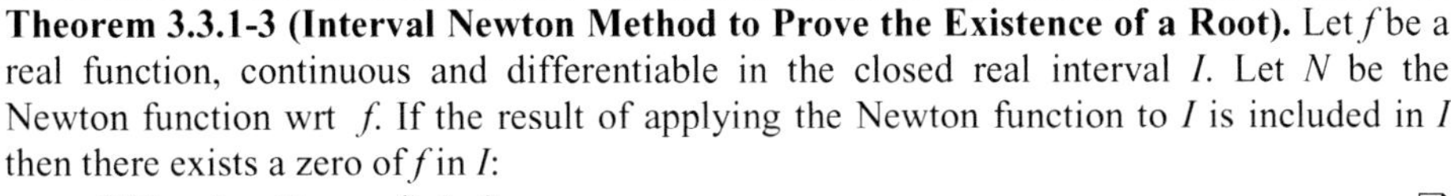

Theorem 3.3.1-3 (Interval Newton Method to Prove the Existence of a Root). Let *f* be a real function, continuous and differentiable in the closed real interval *I*. Let *N* be the Newton function wrt *f*. If the result of applying the Newton function to *I* is included in *I* then there exists a zero of *f* in *I*:
$$N(I) \subseteq I \Rightarrow \exists_{r_0 \in I}\ f(r_0){=}0 \qquad \square$$

The convergence theorem, adapted from [65], assures that the interval arithmetic evaluation of any Newton narrowing function is guaranteed to stop:

Theorem 3.3.1-4 (Convergence of the Interval Newton Method). Let *f* be a real function, continuous and differentiable in the closed real interval *I*. The interval arithmetic evaluation of the Newton narrowing function (*NN*) with respect to *f* will converge (to an *F*-interval or the empty set) in a finite number of Newton steps (*NS*). $\qquad \square$

The following two theorems address the efficiency of the interval Newton method. They are adapted from the original ones, formulated by Moore [99] and Hansen [66] respectively.

Theorem 3.3.1-5 (Efficiency of the Interval Newton Method - Quadratic). Let *f* be a real function, continuous and differentiable in the closed real interval *I*. Let *f'* be the derivative function of *f* and *F'* its interval extension. Let *NS* be the Newton step function with respect to *f*. If *f* has a simple zero r_0 in *I* and $0 \notin F'(I)$ then the Newton narrowing (with infinite precision arithmetic) is asymptotically error-squaring, i.e. there is an interval $I_0 \subseteq I$ containing r_0 and a positive real number *k* such that:
$$width(NS^{(n+1)}(I_0)) \leq k{\times}(width(NS^{(n)}(I_0)))^2 \quad \text{w/ } NS^{(n)}(I){=}NS(NS^{(n-1)}(I)) \text{ and } NS^{(1)}(I){=}NS(I) \quad \square$$

Theorem 3.3.1-6 (Efficiency of the Interval Newton Method - Geometric). Let *f* be a real function, continuous and differentiable in the closed real interval *I*. Let *f'* be the derivative function of *f*. Let *F* and *F'* be respectively interval extensions of *f* and *f'*. Let *c* be the mid value of the interval *I* (*c=center(I)*). Let *NS* be the Newton step function with respect to *f*. If $0 \notin F([c])$ and $0 \notin F'(I)$ then:
$$width(NS(I)) \leq 0.5{\times}width(I) \qquad \square$$

The first of the two efficiency theorems states that convergence may be quadratic for small intervals around a simple zero of the real function whereas the second theorem says that even for large intervals the rate of convergence may be reasonably fast (geometric).

In the above presentation we assume that the interval Newton's method is applied to a real function as defined in 3.2-2. However, the method can be naturally extended to deal as well with real functions that include parametric constants represented by intervals [66]. In this case, the intended meaning is to represent the family of real functions defined by any possible real valued instantiation for the interval constants. The existence of a root means that there is a real valued combination, among the variable and all the interval constants, that zeros the function.

Figure 3.7 shows an example of the application of the interval Newton method for enclosing the zeros of the family of functions defined by $f(x_1){=}x_1{\times}([0.5..1.5]{-}x_1)$. Any instantiation of the interval constant with a real value within [0.5..1.5] produces a function whose graphic lies within the two solid lines represented in the figure (a particular instantiation with the real value 1 is represented in figure 3.3). Consequently, the interval arithmetic evaluations of $F_E \equiv X_I \times([0.5..1.5]-X_I)$ and $F'_E \equiv [0.5..1.5]-2X_I$ define interval extensions for this family of functions and their derivatives, respectively. The zeros of this family of functions are explicitly represented in the figure as a point {0.0} and a thick horizontal line [0.5..1.5]. At the bottom of the figure are represented 4 cases of the application of the Newton narrowing function with different initial intervals (the precision assumptions are the same as in the example of figure 3.6). In the first case the initial interval was [-0.5..0.2] and the unique zero was successfully enclosed within a canonical *F*-interval [0..0.001]. In the second case the initial interval was [0.3..1.0] which could not be narrowed because both $F_E([0.65])$ and $F'_E([0.3..1.0])$ include zero. In the third case the initial interval was [1.1..1.8] and the right bound was updated to 1.554. In the fourth case the initial interval was [1.9..2.6] and it was proven that there is no zero within this interval.

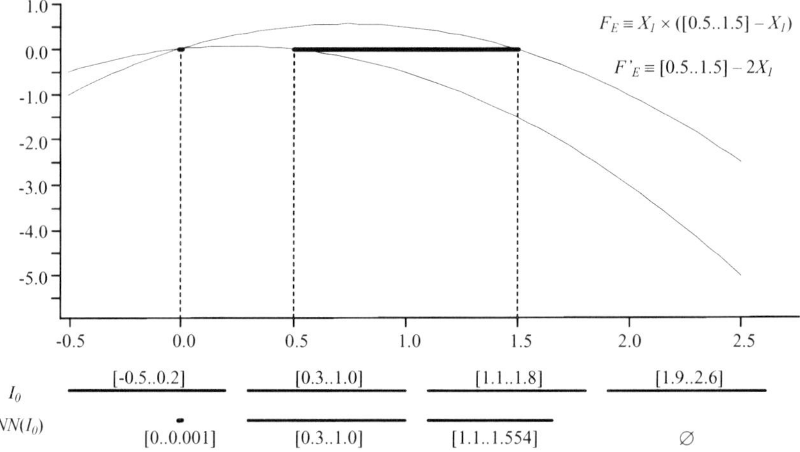

Figure 3.7 The interval Newton method for enclosing the zeros of a family of functions.

3.3.2   *Multivariate Interval Newton Method*

The interval Newton method extended for the multivariate case may be used to solve systems of *n* equations of the form:

$f_i(B)=0$, where $i=1,...,n$ and *B* is an *n*-ary *R*-box.

Similarly to the reasoning for the univariate case, it is possible to conclude that all solutions of the above system must lie within the set *S* defined in vector notation by:

$S = \{y \in B : f(c)+J(x)(y-c)=0, x \in B\}$ where $c \in B$ and *J* is the Jacobian of *f*.

The idea is the same of the univariate method, that is, to define a Newton function whose result, when applied to an *R*-box, includes all the possible solutions of the system within that box. Intersecting the result with the original box and iterating the process, the search space will be eventually reduced.

The formal definitions of the Newton step and Newton narrowing functions (*NS* and *NN*) are almost identical to the univariate case (definitions 3.3.1-2 and 3.3.1-3 respectively) except that their mappings are, in the *n* dimensional case, from *n*-ary *R*-boxes into *n*-ary *R*-boxes. The definition of the Newton function distinguishes the various multidimensional interval Newton methods.

In the first proposal [99] the Newton function was a direct extension of that for the univariate case, written in vector notation as:

$$N(B) = C - V(B)F(C)$$

where: B is an n-ary R-box; C is a degenerate R-box including the mid point of B; F is a vector with elements F_i which are interval extensions of f_i and; $V(B)$ is an interval matrix containing the inverse of every real matrix within $J(B)$ (the extension of the Jacobian matrix of f).

In order to compute the interval matrix $V(B)$ the nonsingularity of the matrices within $J(B)$ was required, implying that the method can only be directly applied (without split) if there is at most one isolated root within B.

Other proposals avoid the accurate computation of the inverse interval matrix which is a quite complex process and not always achievable. Instead, these proposals, use efficient interval approximation methods to solve, for $N(B)$, the linear equation $F(C)+J(B)(N(B)-C)=0$, which defines an enclosing box for the solution set. The resulting Newton function may return less accurate enclosing boxes but it can be applied without restrictions and computed more efficiently. The first proposal, due to Krawczyk [86] and called the Krawczyk method, was later improved by Hansen and Sengupta [67] with a faster method known as the Hansen and Sengupta Gauss-Seidel method.

The properties presented before for the unidimensional case may be extended to the interval Newton methods for the multidimensional case. An exception is made for the efficiency properties of which only the asymptotic quadratic convergence was proven [3] for the special case where the inverse interval matrix $V(B)$ can be obtained. This last property makes the multidimensional interval Newton methods particular effective for narrowing small boxes enclosing a single root, whose existence may eventually be proven during the narrowing process (see theorem 3.3.1-3). However, for large boxes, the narrowing achieved by these methods apparently does not justify the computational costs of their implementation, at least compared with competing constraint propagation techniques.

3.4   Summary

In this chapter, interval analysis techniques relevant for the interval constraints framework were introduced. Interval arithmetic and its main properties were presented. Interval functions and interval expressions were defined and related with the sound evaluation of the range of a real function through the basic concept of an interval extension of the real function. The properties of interval extensions were analysed and several important forms identified. The interval Newton method, widely used in the interval constraints framework, was described and its fundamental properties analysed. The next chapter will address constraint propagation approaches which extensively use interval analysis techniques to guarantee their correctness.

Chapter 4

Constraint Propagation

The filtering algorithms used in Interval Constraints for pruning the variable domains are based on constraint propagation techniques initially developed in Artificial Intelligence for finite domains. They use partial information expressed by a constraint to eliminate some incompatible values from the domain of the variables within the scope of the constraint. Once the domain of a variable is reduced, this information is propagated to all constraints with that variable in their scopes, which must be checked again possibly to further reduce the domains of the other constrained variables. The constraint propagation is terminated when a fixed-point is attained, that is, the variable domains can not be further reduced by any constraint.

In the next section the propagation process is described as successive reduction of variables domains by successive application of narrowing functions associated with the constraints of the CCSP. The properties of the propagation algorithm are derived from the properties of the narrowing functions used for pruning the variable domains.

In the interval constraints framework, a set of narrowing functions is associated to each constraint of the CCSP. The evaluation of each narrowing function is accomplished by algorithms based on Interval Analysis techniques. Section 4.2 discusses the main approaches used to associate narrowing functions to constraints and the algorithms used for their evaluation.

4.1 The Propagation Algorithm

The overall functioning of the propagation algorithm used for pruning the variable domains is based on narrowing functions. A narrowing function is a mapping between elements of the domains lattice where the new element is obtained from the original by eliminating some value combinations incompatible with a particular constraint of the CCSP.

Definition 4.1-1 (Narrowing Function). Let $P=(X,D,C)$ be a CCSP. A narrowing function NF associated with a constraint $c=(s,\rho)$ (with $c \in C$) is a mapping between elements of 2^D (Domain$_{NF} \subseteq 2^D$ and Codomain$_{NF} \subseteq 2^D$)[1] with the following properties (where A is any element of Domain$_{NF}$):

    P1) $NF(A) \subseteq A$ (contractance)

    P2) $\forall_{d \in A}\ d \notin NF(A) \Rightarrow d[s] \notin \rho$ (correctness)   ❏

Property **P1** from the above definition assures that the new element is not larger (wrt set inclusion) than the original element, which guarantees that a fixed-point will be eventually reached when the narrowing functions are successively applied. Property **P2** guarantees the correctness of the application of a narrowing function since every value combination eliminated does not satisfy the constraint c (see definition 2-4) and so, cannot be a solution of the CCSP.

---

[1] We further impose that for any narrowing function the domain and the codomain must be the same subset of the representable elements of 2^D (the reachable sub-lattice - see sub-section 2.2.5).

Monotonicity and idempotency are additional properties common to most of the narrowing functions used in interval constraints. Several authors [114, 12] denominate *narrowing operators* the narrowing functions which satisfy both monotonicity and idempotency (or at least monotonicity [9]).

> **Definition 4.1-2 (Monotonicity and Idempotency of Narrowing Functions).** Let $P=(X,D,C)$ be a CCSP. Let NF be a narrowing function associated with a constraint of C. Let A_1 and A_2 be any two elements of Domain$_{NF}$. NF is respectively monotonic and idempotent iff the following properties hold:
>
> **P3**) $A_1 \subseteq A_2 \Rightarrow NF(A_1) \subseteq NF(A_2)$ (monotonicity)
> **P4**) $NF(NF(A_1)) = NF(A_1)$ (idempotency)   ❑

An important concept related with the narrowing functions is the notion of a fixed-point. For a particular element of the domain of a narrowing function, the set of all fixed-points smaller (wrt set inclusion) than this element may be defined.

> **Definition 4.1-3 (Fixed-Points of Narrowing Functions).** Let $P=(X,D,C)$ be a CCSP. Let NF be a narrowing function associated with a constraint of C. Let A be an element of Domain$_{NF}$. A is a fixed-point of NF iff:
>
> $NF(A) = A$.
>
> The set of all fixed-points of NF within A, denoted Fixed-Points$_{NF}(A)$, is the set:
>
> Fixed-Points$_{NF}(A) = \{ A_i \in \text{Domain}_{NF} \mid A_i \subseteq A \ \wedge \ NF(A_i) = A_i \}$ ❑

The following theorem, based on [114], asserts that the union of all fixed-points of a monotonic narrowing function within an element of its domain is itself a fixed-point which is the greatest fixed-point within the element.

> **Theorem 4.1-1 (Union of Fixed-Points)**[2]. Let $P=(X,D,C)$ be a CCSP. Let NF be a monotonic narrowing function associated with a constraint of C, and A an element of Domain$_{NF}$. The union of all fixed-points of NF within A, denoted $\cup$Fixed-Points$_{NF}(A)$, is the greatest fixed-point of NF in A:
>
> $\cup$Fixed-Points$_{NF}(A) \in$ Fixed-Points$_{NF}(A)$
>
> $\forall_{A_i \in \text{Fixed-Points}_{NF}(A)} \ A_i \subseteq \cup$Fixed-Points$_{NF}(A)$ ❑

From the above theorem it is possible to prove that the contraction resulting from the application of a monotonic narrowing function to an element of its domain is limited by the greatest fixed-point within the element. In other words, no value combination included in the greatest fixed-point may be discarded in the contraction. Moreover, if the monotonic narrowing function is idempotent, the result of the contraction is precisely the greatest fixed-point within the element of application.

> **Theorem 4.1-2 (Contraction Applying a Narrowing Function).** Let $P=(X,D,C)$ be a CCSP. Let NF be a monotonic narrowing function associated with a constraint of C and A an element of Domain$_{NF}$. The greatest fixed-point of NF within A is included in the element obtained by applying NF to A:
>
> $\cup$Fixed-Points$_{NF}(A) \subseteq NF(A)$
>
> In particular, if NF is also idempotent then:
>
> $\cup$Fixed-Points$_{NF}(A) = NF(A)$ ❑

---

[2] The theorems presented in this Chapter are proved in Appendix B.

In the interval constraints framework, each constraint originates several narrowing functions responsible for the elimination of some incompatible value combinations. Function *prune*, implemented in pseudo-code in figure 4.1, describes the overall propagation algorithm which applies successively each narrowing function until a fixed-point is attained. The algorithm is an adaptation of the original propagation algorithm AC3 [98] used for solving CSPs with finite domains.

---

function *prune*(a set Q of narrowing functions, an element A of the domains lattice)

  (1) $S \leftarrow \varnothing$;

  (2)   **while** $Q \neq \varnothing$ **do**

  (3)   choose $NF \in Q$;

  (4) $A' \leftarrow NF(A)$;

  (5)   **if** $A' = \varnothing$ **then return** $\varnothing$;

  (6) $P \leftarrow \{ NF' \in S: \exists_{x \in \mathrm{Relevant}_{NF'}} A[x] \neq A'[x] \}$;

  (7) $Q \leftarrow Q \cup P$; $S \leftarrow S \setminus P$;

  (8)   **if** $A' = A$ **then** $Q \leftarrow Q \setminus \{NF\}$; $S \leftarrow S \cup \{NF\}$ **end if**;

  (9) $A \leftarrow A'$;

  (10)   **end while**

  (11)   **return** A ;

end function

Figure 4.1 The constraint propagation algorithm.

The first argument Q of the function *prune* is a set of narrowing functions initially composed of all the narrowing functions associated with the constraints of the CCSP. The second argument A is initially instantiated with an element of the domains lattice representing the original variable domains (before applying the propagation algorithm). The result is a smaller (or equal) element of the domains lattice.

The algorithm is based on a cycle (lines 2-10) where, in each step of the cycle, A is narrowed by applying a narrowing function NF selected from Q. During the whole process, set S contains the narrowing functions for which A is necessarily a fixed-point and set Q contains the remaining narrowing functions. Initially there are no guarantees on whether the original domains A is a fixed-point of any narrowing function and so S is empty (line 1) and Q contains all the narrowing functions.

The following is an explanation of the sequence of actions executed at each step of the cycle. If Q is the empty set then A is a fixed-point for all the narrowing functions and it cannot be further pruned by them, so the cycle is terminated (line 2) and A is returned (line 11). If Q is not empty then one of its elements NF, chosen accordingly to a selection criterion (line 3), is applied to A with result A' (line 4). If this is the empty set then it is proved that there is no possible value combination within A capable of satisfying the constraint associated with NF and the execution is terminated (line 5)[3] returning $\varnothing$, which means that A does not contain any solution of the CCSP. Otherwise P is defined (line 6) as a subset of S composed of all the elements of S for which A' is no longer guaranteed to be a fixed-point. These elements, which are the narrowing functions with relevant variables[4] whose domains were changed by applying NF to A, must be moved from S to Q (line 7). If A' is a fixed-point of NF (which is guaranteed if $A=A'$)[5] then NF must be moved from Q to S (line 8). Finally, A is updated with the new narrowed set of domains A' (line 9), the current cycle ends (line 10) and a new step cycle restarts (line 2).

---

[3] Actually, line 5 is only written for clarity, in practice it could be dropped since the empty set is a fixed-point of any narrowing function and in the end the result would be the same.

[4] Let NF be a narrowing function associated with a constraint $c=(s,\rho)$. We will say that a variable x is relevant wrt NF ($x \in \mathrm{Relevant}_{NF}$) iff x is an element of s.

[5] Knowing that NF is idempotent, A' is necessarily a fixed-point of NF, and the if condition of line 8 may be dropped.

From the properties of the narrowing functions it is possible to prove that the propagation algorithm terminates and is correct. Moreover, if all the narrowing functions are monotonic then the propagation algorithm is confluent (the result is independent from the selection criteria used in line 3) and computes the greatest common fixed-point included in the initial domains.

Theorem 4.1-3 (Properties of the Propagation Algorithm). Let $P=(X,D,C)$ be a CCSP. Let set S_0 be a set of narrowing functions (obtained from the set of constraints C). Let A_0 be an element of $Domain_{NF}$ (where $NF \in S_0$) and d an element of D ($d \in D$). The propagation algorithm $prune(S_0, A_0)$ (defined in figure 4.1) terminates and is correct:

$$\forall_{d \in A_0} d \text{ is a solution of the CCSP} \Rightarrow d \in prune(S_0, A_0)$$

If S_0 is a set of monotonic narrowing functions then the propagation algorithm is confluent and computes the greatest common fixed-point included in A_0.   ❑

Although, in the case of monotonic narrowing functions, the selection criterion is irrelevant for the pruning obtained by the propagation algorithm, it may be very important for the efficiency of the propagation. In [92, 93] problems of slow convergence of the propagation algorithm are associated with cyclic phenomena (in the successive application of the narrowing functions) and a revised propagation algorithm is suggested for identifying and simplifying such cyclic phenomena dynamically (dynamically adapting the selection criterion).

4.2   Associating Narrowing Functions to Constraints

In the interval constraints framework, a set of narrowing functions associated with a constraint of the CCSP is defined by considering its projections with respect to each variable in the scope of the constraint.

The projection function identifies from a real box B (representing a set of value combinations between the variables of the scope s of a constraint $c=(s,\rho)$) all the possible values of a particular variable $x_i \in s$ for which there is a value combination belonging to the constraint relation ρ.

Definition 4.2-1 (Projection Function). Let $P=(X,D,C)$ be a CCSP. The projection function with respect to a constraint $c=(s,\rho) \in C$ and a variable $x_i \in s$, denoted $\pi_{x_i}^{\rho}$, obtains a set of real values from a real box B (with $B \in 2^{D[s]}$) and is defined by:

$$\pi_{x_i}^{\rho}(B) = \{ d[x_i] \mid d \in \rho \wedge d \in B \} = (\rho \cap B)[x_i]$$ ❑

Clearly, all value combinations within B with x_i values outside $\pi_{x_i}^{\rho}(B)$ are guaranteedly outside the relation ρ and so they do not satisfy the constraint c.

Figure 4.2 gives an example of a CCSP P with a constraint $x_1 \times (x_2-x_1)=0$ and shows the sets obtained by applying its projection functions on the real box $B=<[-0.5..2.5],[0.5..1.5]>$.

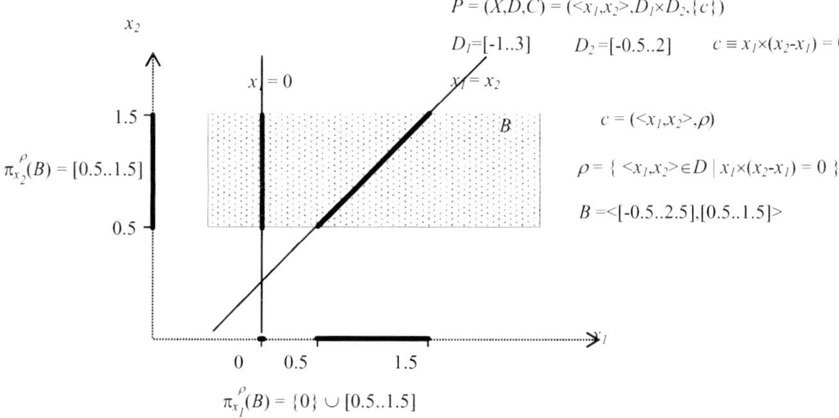

Figure 4.2 An example of a constraint and its projection functions.

Any solution of the constraint must be a point within the line $x_1{=}0$ or within the line $x_1{=}x_2$ (thin solid lines in the figure). These points within the real box B define the segments represented in the figure as thick solid lines within the grey rectangle. The projections of these segments with respect to each variable are the sets obtained by applying the respective projection function to B.

Given the above definition, a box-narrowing function may be defined which narrows the domain of one variable $x_i \in s$ from a box B (representing all the variables of the CCSP) eliminating some values of x_i not belonging to $\pi_{x_i}^{\rho}(B[s])$.

Definition 4.2-2 (Box-Narrowing Function). Let $P{=}(X,D,C)$ be a CCSP (with $X{=}\langle x_1,\ldots,x_i,\ldots,x_n\rangle$). A box-narrowing function with respect to a constraint $(s,\rho)\in C$ and a variable $x_i \in s$ is a mapping, denoted $\mathrm{BNF}_{x_i}^{\rho}$, that relates any F-box $B{=}\langle I_{x_1},\ldots, I_{x_i},\ldots, I_{x_n}\rangle$ ($B\subseteq D$) with the union of m ($1{\leq}m$) F-boxes, defined by:

$$\mathrm{BNF}_{x_i}^{\rho}(\langle I_{x_1},\ldots, I_{x_i},\ldots, I_{x_n}\rangle) = \langle I_{x_1},\ldots, I_1,\ldots, I_{x_n}\rangle \cup \ldots \cup \langle I_{x_1},\ldots, I_m,\ldots, I_{x_n}\rangle$$

satisfying the property:

$$\pi_{x_i}^{\rho}(B[s]) \subseteq I_1 \cup \ldots \cup I_m \subseteq I_{x_i} \qquad \square$$

The box-narrowing functions satisfy both properties of the narrowing functions: contractance follows from the property $I_1\cup\ldots\cup I_m \subseteq I_{x_i}$ (the only changed domain is smaller than the original) and correctness follows from the property $\pi_{x_i}^{\rho}(B[s]) \subseteq I_1 \cup \ldots \cup I_m$ (the eliminated combinations have x_i values outside the projection function). However these properties are insufficient to make them narrowing functions (see definition 4.1-1) since they are not guaranteed to be closed under composition (a box-narrowing function may only be applied to a single F-box but the result may be the union of disjoint F-boxes).

Most approaches solve the above problem by imposing that the result of applying a box-narrowing function to an F-box must be a single F-box. This can always be achieved by substituting the union operations of definition 4.2-2 by union hull operations (see definition 2.2.2-1). With this restriction a box-narrowing function is a narrowing function accordingly to definition 4.1-1 and may be applied within the propagation algorithm. Moreover, the complete set of narrowing functions may be obtained by considering a box-narrowing function associated with each variable of each constraint of the CCSP.

Other approaches [77, 128] that consider structures for representing unions of F-boxes (see subsection 2.2.4), define each narrowing function as a function that applies a box-narrowing function to each F-box represented in the structure. The result is the union of several F-boxes (which is representable by a structure) each one smaller than the original (contractance) and containing the same set of solutions (correctness).

What still remains to be explained is how to obtain an enclosure of a projection function which is necessary to define the associated box-narrowing function. Because the goal is to narrow as much as possible the original domain loosing no solutions, the best possible enclosure is the RF-set approximation (see definition 2.2.3-2) of the set of reals that would be obtained by the projection function. This best possible enclosure cannot be easily obtained for all kinds of constraints so, several interval constraint approaches transform the original CCSP into an equivalent one (see definition 2-7) where all the constraints are in a suitable form for obtaining such enclosure. This method is known as the constraint decomposition method to emphasise the decomposition of a complex constraint into a set of primitive constraints. An alternative approach, denoted here as the constraint Newton method, is to handle directly the original set of constraints and for each one obtain a coarser approximation of the enclosing set by applying an algorithm that alternates bisection with Newton steps (see subsection 3.3.1). Both methods will be described in the next two sub-sections whereas complementary alternatives will be addressed in subsection 4.2.3.

4.2.1   Constraint Decomposition Method

The constraint decomposition method [77, 128, 91, 15] was the original technique used for defining box-narrowing functions capable of narrowing a variable domain into the RF-set approximation (or at least the RF-hull approximation) of the set obtained by the respective projection function. It is based on the transformation of complex constraints into an equivalent set of primitive constraints whose projection functions can be easily computed.

Definition 4.2.1-1 (Primitive Constraint). Let e_c be a real expression with at most one basic operator (see definition 3.2-1) and with no multiple occurrences of its variables. Let e_0 be a real constant or a real variable not appearing in e_c. The constraint c is a primitive constraint iff it is expressed as:

$$e_c \diamond e_0 \quad \text{with} \quad \diamond \in \{\leq,=,\geq\} \qquad \square$$

A set of primitive constraints can be easily obtained from a non-primitive constraint. Recall that by definition 2.2-2 any constraint of a CCSP is expressed in the form $e_c \diamond 0$ (where e_c is a real expression and $\diamond \in \{\leq,=,\geq\}$) and the only reason why it is not a primitive constraint is that the real expression e_c may contain more than one basic operator or may contain multiple occurrences of the same variable. However, if there are n basic operators within expression e_c, then $n-1$ basic operators must be within the expressions that are arguments of the other basic operator (see the recursive definition 3.2-1 of a real expression). Thus, the original constraint may be decomposed by considering new variables (and new equality constraints), one for each of these arguments containing basic operators. The whole set of primitives may be obtained by repeating this process until all constraints contain at most one basic operator. Similarly, for each multiple occurrence of the same variable a new variable may be considered together with a new equality constraint. Alternatively, constraints with multiple occurrences of the same variables could be solved in order to obtain a single occurrence of each variable, which is always possible because, after the previous decomposition, these constraints contain at most a single basic operator.

Consider the CCSP P presented in figure 4.2 which includes the single constraint $c \equiv x_1 \times (x_2 - x_1) = 0$. The constraint is not primitive since it contains two basic arithmetic operators and the variable x_1 occurs twice. Applying the decomposition technique described above, the new variable x_3 is introduced for representing the second argument of the multiplication operator and the primitive constraints $c_1 \equiv x_1 \times x_3 = 0$ and $c_2 \equiv x_2 - x_1 = x_3$ are obtained. The CCSP $P = (\langle x_1, x_2 \rangle, D_1 \times D_2, \{c\})$ may thus be transformed into the equivalent CCSP $P' = (\langle x_1, x_2, x_3 \rangle, D_1 \times D_2 \times [-\infty..+\infty], \{c_1, c_2\})$ where all the constraints are primitive.

The next step of the constraint decomposition method is to solve algebraically each primitive constraint wrt each variable in the scope and to define an interval function enclosing the respective projection function. This is always possible because the constraints are primitive. However an extra care must be taken due to the indefinition of some real expressions for particular real valued combinations (for example: $x_1 \times x_2 = x_3$ is not equivalent to $x_1 = x_3 / x_2$ if x_2 and x_3 are both zero, in which case the expression x_3 / x_2 is not defined). The later problem is naturally handled by considering the natural interval extension of the obtained real function, which includes all the defined real valued combinations (in the above example $X_1 \subseteq X_3 / X_2$).

Definition 4.2.1-2 (Inverse Interval Expression). Let $c = (s, \rho)$ be a primitive constraint expressed in the form $e_c \diamond e_0$ where $e_c \equiv e_1$ or $e_c \equiv \Phi(e_1, \ldots, e_m)$ (Φ is an m-ary basic operator and e_i a variable from s or a real constant). The inverse interval expression of c with respect to e_i, denoted Ψ_{e_i}, is the natural interval expression of the expression obtained by solving algebraically, wrt e_i, the equality $e_c = e_0$ if c is an equality or $e_c = e_0 + k$ if c is an inequality (with $k \leq 0$ for inequalities of the form $e_c \leq e_0$ and $k \geq 0$ for inequalities of the form $e_c \geq e_0$). ❏

Table 4.1 shows the inverse interval expressions of primitive constraints with no operators or with one of the four basic arithmetic operators defined in 3.1-1. The inverse interval expressions associated with primitive constraints including other basic operators could be defined similarly.

Table 4.1 Inverse interval expressions of some primitive constraints.

	Ψ_{e_1}	Ψ_{e_2}	Ψ_{e_3}
$e_1 + e_2 \diamond e_3$	$(E_3 + K) - E_2$	$(E_3 + K) - E_1$	$(E_1 + E_2) - K$
$e_1 - e_2 \diamond e_3$	$(E_3 + K) + E_2$	$E_1 - (E_3 + K)$	$(E_1 - E_2) - K$
$e_1 \times e_2 \diamond e_3$	$(E_3 + K) / E_2$	$(E_3 + K) / E_1$	$(E_1 \times E_2) - K$
$e_1 / e_2 \diamond e_3$	$(E_3 + K) \times E_2$	$E_1 / (E_3 + K)$	$(E_1 / E_2) - K$
$e_1 \diamond e_2$	$(E_2 + K)$	$E_1 - K$	

$\diamond \in \{\leq, =, \geq\}$

e_i is a real variable or a real constant

E_i is the natural interval extension of e_i

$$K = \begin{cases} [-\infty..0] & \text{if } \diamond \equiv \leq \\ [0..0] & \text{if } \diamond \equiv = \\ [0..+\infty] & \text{if } \diamond \equiv \geq \end{cases}$$

Table 4.2 presents the inverse interval expressions associated with the primitive constraints of the decomposed CCSP P' described in the previous example.

Table 4.2 Inverse interval expressions of $c_1 \equiv x_1 \times x_3 = 0$ and $c_2 \equiv x_2 - x_1 = x_3$.

	Ψ_{e_1}	Ψ_{e_2}	Ψ_{e_3}
$x_1 \times x_3 = 0$	$0 / X_3$	$0 / X_1$	$X_1 \times X_3$
$x_2 - x_1 = x_3$	$X_3 + X_1$	$X_2 - X_3$	$X_2 - X_1$

The inverse interval expression wrt a variable allows the definition of the projection function of the constraint wrt to that variable.

Theorem 4.2.1-1 (Projection Function based on the Inverse Interval Expression). Let $P=(X,D,C)$ be a CCSP. Let $c=(s,\rho)\in C$ be an n-ary primitive constraint expressed in the form $e_c \diamond e_0$ where $e_c \equiv e_i$ or $e_c \equiv \Phi(e_1,\ldots,e_m)$ (with Φ an m-ary basic operator and e_i a variable from s or a real constant). Let Ψx_i be the inverse interval expression of c with respect to the variable x_i ($e_i \equiv x_i$). The projection function $\pi_{x_i}^{\rho}$ of the constraint c wrt variable x_i is the mapping:

$$\pi_{x_i}^{\rho}(B) = \Psi x_i(B) \cap B[x_i] \qquad \text{where } B \text{ is an } n\text{-ary real box} \qquad \square$$

An enclosure of the projection function wrt a variable x_i is obtained by firstly applying the evaluation rules of the basic operators (see section 3.1) to obtain the respective inverse interval expression Ψx_i with the real box B, and secondly by intersecting the result with the projection of B wrt x_i.

Table 4.3 shows the projection functions obtained by this method for the primitive constraints of the previous example.

Table 4.3 Projection functions of $c_1 \equiv x_1 \times x_3 = 0$ and $c_2 \equiv x_2 - x_1 = x_3$.

$x_1 \times x_3 = 0$
$\pi_{x_1}^{\rho}(<I_1,I_3>) = (0/I_3) \cap I_1$
$\pi_{x_3}^{\rho}(<I_1,I_3>) = (0/I_1) \cap I_3$

$x_2 - x_1 = x_3$
$\pi_{x_1}^{\rho}(<I_1,I_2,I_3>) = (I_2 - I_3) \cap I_1$
$\pi_{x_2}^{\rho}(<I_1,I_2,I_3>) = (I_3 + I_1) \cap I_2$
$\pi_{x_3}^{\rho}(<I_1,I_2,I_3>) = (I_2 - I_1) \cap I_3$

The soundness of the interval evaluation rules guarantees the enclosure of the projection function. However, the quality of this enclosure depends on the evaluation rules used (as discussed in section 3.1) and the restrictions imposed on the result (a single F-interval or the union of multiple F-intervals). The best possible enclosure (the RF-set approximation of the projection function) is obtainable by using extended interval arithmetic and allowing the result to be composed of multiple F-intervals.

The inclusion monotonicity property of interval arithmetic evaluation guarantees the monotonicity of the box narrowing functions defined by the decomposition method.

In the case of the above example, the interval evaluations will always result in a single F-interval and so these could directly define the narrowing functions that will be considered by the propagation algorithm for pruning the variable domains of the decomposed CCSP P' (see Table 4.4).

Table 4.4 Narrowing functions of CCSP $P'=(<x_1,x_2,x_3>,D_1 \times D_2 \times [-\infty..+\infty],\{x_1 \times x_3=0, x_2-x_1=x_3\})$

$x_1 \times x_3 = 0$	NF_1	$BNF_{x_1}^{\rho}(<I_1,I_2,I_3>) = <(0/I_3) \cap I_1,I_2,I_3>$
	NF_2	$BNF_{x_3}^{\rho}(<I_1,I_2,I_3>) = <I_1,I_2,(0/I_1) \cap I_3>$
$x_2 - x_1 = x_3$	NF_3	$BNF_{x_1}^{\rho}(<I_1,I_2,I_3>) = <(I_2-I_3) \cap I_1,I_2,I_3>$
	NF_4	$BNF_{x_2}^{\rho}(<I_1,I_2,I_3>) = <I_1,(I_3+I_1) \cap I_2,I_3>$
	NF_5	$BNF_{x_3}^{\rho}(<I_1,I_2,I_3>) = <I_1,I_2,(I_2-I_1) \cap I_3>$

If the original goal is, according to the CCSP $P=(<x_1,x_2>,D_1{\times}D_2,\{x_1{\times}(x_2\text{-}x_1)=0\})$, to prune *F*-box $B=<I_1,I_2>\in D_1{\times}D_2$, then, using the decomposition method, the narrowing is executed on the decomposed CCSP $P'=(<x_1,x_2,x_3>,D_1{\times}D_2{\times}[\text{-}\infty..+\infty],\{x_1{\times}x_3=0,\ x_2\text{-}x_1=x_3\})$ by applying the propagation algorithm (*prune*) to the initial *F*-box $B_1=<I_1,I_2,[\text{-}\infty..+\infty]>$ and using the narrow functions defined in table 4.4. From the resulting *F*-box $B_1'=<I_1',I_2',I_3'>$ a box $B'=<I_1',I_2'>$ can be obtained by considering only the domains of the variables appearing in the original CCSP *P*. Table 4.5 summarises the pruning results obtained by this method for four different initial domains.

Table 4.5 Examples of the application of the decomposition method on a CCSP. The CCSP is $P=(<x_1,x_2>,D_1{\times}D_2,\{x_1{\times}(x_2\text{-}x_1)=0\})$ and $Q=\{NF_1,NF_2,NF_3,NF_4,NF_5\}$ (see table 4.4).

| | $B=<I_1,I_2>$ | | $prune(Q,<X_1,X_2,[\text{-}\infty..+\infty]>)$ | $B'=<X_1',X_2'>$ | |
	I_1	I_2		I_1'	I_2'
Case 1:	[-0.5..2.5]	[0.5..1.5]	<[-0.5..2.5],[0.5..1.5],[-2.0..2.0]>	[-0.5..2.5]	[0.5..1.5]
Case 2:	[0.25..1.0]	[0.5..1.5]	<[0.5..1.0],[0.5..1.0],[0..0]>	[0.5..1.0]	[0.5..1.0]
Case 3:	[-1.0..0.25]	[0.5..1.5]	<[0..0],[0.5..1.5],[0.5..1.5]>	[0..0]	[0.5..1.5]
Case 4:	[-1.0..-0.25]	[0.5..1.5]	$\varnothing$	$\varnothing$	

Note that in the first case the original variable domains could not be pruned by the decomposition method. However, in this case, the smallest box within *B* that encloses all the CCSP solutions is <[0.0..1.5],[0.5..1.5]> (as can be checked in figure 4.2). In the other three cases the results obtained were the narrowest possible results without loosing solutions (in particular, in case 4, it was proven that box *B* is inconsistent).

4.2.2   *Constraint Newton Method*

Instead of decomposing each complex constraint into a set of primitive constraints, the constraint Newton method [14, 134] manipulates complex constraints as a whole by using a technique based on the interval Newton's method for searching the zeros of univariate functions (see 3.3.1).

The approach is based on a set of auxiliary functions, that we will denote interval projections, one for each variable of each constraint.

> **Definition 4.2.2-1 (Interval Projection).** Let $P=(X,D,C)$ be a CCSP. Let $c=(s,\rho)\in C$ be an *n*-ary constraint expressed in the form $e_c\diamond0$ (with $\diamond\in\{\leq,=,\geq\}$ and e_c a real expression). Let *B* be an *n*-ary *F*-box. The interval projection of *c* wrt $x_i\in s$ and *B* is the function, denoted $\prod_{x_i}^{\rho B}$, represented by the expression obtained by replacing in e_c each real variable x_j $(x_j{\neq}x_i)$ by the interval constant $B[x_j]$. ❑

The interval projections, $\prod_{x_1}^{\rho B}$ and $\prod_{x_2}^{\rho B}$, associated with the constraint $x_1{\times}(x_2\text{-}x_1)=0$ and an *F*-box $B=<[\text{-}0.5..2.5],[0.5..1.5]>$ (see example of figure 4.2) are the family of univariate real functions represented by the expressions $x_1{\times}([0.5..1.5]\text{-}x_1)$ and $[\text{-}0.5..2.5]{\times}(x_2\text{-}[\text{-}0.5..2.5])$, respectively.

From the properties of the interval projections, a strategy is devised for obtaining an enclosure of the respective projection function.

Theorem 4.2.2-1 (Properties of the Interval Projection). Let $P=(X,D,C)$ be a CCSP. Let $c=(s,\rho)\in C$ be an n-ary constraint and B an n-ary F-box. Let $\prod_{x_i}^{\rho B}$ be the interval projection of c wrt variable $x_i\in s$ and B. The following properties are necessarily satisfied:

    (i)   if $\diamond \equiv$ "=" then $\forall_{r\in B[x_i]}\ r\in\pi_{x_i}^{\rho}(B) \Rightarrow 0\in\prod_{x_i}^{\rho B}([r])$

    (ii)   if $\diamond \equiv$ "$\leq$" then $\forall_{r\in B[x_i]}\ r\in\pi_{x_i}^{\rho}(B) \Rightarrow left(\prod_{x_i}^{\rho B}([r])) \leq 0$

    (iii) if $\diamond \equiv$ "$\geq$" then $\forall_{r\in B[x_i]}\ r\in\pi_{x_i}^{\rho}(B) \Rightarrow right(\prod_{x_i}^{\rho B}([r])) \geq 0$

We will say that a real value r satisfies the interval projection condition if the right side of the respective implication (i), (ii) or (iii) is satisfied.   ❏

Property (i) claims that in equality constraints each element of a projection function wrt a variable must be a zero of the interval projection, that is, zero must be within the interval obtained by its evaluation. Properties (ii) and (iii) claim that in inequality constraints each element of a projection function wrt a variable when evaluated by the complex interval projection will produce an interval where at least some elements are smaller/larger (or equal) than zero.

The key idea of the strategy used in the constraint Newton method is to search for the leftmost and the rightmost elements of the original variable domain satisfying the interval projection condition. The next theorem guarantees that these elements define an interval that contains the projection function.

Theorem 4.2.2-2 (Projection Function Enclosure based on the Interval Projection). Let $P=(X,D,C)$ be a CCSP. Let $c=(s,\rho)\in C$ be an n-ary constraint, B an n-ary F-box and x_i an element of s. Let a and b be respectively the leftmost and the rightmost elements of $B[x_i]$ satisfying the interval projection condition. The following property necessarily holds:

$$\pi_{x_i}^{\rho}(B) \subseteq [a..b]$$
    ❏

From the above theorem, the natural strategy to obtain a new left (right) bound is firstly to verify the interval projection condition in the left (right) extreme of the original variable domain and secondly, only in case of failure, to search for the leftmost (rightmost) zero of the interval projection. This strategy assumes the continuity of the interval projection function since in case of failure of an inequality condition, it assumes that the leftmost (rightmost) element satisfying the interval projection condition must be a zero of the interval projection.

Figure 4.3 presents the pseudocode of function *narrowBounds*, which implements this narrowing strategy. It uses the function *intervalProjCond* for verifying whether the interval projection condition is satisfied at the original interval bounds and uses the *searchLeft* and *searchRight* functions for finding new bounds. The unique argument of the function *narrowBounds* is an F-interval representing the domain of a variable x_i. The result is a smaller F-interval enclosing the projection function (or the empty set if it is proved that the interval projection condition cannot be satisfied).

---

function *narrowBounds*(an F-interval $[a..b]$)

(1)   **if** $a = b$ **then if** *intervalProjCond*($[a]$) **then return** $[a]$ **else return** $\varnothing$; **end if**; **end if**;

(2)   **if not** *intervalProjCond*($[a..a^+]$) **then** $a \leftarrow$ *searchLeft*($[a^+..b]$);

(3)   **if** $a = \varnothing$ **then return** $\varnothing$;

(4)   **if** $a = b$ **then return** $[b]$;

(5)   **if not** *intervalProjCond*($[b^-..b]$) **then** $b \leftarrow$ *searchRight*($[a..b^-]$);

(6)   **return** $[a..b]$;

end function

Figure 4.3 The narrowing algorithm for finding an enclosure of the projection function.

The algorithm works as follows. If the original F-interval $[a..b]$ is degenerate (line 1) then, it either satisfies the interval projection condition and cannot be further narrowed or it doesn't and the empty set is returned because the constraint cannot be satisfied. If $[a..b]$ is not degenerate then the algorithm proceeds (line 2) by inspecting the satisfiability of the interval projection condition in the left bound ($[a..a^+]$) and, in case of failure, a is updated to the left bound of the leftmost canonical interval (within $[a^+..b]$) that zeros the interval projection. If there are no zeros (line 3), the constraint cannot be satisfied and the empty set is returned. If the only zero is the right bound b (line 4) then the degenerate F-interval $[b]$ is returned. Otherwise the algorithm proceeds (line 5) by inspecting the satisfyability of the interval projection condition within the right bound ($[b^-..b]$) and, in case of failure, b is updated to the right bound of the rightmost canonical interval (within $[a..b^-]$) that zeros the interval projection. In this case, the empty set cannot be returned since the left canonical bound is a zero of the interval projection, and so, the F-interval $[a..b]$ is returned (line 6).

Function *intervalProjCond* is described in figure 4.4. It uses the interval projection $\Pi_{x_i}^{\rho B}$ and assumes that the constraint is expressed in the form $e_c \diamond 0$. Its unique argument is a canonical F-interval I and the result is a boolean which is set to false iff it can be proved that the interval projection condition cannot be satisfied by any real value within I.

---

function *intervalProjCond*(a canonical F-interval I)

(1) $[a..b] \leftarrow \Pi_{x_i}^{\rho B}(I)$;

(2)   **case** $\diamond$ **of**

(3)   "=": **return** $0 \in [a..b]$;

(4)   "$\leq$": **return** $a \leq 0$;

(5)   "$\geq$": **return** $b \geq 0$;

(6)   **end case**;

end function

Figure 4.4 The function that verifies if the interval projection condition may be satisfied.

In line 1 the interval projection function is evaluated for the canonical F-interval I and the result is the interval $[a..b]$. Lines 3, 4 and 5 verify if the appropriate interval projection condition (see theorem 4.2.2-1) may be satisfied for a real value within I returning true in that case and false otherwise.

The algorithm for searching for the leftmost zero of an interval projection is specified in function *searchLeft*, implemented in pseudocode in figure 4.5 (function *searchRight* is defined similarly). The algorithm is analogous to the *ShrinkLeft* and *LNAR* algorithms proposed in [14] and [134] respectively. It uses the interval projection function $\Pi_{x_i}^{\rho B}$ and an associated Newton Narrowing function *NN* as described in subsection 3.3.1 (see definition 3.3.1-3).

```
function searchLeft(an F-interval I)
  (1)     Q ← {I};
  (2)     while Q ≠ ∅ do
  (3)         choose I₁ ∈ Q with the smallest left bound (∀_{I∈Q} left(I₁)≤ left(I));
  (4)         Q ← Q \ {I₁};
  (5)         if 0∈∏_{x_i}^{ρA}(I₁) then
  (6)             I₁ ← NN(I₁);
  (7)             if I₁ ≠ ∅ then
  (8)                 I₀ ← cleft(I₁); I₁ ← [right(I₀)..right(I₁)];
  (9)                 if 0∈∏_{x_i}^{ρB}(I₀) then return left(I₀);
  (10)                else Q ← Q ∪ {[left(I₁)..⌊center(I₁)⌋], [⌈center(I₁)⌉..right(I₁)]}; end if;
  (11)            end if;
  (12)        end if;
  (13)    end while;
  (14)    return ∅;
end function
```

Figure 4.5 The algorithm for searching the leftmost zero of an interval projection.

The only argument I of the function *searchLeft* is an F-interval representing the domain of the variable x_i where the search takes place. The returned result is either the empty set if it is proven that there are no zeros of $\prod_{x_i}^{ρB}$ within I, or else is an F-number a satisfying: $0\in\prod_{x_i}^{ρB}([a..a^+])$.

The algorithm is based on a recurring cycle (lines 2 through 13) where, in each cycle, a partition I_1 of the original interval I $(I_1\subseteq I)$ is inspected for its leftmost zero of $\prod_{x_i}^{ρB}$. During the whole process, a set Q will contain partitions of the original interval that might include zeros of $\prod_{x_i}^{ρB}$. This set is initialised with the whole original interval I (line 1). If Q becomes empty then it is proved that there are no zeros of $\prod_{x_i}^{ρB}$ within I and so the cycle terminates and the empty set is returned (line 14). If Q is not empty then its member with the smallest left bound is chosen for inspection (line 3) and is removed from Q (line 4). The inspection of an interval I_1 proceeds as follows. Line 5 verifies if zero is contained in the interval evaluation of $\prod_{x_i}^{ρB}$ with I_1 as argument. In case of failure, I_1 cannot contain a zero of $\prod_{x_i}^{ρB}$ (see theorem 3.2-2) and will not be further considered (a new cycle begins). In case of success, I_1 might contain zeros of $\prod_{x_i}^{ρB}$ and so, the Newton Narrowing function is applied (line 6) for obtaining a smaller interval without losing any existing zero (see theorem 3.3.1-1). Line 7 verifies if the obtained interval is not empty. If it is empty then I_1 cannot contain a zero of $\prod_{x_i}^{ρB}$ (see theorem 3.3.1-2) and a new cycle begins. Otherwise a canonical F-interval I_0 (enclosing the left bound of I_1) is isolated for inspection (line 8). If zero is contained in the interval evaluation of $\prod_{x_i}^{ρB}$ with I_0 as argument then the left bound of I_0 is returned (line 9); else I_1 is split at its mid point and both intervals are added to Q (line 10).

The algorithm, which takes advantage from the efficiency of the Interval Newton method (see theorems 3.3.1-5 and 3.3.1-6), is correct and terminates.

Correctness is guaranteed in the sense that the returned real value is smaller (or equal) than any possible real value within the original domain that is a zero of the interval projection function. On the one hand, the fact that sub-regions of the original domain are only discarded by applying the Newton Narrowing function guarantees that no existing zero is lost. On the other hand, the returned real value is the left bound of a canonical F-interval I_0 that "zeros" $\prod_{x_i}^{\rho B}$ (its interval evaluation contains zero) and this value is smaller than any real value contained in any partition in Q[6].

However, the fact the interval evaluation of $\prod_{x_i}^{\rho B}(I_0)$ contains zero does not guarantee that the left bound of I_0 is a zero of the function (the returned real value may not be the leftmost zero of the function). Moreover, it does not even guarantee the existence of a real value within I_0 that zeros the function (it could be a consequence of the approximate nature of interval arithmetic evaluation).

Termination is guaranteed because at each cycle a sub-region of the original domain is discarded or split into two smaller sub-regions which is a finite process that necessarily ends when canonical sub-regions are obtained.

As a consequence of the above properties of the *searchLeft/searchRight* functions, the interval obtained by applying the *narrowBounds* function to a domain of variable x_i within a box B, necessarily encloses the interval $[a..b]$ whose bounds are the leftmost and the rightmost elements of $B[x_i]$ satisfying the interval projection condition. Hence, and accordingly to theorem 4.2.2-2, the *narrowBounds* function computes an enclosure of the projection function:

$$\pi_{x_i}^{\rho}(B) \subseteq [a..b] \subseteq narrowBounds(B[x_i])$$

In the unrealistic case where the interval arithmetic evaluations were performed with infinite precision, $[a..b]$ is the interval obtained by the *narrowBounds* function. However, in this case, the termination property of the *searchLeft/searchRight* functions is no longer guaranteed.

The box narrowing functions defined by the *narrowBounds* functions are not guaranteed to be monotonic. On the one hand, the Newton Narrowing functions used in the *searchLeft/searchRight* functions are non monotonic since the Newton function (definition 3.3.1-1) is not monotonic (due to its dependence on the centre of an interval). On the other hand, they may be able to prove the non-existence of zeros on entire sub-regions of a domain where the interval evaluation of particular canonical intervals within this region may be insufficient to discard this possibility. Consequently, the interval obtained by the *narrowBounds* function is not necessarily the smallest interval containing all canonical intervals whose evaluation satisfy the interval projection condition. It may be even smaller than this interval because, during the narrowing process, the Newton Narrowing functions may be able to prove that some of these canonical intervals cannot satisfy the interval projection condition.

However, with infinite precision, the monotonicity of the *narrowBounds* functions is guaranteed since the obtained interval is bounded by the leftmost and the rightmost elements of the original interval satisfying the interval projection condition.

Consider the example of the CCSP $P=(<x_1,x_2>,D_1 \times D_2,\{x_1 \times (x_2-x_1)=0\})$ presented in figure 4.2. Using the constraint Newton method, since there is no need to decompose the unique constraint into primitives, the pruning results over the original box $B=<[-0.5..2.5],[0.5..1.5]>$ are much better than the obtained by the previous method. The following is a step by step description of the application of the *narrowBounds* function for narrowing the domain of variable x_1. As in the example of figure 3.6, it is assumed a three digits precision, that is, the distance between two consecutive F-numbers is 0.001.

The *narrowBounds* function is applied over the original interval $[-0.5..2.5]$ with the associated interval projection $\prod_{x_1}^{\rho B}$ represented by the expression $x_1 \times ([0.5..1.5]-x_1)$

---

[6] Due to the splitting strategy (line 9) and the choosing criterion (line 3), all values of the inspected partition I_l are smaller (or equal) than any value of any other partition in Q

(represented in figure 3.2). Since the original interval is not degenerate, the *intervalProjCond* function is applied to its left canonical bound [-0.5..-0.499]. Because $0 \notin \Pi_{x_1}^{\rho B}$ ([-0.5..-0.499])=[-1..-0.499], this bound does not satisfy the interval projection condition and the search for a new left bound within [-0.499..2.5] is accomplished by function *searchLeft*.

Table 4.6 summarises the process, each line illustrating the principal actions executed at each cycle within the searching algorithm described in figure 4.5. The first column shows the set of interval partitions that are considered at the beginning of the cycle. The second column is the verification that the leftmost interval partition I_1 may contain zeros of the interval projection (fig. 4.5, line 5). The third column is the interval obtained by applying the Newton Narrowing function to this partition (fig. 4.5, line 6) where the Newton function is defined accordingly definition 3.3.1-1 with $F_E \equiv X_1 - X_1^2$ and $F'_E \equiv 1 - 2X_1$ (see example of figure 3.6). The fourth column is the verification if the left canonical bound of the previously obtained interval may contain zeros of the interval projection (figure 4.5, line 9).

Table 4.6 Searching a new left bound for x_1 within the interval [-0.499..2.5].

searchLeft([-0.499..2.5])			
$Q=\{I_1,...,I_n\}$	$0 \in \Pi_{x_i}^{\rho B}(I_1)$	$NN(I_1)$	$0 \in \Pi_{x_i}^{\rho B}(I_0)$
{[-0.499..2.5]}	$0 \in [-5..4.998]$	[-0.499..2.5]	$0 \notin [-0.998..-0.497]$
{[-0.498..1.001],[1.001..2.5]}	$0 \in [-0.995..2]$	[-0.498..1.001]	$0 \notin [-0.997..-0.496]$
{[-0.497..0.252],[0.252..1.001],[1.001..2.5]}	$0 \in [-0.992..0.504]$	[0..0.001]	$0 \in [0..0.002]$

return 0

In the first two cycles, the Newton Narrowing function was unable to reduce the leftmost interval and the algorithm proceeded by considering smaller partitions. However, in the third cycle, the Newton Narrowing function was powerful enough to isolate a canonical zero of the interval projection. The final returned value is the left bound (0) of this canonical interval.

After verifying that $0 \notin \Pi_{x_1}^{\rho B}$ ([2.499..2.5])=[-5..-2.497], the search for a new right bound was performed similarly, by applying function *searchRight* to the interval [0..2.499], obtaining the new bound 1.501. Consequently, the final result obtained by the *narrowBounds* function for narrowing the domain [-0.5..2.5] of variable x_1 is the new and smaller interval [0..1.501].

The pruning achieved by using the constraint Newton method to solve any of the cases presented in table 4.5 is identical to the pruning achieved by the decomposition method. An exception is the first case, presented above, where the x_1 domain was narrowed into a smaller interval [0..1.501] which is a fairly good approximation of [0..1.5] (the smallest interval enclosing the projection function –see figure 4.2).

4.2.3   Complementary Approaches

Several variations of the two basic methods for obtaining box-narrowing functions have been considered. The idea is to take advantage of the properties of these methods when applied to constraints that are expressed in a particular form.

A modification of the Newton's method, firstly presented in [134], is to use other interval extensions of the interval projection function associated with a constraint.

As defined in the previous subsection, the interval projection is a univariate function obtained from the original constraint expression by a process similar to the one presented in the definition of the natural interval extension (see definition 3.2-6) except that some variables are replaced by their interval values. Consequently, the interval arithmetic evaluation of this function with an interval argument, which computes an enclosure of the function range, corresponds to the evaluation of its natural interval extension. If instead of this natural interval extension, other interval extension (expressed in some other form) is considered, then its interval arithmetic evaluation would still compute an enclosure of the function range (see section 3.2). Moreover, the quality of this enclosure is dependent on the form of the interval extension.

Pascal V. Hentenryck et al propose in [134] the use of the Distributed and the Taylor[7] interval extensions together with the natural interval extension for obtaining different enclosures of the interval projection function. This way, different box-narrowing functions are simultaneously defined wrt the same variable of the same constraint and may be applied at different stages of the pruning process accordingly to their specific properties.

A modification of the decomposition method, presented in [77] and known as global tolerance propagation, does not require the complete decomposition of the whole set of constraints into primitive constraints. It transforms the original set of constraints into an equivalent one where for each constraint (not necessarily primitive) the inverse interval expressions can be easily computed by interval arithmetic evaluations. Moreover, for enforcing global consistency it is sufficient to obtain a set of constraints whose variables are not connected circularly to each other by a chain of mutually different constraints [77].

In practice, this equivalent set of constraints is often impossible to obtain, either due to algebraic limitations or the imprecision of the interval arithmetic evaluations, and the approach may only be applied in a few special situations.

Another modification to the general basic methods is the introduction of a pre-processing phase preceding the definitions of the box-narrowing functions. The goal is to define an equivalent CCSP by applying symbolic rewriting techniques over the original set of constraints. The obtained equivalent CCSP will be expressed in a more suitable form for applying efficiently the narrowing propagation algorithm.

Benhamou in [9] characterised the pre-processing techniques in terms of constraint rewriting operators. Practical proposals for applying these techniques aim at improving propagation efficiency by introducing redundant constraints. In particular, for CCSPs addressing the solution of multivariate polynomials over the reals, rewriting approaches were defined where Grobner bases are computed [12] (or partially computed [13]) from the original set of constraints.

Another variation based on the two basic methods for obtaining box-narrowing functions was presented in [11]. In this work, the authors developed an algorithm, denoted *HC4revise*, capable of implementing a narrowing function associated with any complex constraint without decomposing it. Moreover, the narrowing results achieved by *HC4revise* are the same as those that would be obtained if the decomposition method were applied to this constraint and the narrowing propagation executed with the resulting box-narrowing functions. This allowed the implementation of a more efficient algorithm, denoted *HC4*, with the same results as the decomposition method, which does not require the decomposition of complex constraints into primitives.

---

[7] The Taylor interval extension does not require the usage of the *narrowBounds* function because it can be solved wrt the variable. However it requires that the constraint must be of the form $E_c=0$ where E_c denotes a function which has continuous partial derivatives [VMK97].

Based on the *HC4revise* algorithm, a complementary approach was proposed in [11], which may take advantage of the way that a complex constraint is expressed. The idea is that having an algorithm, such as the *HC4revise* algorithm, that does not require decomposing complex constraints, makes it possible to combine the essence of both basic methods, and choose either one or the other, according to the form of the expression of the interval projection.

The evaluation error of the interval projection function is a consequence of the dependency problem (see definition 3.2-7) and so, when there are no multiple occurrences of the same variable (the unique variable) the *HC4revise* may be applied without introducing errors, otherwise, the Newton's method may be preferable. The resulting algorithm, denoted *BC4*, integrate the efficiency of the *HC4revise* algorithm (and efficacy without dependency) with the efficacy of the Newton's method for the narrowing propagation of complex constraints.

Finally, some approaches [125], complementary to the box-narrowing functions associated with each variable of each constraint, consider narrowing functions capable of narrowing several variable domains simultaneously.

These functions are based on the multivariate interval Newton's method (see subsection 3.3.2) and require the grouping of constraints into a square subsystem (the number of considered constraints equals the total number of variables within their scopes) which can be seen as a single complex constraint. Despite the inherent complexity of this multivariate approach, it may be particularly effective where the projection approaches may fail, namely, in pruning space regions in the neighbourhood of a root [125].

4.3   Summary

In this chapter the generic constraint propagation algorithm was described in terms of narrowing functions associated with the constraint set. Its properties were derived from the properties of the narrowing functions. The main methods used in the interval constraint framework for associating narrowing functions to constraints were presented. Their extensive use of interval analysis techniques for guaranteeing correctness of the resulting narrowing functions was emphasised. The next chapter defines local consistency as a property that depends exclusively on the narrowing functions associated with the constraint set, and overviews the main consistency criteria used in continuous domains.

Chapter 5

Partial Consistencies

The fixed-points of a set of narrowing functions associated with a constraint characterize a local property enforced among the variables of the constraint scope. Such property is called local consistency since it depends exclusively on the narrowing functions associated with a single (local) constraint and defines the value combinations that are not pruned by them (consistent). Section 5.1 characterizes the main types of local consistency types used in continuous domains.

Local consistency is a partial consistency, in the sense that imposing it on a CCSP is not sufficient to remove all inconsistent value combinations among its variables. Stronger higher order consistency requirements may be subsequently imposed establishing global properties over the variable domains. Higher order consistencies will be discussed in section 5.2.

5.1　Local Consistency

Local consistencies used for solving CCSPs are approximations of arc-consistency, a local consistency developed in Artificial Intelligence [97, 98] for solving CSPs with finite domains. A constraint is said to be arc-consistent wrt a set of value combinations iff, within this set, for each value of each variable of the scope there is a value combination of these variables that satisfy the constraint.

Definition 5.1-1 (Arc-Consistency). Let $P=(X,D,C)$ be a CSP. Let $c=(s,\rho)$ be a constraint of the CSP ($c\in C$). Let A be an element of the power set of D ($A\in 2^D$). The constraint c is arc-consistent wrt A iff:

$$\forall_{x_i\in s}\ \forall_{d_i\in A[x_i]}\ \exists_{d\in A[s]}\ (d[x_i]=d_i\ \wedge\ d\in\rho)$$

which, extending the definition of a projection function to any element of 2^D, is equivalent

to:　$\forall_{x_i\in s}\ A[x_i]=\{\ d[x_i]\mid d\in\rho\cap A[s]\ \}=\pi_{x_i}^{\rho}(A[s])$　❑

Consider the example of figure 4.2 with box $B_1=<[-0.5..2.5],[0.5..1.5]>$, box $B_2=<[0..1.5],[0.5..1.5]>$ and the element $A=<[0..0],[0.5..1.5]>\cup<[0.5..1.5],[0.5..1.5]>$. In this example the variables of the CCSP are all represented in the constraint scope $s=<x_1,x_2>$ thus $B_1[s]=B_1$, $B_2[s]=B_2$ and $A[s]=A$. The boxes B_1 and B_2 are not arc-consistent since, in both, there are x_1 values (for example $x_1=0.25$) without any corresponding x_2 value satisfying the constraint ($\pi_{x_1}^{\rho}(B_1)=\pi_{x_1}^{\rho}(B_2)=\{0\}\cup[0.5..1.5]$ is different from their respective domains $B_1[x_1]=[-0.5..2.5]$ and $B_2[x_1]=[0..1.5]$). However, element A is arc-consistent because $\pi_{x_1}^{\rho}(A)=A[x_1]=\{0\}\cup[0.5..1.5]$ and $\pi_{x_2}^{\rho}(A)=A[x_2]=[0.5..1.5]$.

In continuous domains, arc-consistency cannot be obtained in general due to machine limitations for representing real numbers. In practice, each real value must be approximated by a canonical F-interval and so, the best possible approximation of arc-consistency wrt a set of real valued combinations is the RF-set approximation of each variable domain within this set.

This is the idea of interval-consistency, which can be defined by replacing, in the definition of arc-consistency, the notion of a real value by the notion of a canonical F-interval. A constraint is said to be interval-consistent wrt a set of real valued combinations iff for each canonical F-interval representing a sub-domain of a variable there is a real valued combination of the variables of the scope satisfying the constraint.

Restricting F-intervals to closed form, a non degenerate canonical F-interval will only be considered within a variable domain if there is a real valued combination satisfying the constraint in its interior. However, due to the closed form imposition, a degenerate canonical F-interval will be considered either if there is a real valued combination satisfying the constraint with this value for that variable or if this real valued combination is within the interior of the adjacent canonical F-intervals.

Definition 5.1-2 (Interval-Consistency). Let $P=(X,D,C)$ be a CCSP. Let $c=(s,\rho)$ be a constraint of the CCSP ($c \in C$). Let A be an element of the power set of D ($A \in 2^D$). The constraint c is interval-consistent wrt A iff:

$$\forall_{x_i \in s} \ \forall_{[a..a^+] \subseteq A[x_i]} \exists_{d \in A[s]} \ (d[x_i] \in (a..a^+) \ \wedge \ d \in \rho) \ \wedge$$

$$\forall_{[a] \subseteq A[x_i]} \exists_{d \in A[s]} \ (d[x_i] \in (a^-..a^+) \ \wedge \ d \in \rho) \qquad \text{(where } a \text{ is an } F\text{-number)}$$

which, extending the definition of a projection function to any element of 2^D, is equivalent

to: $\forall_{x_i \in s} A[x_i] = S_{apx}(\{\ d[x_i]\ |\ d \in \rho \cap A[s]\ \}) = S_{apx}(\pi_{x_i}^{\rho}(A[s]))$ ❑

Consider again the example of figure 4.2 and a three digits machine precision. The boxes B_1 and B_2 of the previous example, which were not arc-consistent, are also not interval-consistent since they include, in the domain of x_1, non degenerate canonical F-intervals with no corresponding x_2 value in its interior satisfying the constraint (for example $[0.250..0.251] \subseteq B_2[x_1] \subseteq B_1[x_1]$ and if $x_1 \in (0.250..0.251)$ there is no x_2 value satisfying the constraint). The element A of the previous example, which was arc-consistent, is also interval-consistent since it is representable by a three digits machine precision and so $S_{apx}(\pi_{x_1}^{\rho}(A))=\pi_{x_1}^{\rho}(A)=A[x_1]$ and $S_{apx}(\pi_{x_2}^{\rho}(A))=\pi_{x_2}^{\rho}(A)=C[x_2]$.   However,   the   box $B_3=<[0.5..\pi/2],[0.5..\pi/2]>$, which is arc-consistent, is not representable and the smallest interval-consistent F-box including B_3 is $B_3'=<[0.5..1.571],[0.5..1.571]>$.

Interval-consistency was one of the first local consistency types used in continuous domains [77, 128]. It can only be enforced on primitive constraints (decomposition method) where the RF-set approximation of the projection function can be obtained by using extended interval arithmetic. Structures, like divisions in [77] and taxonomies in [128], must be considered for representing each variable domain as a non-compact set of real values. The narrowing functions are defined from the application of a box-narrowing function to each F-box obtained by all possible F-interval combinations between the domains represented in each structure.

In practice, the enforcement of interval-consistency can be applied only to small problems [77]. In order to maintain the RF-set approximation of the projection functions, the number of non-contiguous F-intervals represented within each structure may grow exponentially, requiring an unreasonable number of computations for each box-narrowing function.

Because it may be computationally too expensive to keep a structure for representing multiple F-intervals, the approximations of arc-consistency most widely used in continuous domains assume the convexity of the variable domains, in order to represent them by single F-intervals.

Hull-consistency (or 2B-consistency), firstly introduced by Lhomme in [91] and extensively used in continuous domains [8, 74, 15], is a coarser approximation of arc-consistency than interval-consistency, which requires the satisfaction of the arc-consistency property only at the bounds of the F-intervals that represent the variable domains.

A constraint is said to be hull-consistent wrt an F-box iff, for each bound of the F-interval representing the domain of a variable there is a real valued combination of the variables of the scope satisfying the constraint. Due to machine limitations for representing real numbers, the notion of a bound of an F-interval must be extended to a canonical bound (an extreme canonical F-interval) which also includes all non-representable real values within two consecutive F-numbers.

The definition of hull-consistency can be derived from the definition of interval-consistency by simply considering, for each variable domain, the two extreme canonical F-intervals instead of all possible canonical intervals. Consequently, the hull-consistency approximation of arc-consistency wrt a set of real valued combinations is the RF-hull approximation of each variable domain within this set.

Definition 5.1-3 (Hull-Consistency). Let $P=(X,D,C)$ be a CCSP. Let $c=(s,\rho)$ be a constraint of the CCSP ($c \in C$). Let B be an F-box which is an element of the power set of D ($B \in 2^D$). The constraint c is hull-consistent wrt B iff:

$$\forall_{x_i \in s} \ \exists_{d_l \in B[s]} \ (d_l[x_i] \in [a..a^+) \wedge d_l \in \rho) \ \wedge$$

$$\exists_{d_r \in B[s]} \ (d_r[x_i] \in (b^-..b] \wedge d_r \in \rho) \qquad \qquad (\text{where } B[x_i]=[a..b])$$

which is equivalent to:

$$\forall_{x_i \in s} \ B[x_i] = I_{hull}(\{ \ d[x_i] \mid d \in \rho \cap B[s] \ \}) = I_{hull}(\pi_{x_i}^{\rho}(B[s])) \qquad \qquad \square$$

In the previous example (with three digits machine precision) box B_1 is not hull-consistent because within the x_1 bounds there are no corresponding x_2 values satisfying the constraint (for example, if $x_1 \in [-0.500..-0.499)$ there is no x_2 value satisfying the constraint). However, box B_2 is hull-consistent since, for example, if $x_1=0.000 \in [0.000..0.001)$, any x_2 value satisfies the constraint and if $x_1=1.500 \in (1.499..1.500]$, then $x_2=1.500$ satisfies the constraint (and similarly wrt the domain of the other variable). Element A and real box B_3 are not F-boxes, so the hull-consistency criterion is not applicable in these cases. Box B_3', which was interval-consistent, is also hull-consistent since any interval-consistent box is hull-consistent. In this case, $\pi_{x_i}^{\rho}(B[s])$ must be a single F-interval and so $S_{apx}(\pi_{x_i}^{\rho}(B[s]))=I_{hull}(\pi_{x_i}^{\rho}(B[s]))$.

The existing approaches to enforce hull-consistency are all based on the constraint decomposition method where the RF-hull approximation of the projection function of each primitive constraint is obtained by using extended interval arithmetic complemented with union hull operations to avoid multiple disjoint F-intervals.

The major drawback of this decomposition approach is the worsening of the locality problem, which is a direct consequence of the dependency problem (see definition 2.2.2-7). The existence of intervals satisfying a local property on each constraint does not imply the existence of value combinations satisfying simultaneously all of them. When a complex constraint is subdivided into primitive constraints this will only worsen this problem due to the addition of new variables and the consequent loss of dependency between values of related variables. Hull-consistency enforcement is particularly ineffective if the original constraints contain multiple occurrences of the same variables.

An example of the bad results obtained by the decomposition approach was given in subsection 4.2.1 for pruning the domains of box $B=<[-0.5..2.5],[0.5..1.5]>$ (figure 4.2). As seen above, box B is not hull-consistent. However, the enforcement of hull-consistency in the decomposed CCSP did not prune its domains ($<[-0.5..2.5],[0.5..1.5],[-2.0..2.0]>$ is hull-consistent in the decomposed CCSP – see table 4.5, case 1).

The drawbacks of the decomposition approach motivated the constraint Newton method, which can be applied directly to complex constraints. The local consistency achieved by this method, known as box-consistency and firstly characterized in [14], has been successfully used in many applications on continuous domains [134, 135]. It was developed with the goal of providing a better trade-off between efficiency (of the enforcing algorithm) and pruning (of the variable domains).

Box-consistency is a coarser approximation of arc-consistency than hull-consistency. Instead of requiring the existence of a consistent real valued combination within each bound of each scope variable, it replaces the real valued combination by an enclosing box and requires a weaker form of consistency. The box is formed by the respective bound together with the F-intervals of the other variables of the scope. The weaker form of consistency is associated with a particular interval extension of the left side (e_c) of the constraint ($e_c \diamond 0$) and is satisfied iff the F-interval obtained by applying this interval extension to the box contains at least a real value satisfying the constraint[1].

Definition 5.1-4 (Box-Consistency). Let $P=(X,D,C)$ be a CCSP. Let $c=(s,\rho)$ be a constraint of the CCSP ($c \in C$) expressed in the form $e_c \diamond 0$ (with $\diamond \in \{\leq,=,\geq\}$ and e_c a real expression). Let F_E be an interval expression representing an interval extension F of the real function represented by e_c. Let B be an F-box which is an element of the power set of D ($B \in 2^D$). c is box-consistent wrt B and F_E iff:

$$\forall x_i \in s \ \exists r_1 \in F_E(B_1) \ r_1 \diamond 0 \ \land \ \exists r_2 \in F_E(B_2) \ r_2 \diamond 0$$
where B_1 and B_2 are two F-boxes such as:
$$B_1[x_i]=cleft(B[x_i]), B_2[x_i]=cright(B[x_i]) \text{ and } \forall_{x_j \in s} (x_j \neq x_i \Rightarrow B_1[x_j]=B_2[x_j]=B[x_i]). \quad \square$$

With three digits machine precision, the constraint $c \equiv x_1 \times (x_2 - x_1) = 0$ is not box-consistent wrt $B_1 = <[-0.5..2.5],[0.5..1.5]>$ and the interval extension represented by $X_1 \times (X_2 - X_1)$ since $0 \notin [-0.5..-0.499] \times ([0.5..1.5]-[-0.5..-0.499])=[-1..-0.498]$. However, the constraint is box-consistent wrt $B_2 = <[0..1.5],[0.5..1.5]>$ and the same interval extension because, wrt variable x_1, $0 \in [0..0.001] \times ([0.5..1.5]-[0..0.001])$ and $0 \in [1.499..1.5] \times ([0.5..1.5]-[1.499..1.5])$, and wrt variable x_2, $0 \in [0..1.5] \times ([0.5..0.501]-[0..1.5])$ and $0 \in [0..1.5] \times ([1.499..1.5]-[0..1.5])$.

Notice that the notion of box-consistency is always associated with a particular interval extension of the left hand side of the constraint. Enforcing box-consistency with different interval extensions may lead to different pruning results. However, if a constraint is hull-consistent wrt an F-box, it must also be box-consistency wrt the same F-box for any possible interval extension. The reason is that, independently from the interval extension used, the weaker form of consistency required by box-consistency is always satisfied when consistency required by hull-consistency is satisfied.

Although box-consistency is weaker than hull-consistency for the same constraint, in practice, the enforcement of box-consistency may achieve better pruning results since it may be directly applied to complex constraints whereas hull-consistency is only enforced in primitive constraints implying the previous decomposition of a complex constraint (see, in subsection 4.2.2, the better results obtained by the constraint Newton method for pruning the domains of box $B = <[-0.5..2.5],[0.5..1.5]>$).

For primitive constraints box-consistency and hull-consistency are equivalent if the interval extension used in box-consistency does not contain multiple occurrences of the same variable and its evaluation is computed with infinite precision. This was proved in [28] and is a consequence of the absence of the dependency problem in the evaluation of

---

[1] The satisfaction of this weaker form of consistency for a consistent real valued combination is guaranteed by the soundness properties of interval extensions and their evaluations (theorem 3.2-2).

the interval extension, which guarantees that no overestimation error is made with infinite precision.

For complex constraints box-consistency is stronger than hull-consistency applied on the primitive constraints obtained by decomposition [28]. This is a consequence of the amplification of the locality problem induced by the constraint decomposition. Enforcement of a consistency criterion directly to a complex constraint is necessarily stronger (or equal) than its enforcement on the primitive constraints obtained by the decomposition of the constraint. In particular, this is the case for the box-consistency criterion, and because for primitive constraints box-consistency and hull-consistency are equivalent, box-consistency on complex constraints must be stronger (or equal) than hull-consistency on the respective primitive constraints.

Nevertheless, the pruning obtained by box-consistency is often insufficient for non linear constraints. If there are several uncertain variables, the Newton method to enforce box-consistency aims at tightening the bounds of each one substituting the other variables by their F-interval domains. Hence, if there are n uncertain variables, it is still necessary to work with n univariate functions with n-1 interval values. Depending on the complexity of the constraint, the uncertainty of the n-1 interval values may cause a wide range of possible values for the univariate functions, preventing possible domain reduction.

Consider again constraint $c \equiv x_1 \times (x_2 - x_1) = 0$ and a different interval extension of its left side represented by $F_{E'} \equiv X_1 \times (2X_2 - (X_1 + X_2))$. Note that the real function f represented by $f_E \equiv x_1 \times (x_2 - x_1)$ is the same represented by $f_{E'} \equiv x_1 \times (2x_2 - (x_1 + x_2))$ (the real expressions are equivalent) and $F_{E'}$ is the natural interval expression of $f_{E'}$ and consequently an interval extension of f. Constraint c, which was not box-consistent wrt $B_1 = <[-0.5..2.5],[0.5..1.5]>$ and F_E, is box-consistent wrt B_1 and $F_{E'}$. Using the interval expression $F_{E'}$ (instead of F_E) for verifying the criterion for each bound of each variable, the uncertainty of the interval values allows a wider range of possible values. For example, the condition on the left bound of variable x_1 is now satisfied since $0 \in F_{E'}(<[-0.5..-0.499],[0.5..1.5]>) = [-1.5..0.001]$.

Generalising the concept of local consistency from a constraint to the set of constraints of a CCSP, we can say that a CCSP $P = (X,D,C)$ is locally consistent (interval, hull or box-consistent) wrt a set of real valued combinations $A \in 2^D$ iff all its constraints are locally consistent wrt A. Since the propagation algorithm obtains the greatest common fixed-point (of the monotonic narrowing functions) included in the original domains, then the result of applying the propagation algorithm to a set $A \in 2^D$ is the largest subset $A' \subseteq A$ for which each constraint is locally consistent.

Definition 5.1-5 (Local-Consistency). Let $P = (X,D,C)$ be a CCSP. Let A be an element of the power set of D $(A \in 2^D)$. P is locally-consistent wrt A iff:

$\forall_{c \in C}$ c is locally-consistent wrt A

Let S be a set of monotonic narrowing functions associated with the constraints in C which enforce a particular local consistency by constraint propagation:

P is locally-consistent wrt prune(S,A)

$\forall_{A' \subseteq A}$ (P is locally-consistent wrt $A' \Rightarrow A' \subseteq$ prune(S,A)) ❑

When only local consistency techniques are applied to non-trivial problems the achieved reduction of the search space is often poor (a problem called early quiescence in [44]).

Consider the CCSP represented in figure 5.1 where there are two real variables, x_1 and x_2 with values ranging within [-5..5] and two constraints, $c_1 \equiv x_1^2 + x_2^2 - 2^2 \leq 0$ and $c_2 \equiv (x_1 - 1)^2 + (x_2 - 1)^2 - 2.5^2 \geq 0$. The thick solid square is the initial domain box. The two circumferences represent the two constraints. The grey area represents the complete set of solutions. The thin solid square is the box obtained by enforcing a local consistency, either box-consistency (with the natural interval extensions represented by $F_{E_1} \equiv X_1^2 + X_2^2 - 2^2$ and $F_{E_2} \equiv (X_1-1)^2 + (X_2-1)^2 - 2.5^2$ respectively) or hull-consistency (on the decomposed problem). The dashed square is the smallest F-box enclosing all solutions within the initial box.

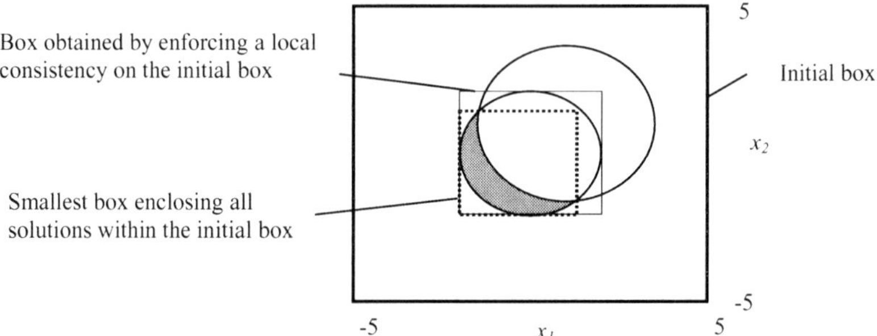

Figure 5.1 Insufficient pruning achieved by local consistency enforcement.

The figure shows that the local consistency criterion cannot prune the search space inside the smaller circumference – the pruning is the same as it would be without the constraint associated with the larger circumference. Depending on the decision problem to solve, this may be irrelevant or, on the contrary, it may justify the enforcement of a stronger consistency.

5.2   Higher Order Consistency

Better pruning of the variable domains may be achieved if, complementary to a local property, some (global) properties are also enforced on the overall constraint set.

As in local consistency, higher order consistency types used in continuous domains are approximations of similar concepts originally developed for solving CSPs over finite domains. The most general concept to capture a global property among the overall constraint set of a CSP is the definition of strong k-consistency given by Freuder in [55].

A CSP is k-consistent ($k \geq 2$) iff any consistent instantiation of k-1 variables can be extended by instantiating any of the remaining variables. A CSP is strongly k-consistent if it is i-consistent for all $i \leq k$.

In particular, strong 2-consistency corresponds to arc-consistency (see definition 4.3-1) and hull-consistency (see definition 4.3-3) can be seen as an approximation of strong 2-consistency restricted to the bounds of the variable domains (that is why the original denomination was 2B-consistency: B for bounds).

Similarly, higher order consistency types used in continuous domains are approximations of strong k-consistency (with $k>2$) restricted to the bounds of the variable domains.

Strong 3-consistency adapt path-consistency, a higher order extension of arc-consistency [97, 98], to continuous domains. Specifically, 3B-consistency [91] and Bound-consistency [119], are generalisations of hull and box-consistency respectively. In both, the property enforced on the overall constraint set is the following: if the domain of one variable is reduced to one of its bounds then the obtained F-box must contain a sub-box for which the CCSP is locally consistent (hull or box-consistency respectively).

The following is a generic definition for the consistency types used in continuous domains. Accordingly to this definition local consistency is just a special case of the generic kB-Consistency with $k=2$.

Definition 5.2-1 (kB-Consistency). Let $P=(X,D,C)$ be a CCSP. Let A be an element of the power set of D ($A \in 2^D$) and k an integer number.

P is 2B-Consistent wrt A iff P is locally-consistent wrt A

$\forall_{k>2}$ P is kB-Consistent wrt A iff

$\quad \forall_{x_i \in X}$ $(\exists_{A_1 \subseteq B_1}$ P is (k-1)B-Consistent wrt $A_1 \wedge \exists_{A_2 \subseteq B_2}$ P is (k-1)B-Consistent wrt A_2)

$\qquad$ where B_1 and B_2 are two elements of the power set of D such that:

$\qquad B_1[x_i]=cleft(B[x_i])$, $B_2[x_i]=cright(B[x_i])$ and $\forall_{x_j \in X} (x_j \neq x_i \Rightarrow B_1[x_j]=B_2[x_j]=B[x_i])$. $\quad \square$

In the rest of this work we will denote by kB-Hull-consistency and kB-Box-consistency the cases where the local consistency enforced is respectively Hull- and Box-consistency. If $k=2$ the designation kB may be omitted and the generic term Local-consistency may be used to designate an unspecified type of local consistency. kB-Hull-consistency corresponds to the notion of kB-consistency proposed by Lhomme in [91] and 3B-Box-consistency corresponds to the notion of Bound-consistency introduced in [119].

The algorithms to enforce these stronger consistencies interleave constraint propagation with search techniques. The price to pay for stronger consistency is thus the growth in computational cost of the enforcing algorithm, limiting the practical applicability of such criteria.

Figures 5.2 and 5.3 present an algorithm for enforcing kB-Consistency, either kB-Box-consistency or kB-Hull-consistency. The algorithm is a generalisation of the 3B-consistency algorithm proposed in [91] with some improvements suggested in [20]. The input is the order k, a CCSP $P=(X,D,C)$ and an F-box $B \subseteq D$. The output is the largest kB-Consistent F-box within B or the empty set if it is proved that there is no such box.

The main function named *kB-consistency* uses an auxiliary function with the same name but an extra parameter (*size*). This auxiliary function computes the largest F-box within F-box B with the following property: if the domain of one variable is reduced to its leftmost/rightmost subinterval (with width not exceeding *size*) then the obtained F-box must contain a sub-box for which the CCSP is (k-1)B-Consistent[2]. Therefore, when *size* is small enough to force such subintervals to be canonical, the auxiliary function computes the largest kB-Consistent F-box within B.

---

function *kB-consistency*(an integer $k \geq 2$, a CCSP $P=(X,D,C)$, an F-box B)
(1)   **if** $k=2$ **then return** *prune*(set of NF from C,B);
(2)   *size* $\leftarrow$ *largestWidth*(B);
(3)   **repeat**
(4)   *size* $\leftarrow$ *size*/2;
(5) $B \leftarrow$ *kB-consistency*(*size*,k,P,B);
(6)   **if** $B = \varnothing$ **then return** $\varnothing$;
(7)   **until** *canonical*(*size,B*);
(8)   **return** B;
end function

Figure 5.2 The generic *kB-consistency* algorithm.

The main *kB-consistency* function (figure 5.2) first checks (line 1) whether the enforcement of a local consistency (if $k=2$) is sufficient and, in that case, it calls the propagation algorithm with the appropriate set of narrowing functions for enforcing the local criterion (Box-consistency or Hull-consistency). Otherwise, if the enforcement of an higher order consistency is required, the *size* value is initialised with the largest domain width within the

---

[2] In the following this property will be denoted kB(size)-Consistency.

original *F*-box (line 2) and smaller values are subsequently considered (line 4) for improving the domain reduction computed by the auxiliary *kB-consistency* function (line 5). The procedure terminates when the box is proved to be inconsistent returning the empty set (line 6) or when the *size* value is smaller enough to assure the *kB-consistency* wrt the current *F*-box *B* (line 7), in which case this box is returned (line 8).

The generic *kB-consistency* algorithm is correct and terminates if the *kB-consistency* function terminates and computes the largest kB(size)-Consistent *F*-box within *B*.

```
function kB-consistency(F-number size, an integer k≥2, a CCSP P=(<x_1,…,x_n>,D,C), an F-box B)
 (1)   if k=2 then return prune(set of NF from C,B);
 (2)   for j = 1 to 2n do mem[j] ← universalBox;
 (3)   j ← 1; unfixedBounds ← 2n;
 (4)   while unfixedBounds>0 do
 (5)       if mem[j] ⊈ B then
 (6)           fixed ←FALSE;
 (7)           repeat
 (8)               if isOdd(j) then
 (9)                   i ← (j+1)/2;
(10)                   I_1 ← [left(B[x_i])..min(right(B[x_i]),⌈left(B[x_i])+size⌉)];
(11)                   I_2 ← [min(right(B[x_i]),⌈left(B[x_i])+size⌉)..right(B[x_i])];
(12)               else
(13)                   i ← j/2;
(14)                   I_1 ← [max(left(B[x_i]),⌊right(B[x_i])-size⌋)..right(B[x_i])];
(15)                   I_2 ← [left(B[x_i])..max(left(B[x_i]),⌊right(B[x_i])-size⌋)];
(16)               end if;
(17)               B[x_i] ← I_1;
(18)               mem[j] ← kB-consistency(size,k-1,P, mem[j] ∩ B);
(19)               B[x_i] ← I_1 ∪ I_2;
(20)               if mem[j] ≠ ∅ then
(21)                   fixed ←TRUE;
(22)                   adjustBound(j,mem[j],B);
(23)               else
(24)                   if left(I_2) = right(I_2) then return ∅;
(25)                   B[x_i] ← I_2;
(26)                   mem[j] ← universalBox;
(27)                   unfixedBounds ← 2n;
(28)                   B' ← kB-consistency(size,k-1,P,B);
(29)                   if B'= ∅ then return ∅;
(30)                   for l = 1 to 2n do adjustBound(l,B',B);
(31)               end if;
(32)           until fixed
(33)       end if;
(34)       unfixedBounds ← unfixedBounds-1;
(35)       if j = 2n then j ← 1; else j ← j +1; end if;
(36)   end while;
(37)   return B;
end function
```

Figure 5.3 The auxiliary *kB-consistency* function used by the generic algorithm.

The auxiliary *kB-consistency* function (figure 5.3) also calls the propagation algorithm if local consistency is all that is required (line 1). Otherwise, all *2n* bounds of the *n* variables must be fixed (narrowed) in a round robin fashion until the kB(size)-Consistency property is achieved (while cycle from line 4 to line 36).

In order to narrow a particular variable bound (lines 6-32), a sub-box is considered (lines 9-11,17 for the left bound and 13-15,17 for the right bound) where the domain of that variable is reduced to its leftmost/rightmost subinterval with width equal to *size* (if the original domain is smaller than *size* then the whole box is considered). The possibility of (k-1)B(size)-Consistency is verified for this sub-box (line 18) and if it succeeds (line 20) the bound j may have to be adjusted in line 22 (the procedure *adjustBound(j,B',B)* verifies if the value of the bound j in B' is different from the one in B, in which case the later is updated and *mem[j]* is reinitialized into the *universalBox*). In case of failure the sub-box is discarded from the original box (line 25) and, after narrowing the remaining *F*-box (lines 28-30), the next leftmost/rightmost subinterval is considered for the same variable bound until its satisfaction (line 20) or the complete elimination of the original box (line 24).

Once a particular variable bound is changed, all the other variable bounds must be checked again (line 27) to guarantee that the while cycle (lines 4-36) only terminates when (k-1)B(size)-Consistency is satisfied simultaneously for all bounds. The above procedure assures the correctness of the algorithm and termination is guaranteed due to the fact that after any $2n$ steps of the while cycle either all the bounds are fixed (and the algorithm terminates) or at least one bound is reduced (and after a finite number of such reductions the original box will be completely discarded).

The *mem* vector of *F*-boxes, one for each variable bound (initialized with the *universalBox* where every variable domain ranges between $-\infty$ and $+\infty$), implements the improvements suggested in [20] to the original algorithm. The idea is to memorize the result of the previous bound reduction (line 18) and use it in the next reduction attempt for the same bound, either by guaranteeing the property satisfaction without checking (in line 5, if $mem[j] \subseteq B$ the bound is not narrowed) or, by reducing the subbox that is checked (in line 18, (k-1)B(size)-Consistency is verified in $mem[j] \cap B$ instead of in B).

All the consistency criteria used in continuous domains, either local or higher order consistencies, are partial consistencies. The adequacy of a partial consistency for a particular CCSP must be evaluated taking into account the trade-off between the pruning it achieves and its execution time. Moreover, it is necessary to be aware that the filtering process is performed within a larger procedure for solving the CCSP and it may be globally advantageous to obtain faster, if less accurate, results.

5.3   Summary

In this chapter Interval-, Hull- and Box-consistency were identified as the main local consistency criteria used in continuous domains. Their definitions were presented and the methods for enforcing them discussed. Higher order consistency criteria were defined as generalisations of the local consistency criteria, and a generic enforcing algorithm was presented. In the next chapter Global Hull-consistency is proposed as an alternative consistency criterion in continuous domains. Several alternative enforcing algorithms are suggested and their properties derived.

Chapter 6

Global Hull-Consistency

The pruning of the search space achieved by local consistency techniques on non-trivial problems is often poor. Nevertheless, the computational cost of enforcing stronger consistencies may limit their practical applicability.

In this context we propose a strong consistency criterion, Global Hull-consistency, and show that its use in some such problems has reasonable computational costs. The need for a strong consistency requirement originated on solving constraint problems which include parametric ordinary differential equations [36, 37, 38], which will be addressed in part II.

The key idea of Global Hull-consistency is to generalise local Hull-consistency criterion to a higher level, by considering the set of all constraints as a single global constraint. Hence, it must guarantee arc-consistency at the bounds of the variable domains for this single global constraint: if a variable is instantiated with the value of one of its bounds then there must be a consistent instantiation of the other variables, and this complete instantiation is a solution of the CCSP.

If real values could be represented with infinite precision, Global Hull-consistency would be similar to the notion of e-consistency (e- for external) presented in [30]. In this case, enforcing Global Hull-consistency (or e-consistency) on a real box corresponds to obtaining the smallest external real box enclosing all solutions of a CCSP within the original box.

However, due to limitations of the representation of real values, the result of enforcing Global Hull-consistency on a box of domains must be an F-box, enclosing the real box obtained by enforcing e-consistency. These limitations prevent the enforcing algorithms from dealing directly with real valued instantiations, requiring them to operate with their canonical F-box approximations. Because within a canonical solution there might be a solution of the CCSP (and this possibility cannot be discarded due to the system limitations), the best thing that can be done is to guarantee that for each bound of each variable there is a canonical F-box instantiation which is a canonical solution. This is formalised in the following definition of Global Hull-consistency.

Definition 6-1 (Global Hull-Consistency). Let $P=(X,D,C)$ be a CCSP. Let B be an F-box which is an element of the power set of D ($B \in 2^D$). P is Global Hull-consistent wrt B iff:

$$\forall_{x_i \in X} \, \exists_{B_l \subseteq B} \, (B_l[x_i] = cleft(B[x_i]) \land B_l \text{ is a canonical solution of } P) \, \land$$

$$\exists_{B_r \subseteq B} \, (B_r[x_i] = cright(B[x_i]) \land B_r \text{ is a canonical solution of } P) \qquad \square$$

Any strategy to enforce Global Hull-consistency must be able to localise the canonical solutions within a box of domains that are extreme with respect to each bound of each variable domain. Global Hull-consistency is the strongest criterion for narrowing a box of domains into a single smaller F-box that looses no possible solution. Narrowing the obtained F-box further would necessarily exclude one extreme canonical solution, possibly discarding a solution.

Figure 6.1 shows the box obtained by enforcing Global Hull-consistency on the example presented in figure 5.1. The small boxes (1, 2, 3 and 4) represent the extreme canonical solutions (wrt each variable bound) that were found by the enforcing algorithm.

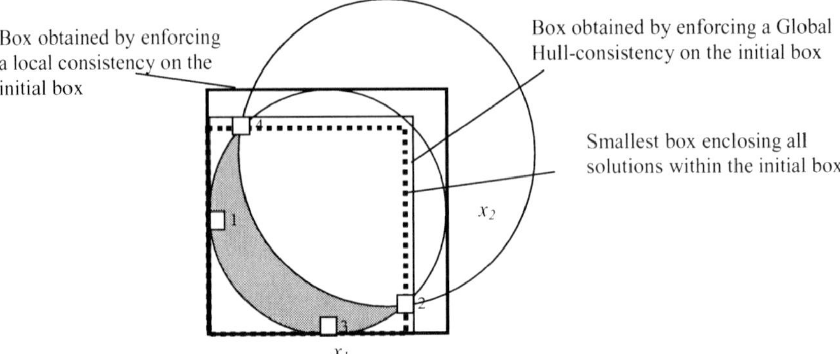

Figure 6.1 Pruning achieved by enforcing Global Hull-consistency.

The obtained F-box is an approximation of the smallest real box enclosing the solution space and the quality of the enclosure depends on the width of the canonical F-boxes (the available precision for the representation of two consecutive real values). Narrowing this F-box further would result in the elimination of at least one of the extreme canonical solutions (1, 2, 3 or 4) and all the real solutions that it might contain.

The existing constraint systems developed for continuous domains are able to enforce some kind of partial consistency (usually Local-consistency or eventually 3B-Consistency) and use it for isolating solutions of a CCSP through a branch and bound strategy, implementing a backtrack search of the space of possibilities (see chapter 2). Thus, the ultimate goal of these systems is to find individual solutions and not to enclose the complete solution space within a single box.

In CCSPs where the number of solutions is small the strategy aiming at identifying them all could be used for enforcing Global Hull-consistency, and the resulting box should enclose the complete set of solutions. However, in the case of under-constrained CCSPs, the huge number of solutions (usually infinite in continuous domains) makes this strategy inadequate and specialised algorithms are needed for enforcing Global Hull-consistency within reasonable computational costs.

In the rest of this chapter several algorithms will be presented for enforcing Global Hull-consistency. The next section discusses the relations between Global Hull-consistency and kB-Consistency and shows how the former can be obtained by an algorithm that enforces an appropriate higher order consistency. Section 6.2 addresses algorithms which can be easily implemented by the existing backtrack search systems without significant modifications of their propagation mechanisms. Section 6.3 proposes the substitution of backtrack search by ordered search and discusses which of the previous algorithms may profit from it. Section 6.4 presents a specialised algorithm which uses a binary tree for the representation of the search space and includes a local search procedure for anticipating the localisation of extreme canonical solutions.

6.1   The Higher Order Consistency Approach

If a CCSP contains only a single variable then enforcing a Local-consistency (2B-Consistency), either Hull-consistency or Box-consistency, is sufficient to guarantee that each bound is a canonical solution of the CCSP and so, the resulting box is Global Hull-consistent. The reason for this is that according to the definitions of Hull and Box-consistency (definitions 5.1-3 and 5.1-4) if the variable is instantiated with its leftmost/rightmost canonical subinterval then each constraint must be satisfied (in the sense that within the interval obtained by its interval arithmetic evaluation there is a real value satisfying the constraint). Because there is only one variable in the CCSP, its instantiation is a complete instantiation which satisfies all the constraints and, consequently, is a canonical solution of the CCSP.

When a CCSP contains two variables, enforcing 3B-Consistency guarantees that if the domain of one variable is reduced to one of its (canonical) bounds the resulting *F*-box (with at most one non-canonical domain) must contain a sub-box for which the CCSP is locally consistent (see definition 5.2-1). But if this sub-box is locally consistent, the instantiation of the non-canonical domain to one of its bounds satisfies all the constraints and, consequently, there is at least a canonical solution within the sub-box. Therefore, enforcing 3B-Consistency on a CCSP with two variables guarantees Global Hull-consistency.

The above property between Global Hull-consistency and kB-Consistency may be generalised for any number of variables occurring in a CCSP and is formalised in the following theorem.

Theorem 6.1-1 (Equivalence between Global Hull-consistency & (n+1)B-Consistency).
Let $P=(X,D,C)$ be a CCSP with n variables ($X=<x_1,\ldots,x_n>$). Let B be an F-box which is an element of the power set of D ($B \in 2^D$). The following property necessary holds:
P is Global Hull-consistent wrt B iff P is (n+1)B-Consistent wrt B ❑

6.1.1   The *(n+1)B-consistency Algorithm*

Given theorem 6.1-1, a straightforward approach for enforcing Global Hull-consistency is to use the *kB-consistency* algorithm (Chapter 5, section 5.2) and choose an appropriate value for the order k (it must be equal to the number of variables plus one).

Figure 6.2 illustrates this approach whose algorithm will be denoted *(n+1)B-consistency* to emphasise its dependency on the number of variables of the CCSP.

---

function *(n+1)B-consistency*(a CCSP $P=(<x_1,\ldots,x_n>,D_1\times\ldots\times D_n,C)$, an F-box B)
  (1) **return** *kB-consistency*(n+1,P,B);
end function

Figure 6.2 The *(n+1)B-consistency* algorithm.

The input is a CCSP P with n variables and an F-Box B (with $B \subseteq <I_{apx}(D_1),\ldots, I_{apx}(D_n)>$). The result (line 1) is the largest $(n+1)$B-Consistent F-box within B (which is equivalent to the largest Global Hull-consistent F-box within B) or the empty set if it is proved that there is no such box. The correctness of the algorithm is guaranteed by theorem 6.1-1.

6.2   Backtrack Search Approaches

Most interval constraint systems provide a search mechanism (alternating pruning and branching steps) which implements a backtracking search for obtaining canonical solutions. The pruning is normally achieved by enforcing some Local-consistency (Hull/Box-consistency or eventually 3B-Consistency) but it can be generalised for any kB-Consistency criterion (with $1<k\leq n+1$, n being the number of variables). The branching is normally achieved by choosing a variable domain (according to a split strategy) and to separately consider the two boxes obtained by dividing this domain at the mid point. In order to keep track of search space regions that may contain canonical solutions, a stack of F-boxes is maintained and explored throughout backtracking.

Figure 6.3 presents the function *backtrackSearch* which implements this generic algorithm. It assumes that the pruning is achieved by the *kB-consistency* function with some predefined value for the order k and a predefined type of local consistency (Hull or Box-consistency). It also assumes that the branching is implemented by procedure *splitBox* according to some, not shown, split strategy (LW, RR or x_i) and a side (LEFT or RIGHT). The procedure receives the F-box to split and returns the branches that will be explored subsequently. The split strategy determines the variable to split: if LW, the chosen variable is that with the largest domain width; if RR, each variable is chosen accordingly to a round-robin strategy; and if x_i, variable x_i is chosen if its domain is not canonical, otherwise one of the other split strategies is used. The side determines which branch will be explored first.

Function *backtrackSearch* has two arguments, the first is a CCSP P and the second a stack S of F-boxes where the canonical solutions of P will be searched from. The second argument is also an output argument returning the remaining values of S, which account for the reduction of the search space achieved during the search. The result of the function is a canonical solution of P within one of the F-boxes of S or the empty set if no such canonical solution exists.

function *backtrackSearch*(a CCSP $P=(X,D,C)$, **inout** a stack S of F-boxes)

(1)   **while** $S.size()>0$ **do**

(2) $B \leftarrow S.pop()$;

(3) $B \leftarrow kB\text{-}consistency(P,B)$;

(4)   **if** $B \neq \varnothing$ **then**

(5)   **if** *isCanonical*(B) **then return** B;

(6) $splitBox(B,B_1,B_2)$;

(7) $S.push(B_2)$;

(8) $S.push(B_1)$;

(9)   **end if**;

(10) **end while**;

(11) **return** $\varnothing$;

end function

Figure 6.3 The generic backtrack search algorithm for finding canonical solutions.

During the execution of the *backtrackSearch* function the stack S of F-boxes representing the remaining search space is maintained by a set of functions which implement the usual stack operations: *size* returns the number of elements; *pop* returns the top element and removes it from the stack; *push* adds a new top element.

The function is implemented as a *while* cycle (lines 1-10) executed as long as there are F-Boxes in the stack (line 1) and no canonical solution was found. If the stack becomes empty then there are no more canonical solutions and the empty set is returned (line 11). Otherwise the top element is removed from the stack (line 2) and narrowed by the pruning function (line 3). If it is inconsistent then nothing is done in this cycle and a new one is started with the next top element. If the resulting box is canonical then a canonical solution has been found and the box is returned (line 5). Otherwise, if the pruning function could not discard the F-box obtaining a smaller F-box which is not canonical then the box is split by the *splitBox* function (line 6) and the two resulting F-boxes are added to the top of the stack (lines 7 and 8).

The algorithm is correct and terminates. On the one hand, the narrowing of the search space is achieved by enforcing a partial consistency requirement (kB-Consistency) which does not eliminate canonical solutions. On the other hand, the split of the boxes that cannot be further pruned guarantee that canonical F-boxes are eventually analysed, either reducing the search space (if such boxes violate the partial consistency criterion), or returning the canonical solution found. The finite number of canonical F-boxes within any search space guarantees termination of the algorithm.

The following subsections present four different algorithms, based on the above generic backtrack search mechanism, for enforcing Global Hull-consistency.

6.2.1 The BS_0 Algorithm

The simplest backtracking algorithm, BS_0, for enforcing Global Hull-consistency on an F-box uses *backtrackSearch* for finding all the solutions within the box and returns the smallest F-box enclosing them.

The algorithm is presented in Figure 6.4. The input is a CCSP P and an F-Box B (with $B \subseteq <I_{apx}(D_1),\ldots, I_{apx}(D_n)>$). The result is the smallest F-box containing all the canonical solutions within B (which is equivalent to the largest Global Hull-consistent F-box within B) or the empty set if B does not contain any canonical solutions.

function BS_0(a CCSP $P=(X,D_1\times\ldots\times D_n,C)$, an F-box B)

(1) $S \leftarrow B$;

(2) $B_{sol} \leftarrow backtrackSearch(P,S)$;

(3)   **if** $B_{sol} = \varnothing$ **then return** $\varnothing$ **else** $B_{in} \leftarrow B_{sol}$;

(4)   **repeat**

(5) $B_{sol} \leftarrow backtrackSearch(P,S)$;

(6)   **if** $B_{sol} \neq \varnothing$ **then** $B_{in} \leftarrow B_{in} \uplus B_{sol}$;

(7)   **until** $B_{sol} = \varnothing$;

(8)   **return** B_{in};

end function

Figure 6.4 The BS_0 algorithm.

The algorithm maintains an inner box B_{in} that is the smallest box enclosing all the currently found canonical solutions. When all the canonical solutions have been found this inner box must be the largest Global Hull-consistent F-box within the original domains and is returned as the final result of the algorithm (line 8).

A stack S of F-boxes is initialised with the single box B (line 1) and a canonical solution is searched through backtracking on this stack (line 2). If no canonical solution is found the empty set is returned; otherwise, the inner box is initialised to the obtained canonical solution (line 3). The algorithm proceeds with a *repeat* cycle, where new solutions are searched within the remaining search space (line 5) and the inner box is enlarged to include them (line 6)[1], until there are no more solutions (line 7).

The correcteness of the algorithm is guaranteed by the correcteness of the *backtrackSearch* algorithm and by definition 6-1 which implies that the smallest box containing all the canonical solutions is the largest Global Hull-consistent F-box.

Termination is guaranteed by the termination of the *backtrackSearch* function that reduces the finite search space in each invocation (at least the box associated with the new canonical solution is removed from S) thus requiring a finite number of steps of the *repeat* cycle.

---

[1] The symbol $\uplus$ represents the union hull operation - see definition 2.2.2-1.

6.2.2   The BS_1 Algorithm

The previous brute-force algorithm blindly searches for new canonical solutions ignoring those already found. Consequently, many space regions which are searched are irrelevant to Global Hull-consistency. In particular, canonical solutions inside the inner box (which encloses all known solutions) are useless regarding Global Hull-consistency.

The backtrack search algorithm, BS_1, thus avoids searching regions within the inner box. To achieve this, it separately searches extreme canonical solutions with respect to each variable bound and adds a new constraint whenever a new canonical solution is found to narrow the relevant search space.

Figure 6.5 presents the algorithm. The input is a CCSP P and an F-Box $B \subseteq <I_{apx}(D_1),\ldots,$ $I_{apx}(D_n)>$, and the result is the largest Global Hull-consistent F-box within B or the empty set if it does not contain any canonical solutions.

---

function BS_1(a CCSP $P=(<x_1,\ldots,x_n>,D_1\times\ldots\times D_n,C)$, an F-box B)

(1) $S \leftarrow B$; $splitSide \leftarrow$ LEFT;

(2) $B_{sol} \leftarrow backtrackSearch(P,S)$;

(3)   **if** $B_{sol} = \varnothing$ **then return** $\varnothing$ **else** $B_{in} \leftarrow B_{sol}$;

(4)   **for** $i=1$ **to** n **do**

(5)   **while** $left(B[x_i]) < left(B_{in}[x_i])$ **do**

(6) $B_{sol} \leftarrow backtrackSearch((<x_1,\ldots,x_n>,D_1\times\ldots\times D_n,C\cup\{x_i \leq left(B_{in}[x_i])\}),S)$;

(7)   **if** $B_{sol} = \varnothing$ **then** $B[x_i] \leftarrow [left(B_{in}[x_i])..right(B[x_i])]$;

(8)   **else** $B_{in} \leftarrow B_{in} \uplus B_{sol}$; **end if**;

(9)   **end while**;

(10) $S \leftarrow B$; $splitSide \leftarrow$ RIGHT;

(11)   **while** $right(B[x_i]) > right(B_{in}[x_i])$ **do**

(12) $B_{sol} \leftarrow backtrackSearch((<x_1,\ldots,x_n>,D_1\times\ldots\times D_n,C\cup\{x_i \geq right(B_{in}[x_i])\}),S)$;

(13)   **if** $B_{sol} = \varnothing$ **then** $B[x_i] \leftarrow [left(B[x_i])..right(B_{in}[x_i])]$;

(14)   **else** $B_{in} \leftarrow B_{in} \uplus B_{sol}$; **end if**;

(15)   **end while**;

(16) $S \leftarrow B$; $splitSide \leftarrow$ LEFT;

(17) **end for**;

(18) **return** B_{in};

end function

Figure 6.5 The BS_1 algorithm.

Similarly to BS_0, function BS_1 maintains an inner box B_{in} that is returned as the final result when a solution exists (line 18).

The initialisation of the stack S of F-boxes (line 1), the search for a first canonical solution (line 2) and the initialisation (line 3) of the inner box (or returning of the empty set) are identical to BS_0. However, BS_1 proceeds by considering each variable domain separately (the *for* cycle between lines 4 and 17), to search the extreme canonical solutions with respect to its left (lines 5-9) and right (lines 11-15) bounds.

The search for the leftmost canonical solution for a variable is executed in a *while* cycle (lines 5-9) within the regions of the current search space where this variable has values which are less than any known solution (line 6). While new solutions are found, the inner box is updated to enclose them (line 8) and the cycle continues. Eventually, the last canonical solution found is the leftmost solution and the original search box is updated accordingly (line 7) terminating the cycle. The search for the rightmost solution is similar.

Before every search for an extreme canonical bound, the search space S must be reinitialised with the current domains box (lines 1, 10 and 16) in order to disregard all the previous additional constraints that were considered for restraining the search space in other search contexts and the side branching option used by function *backtrackSearch* is redefined towards the appropriate direction (variable *splitSide*).

The algorithm is correct since each *while* cycle obtains an extreme canonical solution wrt the variable under consideration. This is achieved in a finite number of steps because the backtrack search either fails, terminating the algorithm at once with the correct result, or finds a new solution and reduces the finite search space for the following iteration.

6.2.3   The BS_2 Algorithm

The key idea of the next algorithm, BS_2, (suggested by Frédéric Benhamou in a personal communication), is to control the branching strategy to direct the search towards the extreme canonical solutions. Instead of constraining the search space whenever a new canonical solution is found, the branching strategy guarantees that the first solution found is already an extreme canonical solution with respect to some variable bound. The search is separately executed for each bound of each variable domain and the branching options, of strategy and side, adopted by function *backtrackSearch*, are explicitly controlled by variables *splitStrategy* and *splitSide*, respectively

Figure 6.6 illustrates the BS_2 algorithm. The input is a CCSP P and an F-Box B ($B \subseteq <I_{apx}(D_1),...,I_{apx}(D_n)>$), and the result is the largest Global Hull-consistent F-box within B or the empty set if it does not contain any canonical solutions.

```
function BS₂(a CCSP P=(<x₁,…,xₙ>,D₁×…×Dₙ,C), an F-box B)
  (1)   S ← B;
  (2)   splitStrategy ← x₁; splitSide ← LEFT; B_sol ← backtrackSearch(P,S);
  (3)   if B_sol = ∅ then return ∅ else B_in ← B_sol;
  (4)   for j=2 to 2n do
  (5)      if isOdd(j) then
  (6)         splitSide ← LEFT; i ← (j+1)/2;
  (7)         B[x_{i-1}] ← [left(B[x_{i-1}])..right(B_in[x_{i-1}])];
  (8)      else
  (9)         splitSide ← RIGHT; i ← j/2;
  (10)        B[x_i] ← [left(B_in[x_i])..right(B[x_i])];
  (11)     end if;
  (12)     S ← B;
  (13)     splitStrategy ← x_i; B_sol ← backtrackSearch(P,S);
  (14)     B_in ← B_in ⊎ B_sol;
  (15)  end for;
  (16)  return B_in;
end function
```

Figure 6.6 The BS_2 algorithm.

When the *splitStrategy* and the *splitSide* are respectively x_i and LEFT/RIGHT, the *backtrackSearch* function returns the leftmost/rightmost canonical solution wrt the variable x_i. Consider the stack S of F-boxes, representing the search space at a given moment, as a sequence of k F-boxes $B_1,...,B_k$, with B_k its top element. If S is obtained from an original F-box B after several iterations of the *while* cycle of the *backtrackSearch* function and during the process only variable x_i was split then for any two consecutive F-boxes within S, B_j and B_{j+1}, $right(B_{j+1}[x_i]) \leq left(B_j[x_i])$ if *splitSide*=LEFT and $right(B_j[x_i]) \leq left(B_{j+1}[x_i])$ if *splitSide*=RIGHT. Thus, if there are no other split variables, when a solution is found, it must be the result of narrowing the top element into a canonical box and so any x_i value must be smaller/larger or equal than any other x_i value of any other F-box implying that the canonical solution is the leftmost/rightmost canonical solution wrt the variable x_i. With an x_i split strategy the only possible cases where the split variable is not x_i is when the domain of x_i is already canonical and so both subboxes will have the same x_i domain which must contain the smallest/largest possible value for x_i within S. Therefore, with an x_i split strategy, the search space at any given moment may be represented as a sequence of $k+m$ F-boxes $B_1,\ldots,B_k,B_{k+1},\ldots,B_{k+m}$, with B_{k+m} its top element and where all the $B_{k+1},\ldots,B_{k+m}$ boxes have the same smallest/largest x_i canonical domain (for some $m \geq 0$) and for any two consecutive F-boxes B_j and B_{j+1}, with $1 < j \leq k$, $right(B_{j+1}[x_i]) \leq left(B_j[x_i])$ if $splitSide$=LEFT and $right(B_j[x_i]) \leq left(B_{j+1}[x_i])$ if $splitSide$=RIGHT. Thus, in any case, if a solution is found within the top element it must be the leftmost/rightmost canonical solution wrt variable x_i.

During the whole execution of the BS_2 algorithm, exactly $2n$ canonical solutions are found (if there are any canonical solutions, otherwise the empty set is returned in line 3), one for each variable bound, and the enclosing inner box B_{in} that is maintained (line 14) is returned in the end (line 16).

After the initialisation of the stack S of F-boxes with the original box (line 1), the branching options set the search for the leftmost canonical solution of variable x_1 (line 2). Then, the algorithm proceeds by separately searching the extreme canonical solutions of the remaining $2n-1$ variable bounds (*for* cycle, in lines 4-15).

The search for the leftmost (rightmost) canonical solution wrt a particular variable is executed by a single function call to the *backtrackSearch* algorithm over the current domains box (line 13) whose branching options (variable and side) must be appropriately set (lines 13 and 6/9).

Before every new search for an extreme canonical bound, the current search space S must be set to the current domains box (line 12) that considers the latest update on its bounds (lines 7/10).

Clearly the algorithm terminates (it does not contain any possibly infinite cycles and the *backtrackSearch* function terminates) and is correct. As proved before, with an x_i split strategy for a particular split side, left or right, the first canonical solution found by the generic backtrack search algorithm is the leftmost/rightmost.

6.2.4   The BS_3 Algorithm

All previous approaches independently search for each extreme variable bound. In BS_1, the addition of new constraints improve the search by preventing search on irrelevant space regions. In BS_2, the branching strategy exploits a partition on the search space that guarantees that the canonical solutions found are indeed extreme canonical solutions.

The last of the backtrack search algorithms, BS_3, tries to reuse the computational effort as much as possible within the limitations imposed by a stack data structure for representing the search space. It is a branch and bound algorithm where the extreme canonical solutions with respect to each bound of each variable domain are searched simultaneously within a round robin scheme.

The basic idea is to maintain an outer box that includes all possible canonical solutions and an inner box that is the smallest box enclosing all currently found canonical solutions. The search for extreme canonical bounds is restricted to the region between these two boxes.

The algorithm stops when the inner box equals the outer box returning it as the largest box within the original domains satisfying the Global Hull-consistency criterion.

Figure 6.7 shows BS_3. From a CCSP P and an F-Box B, the function BS_3 computes the largest Global Hull-consistent F-box within B or the empty set if it does not contain any canonical solutions.

```
function BS₃(a CCSP P=(<x₁,...,xₙ>,D₁×...×Dₙ,C), an F-box B)
(1)    S ← B; B_sol ← backtrackSearch(P,S);
(2)    if B_sol = ∅ then return ∅ else B_in ← B_sol; B_out ← B; end if;
(3)    repeat
(4)        fixed ← TRUE;
(5)        for j=1 to 2n do
(6)            if isOdd(j) then
(7)                i ← (j+1)/2;
(8)                splitSide ← LEFT;
(9)                I₁₂ ←[left(B_out[xᵢ])..left(B_in[xᵢ])]; I₃ ←[left(B_in[xᵢ])..right(B_out[xᵢ])];
(10)               I₁ ←[left(I₁₂)..⌈center(I₁₂)⌉];I₂ ←[⌈center(I₁₂)⌉..right(I₁₂)];
(11)           else
(12)               i ← j/2;
(13)               splitSide ← RIGHT;
(14)               I₁₂ ←[right(B_in[xᵢ])..right(B_out[xᵢ])]; I₃ ←[left(B_out[xᵢ]).. right(B_in[xᵢ])];
(15)               I₁ ←[⌊center(I₁₂)⌋..right(I₁₂)]; I₂ ←[left(I₁₂)..⌊center(I₁₂)⌋];
(16)           end if;
(17)           if width(I₁₂) > 0 then
(18)               fixed ← FALSE;
(19)               B₁ ← B_out; B₁[xᵢ]← I₁; S ← B₁; B_sol ← backtrackSearch(P,S);
(20)               if B_sol = ∅ then
(21)                   B_out[xᵢ] ← I₂ ∪ I₃;
(22)                   if width(I₂) > 0 then
(23)                       B₂ ← B_out; B₂[xᵢ]← I₂; S ← B₂; B_sol ← backtrackSearch(P,S);
(24)                       if B_sol = ∅ then B_out[xᵢ] ← I₃; else B_in ← B_in ⊎ B_sol end if;
(25)                   end if;
(26)               else B_in ← B_in ⊎ B_sol end if;
(27)           end if;
(28)       end for;
(29)   until fixed;
(30)   return B_in;
end function
```

Figure 6.7 The *BS₃* algorithm.

Initially, the generic backtrack search mechanism is applied to the original box to obtain a first canonical solution (line 1). If no canonical solution is found then the CCSP has no solutions and the empty set is returned. Otherwise the inner box is initialised to the canonical solution and the outer box is initialised to the original box (line 2).

The algorithm searches extreme canonical solutions for each variable bound in round robin until all of them are found. This is implemented as a *for* cycle (lines 5-28) (with $2n$ iterations, one for each bound) within a *repeat* cycle (lines 3-29) which terminates when the inner box equals the outer box, returning it (line 30).

In each iteration for searching the left/right bound of a particular variable x_i the following steps are accomplished.

Firstly, the branching option of side (*splitSide*) is set appropriately (lines 8/13).

Secondly two *F*-intervals, I_{12} and I_3, are considered, where the first defines the portion with values smaller/larger than the left/right bound of the inner box and the second contains the remaining values (lines 9/14). Interval I_{12} is subsequently split at its mid point originating two subintervals I_1 and I_2 (lines 10/15).

If the width of I_{12} is zero, the inner and outer boxes have equal bounds, that of the extreme canonical solution already found, and nothing else needs to be done in the iteration. Otherwise (line 17), the relevant search space is set to that between the inner and the outer boxes. More precisely, to an *F*-box equal to the outer box but for the domain of x_i. To improve efficiency, a canonical solution is searched within this search space, with the domain of x_i firstly set to I_1 (line 19) and subsequently set to I_2 (line 23). If no canonical solution is found, the x_i domain of outer box is updated to the remaining interval (lines 21 and 24 respectively). Otherwise the inner box is updated to include the newly found canonical solution (lines 24 and 26 respectively).

Notice that each iteration for a left/right bound of a particular variable x_i eliminates at least half of the relevant search space. In fact, the search space is divided into two halves, the *F*-boxes B_1 and B_2. If a canonical solution is found within B_1 then the inner box is enlarged to include it and *F*-box B_2 will be considered no further. Otherwise, B_1 is eliminated from the search space.

The algorithm is correct and terminates. Correcteness is guaranteed by the correcteness of the *backtrackSearch* algorithm and because the termination of the *repeat* cycle implies that for every bound it is proved that there are no more canonical solutions outside the inner box (which is returned in the end). Termination is guaranteed by the reduction of the finite search space performed at each iteration.

6.3   Ordered Search Approaches

All the previous algorithms use a generic backtrack search mechanism for finding individual canonical solutions, in order to enforce Global Hull-consistency. Thus the ultimate goal is not merely finding canonical solutions but among them the most extreme with respect to a variable bound. However, the simple generic backtrack search mechanism makes no preferences on the order in which the boxes of the search space are explored, following the natural backtracking sequence.

If an ordered search is performed instead, the Global Hull-consistency enforcing algorithms might anticipate the exploration of preferable space regions, thus compensating the extra computational cost of such strategy.

The algorithms in the following subsections are variations of the backtrack searching algorithms that use instead a generic ordered search mechanism for finding canonical solutions.

The generic ordered search mechanism is implemented by the function *orderedSearch* represented in figure 6.8. Its overall functioning is similar to the *backtrackSearch* function except that in this case the search space is represented by a different data structure, which is a list of *F*-boxes sorted by a particular variable bound. It is assumed that the variable and the bound are determined by the global parameters *orderedVariable* and *orderedSide* respectively.

```
function orderedSearch(a CCSP P=(X,D,C), inout an ordered list L of F-boxes)
 (1)   while L.size()>0 do
 (2)       B ← L.pop_front();
 (3)       B ← kB-consistency(P,B);
 (4)       if B ≠ ∅ then
 (5)           if isCanonical(B) then return B;
 (6)           splitBox(B,B₁,B₂);
 (7)           L.insertOrdered(B₂);
 (8)           L.insertOrdered(B₁);
 (9)       end if;
(10)   end while;
(11)   return ∅;
end function
```

Figure 6.8 The generic ordered search algorithm for finding canonical solutions.

The function *orderedSearch* has two arguments, the first is a CCSP P and the second an ordered list L of F-boxes where the canonical solutions of P will be searched. The second argument is also an output argument allowing multiple calls to this function with the side effect of reducing the remaining search space. If the *orderedSide* parameter is set to LEFT/RIGHT, then the F-boxes within L are in ascending/descending order of their left/right bounds of the domain of the variable defined in the *orderedVariable* parameter. The result of the *orderedSearch* function is a canonical solution of P within one of the F-boxes of L or the empty set if there are no such canonical solutions.

During the execution of the *orderedSearch* function the ordered list L of F-boxes representing the remaining search space is maintained by a set of functions which implement the usual list operations: *size* returns the number of elements; *pop_front* returns the first element and removes it from the list; *insertOrdered* adds a new element in its appropriate position with respect to the order determined by the *orderedVariable* and *orderedSide* parameters.

The only differences between this algorithm and the *backtrackSearch* algorithm result from the maintenance of the different data structures for representing the search space. The *pop* and *push* functions, which are typical for stack data structures, are replaced, in the *orderedSearch* algorithm, by the *pop_front* and *insertOrdered* functions, which are common in list data structures.

The algorithm is correct and terminates due to the same reasons that guarantee the correcteness and termination of the *backtrackSearch* algorithm. The only difference between the algorithms is the order in which the F-boxes are explored and this cannot affect their correcteness or termination properties.

The following two subsections present ordered search algorithms which are variations of the BS_1 and BS_3 backtrack search algorithms. Algorithms BS_0 and BS_2 would not profit from an ordered search. In algorithm BS_0 every F-box must be explored and so it is indifferent the order by which this is accomplished. In algorithm BS_2, the stack of F-boxes used in each *backtrackSearch* function remains ordered due to the branching strategy, and no specialised data structure for that purpose is needed.

6.3.1 The OS_1 Algorithm

BS_1 may benefit from an ordered search since finding better (wrt to a variable bound) canonical solutions improves the pruning of the search space.

The modified algorithm, OS_1, is obtained from the original BS_1 function (see figure 6.5) by calling function *orderedSearch* (lines 2, 6 and 12) whose parameters must be previously defined. The *orderedSide* parameter must be set instead of the now useless *splitSide* parameter (lines 1, 10 and 16). The *orderedVariable* parameter is initialised to variable x_1 (line 1) and updated to the current variable x_i at the beginning of each iteration of the *for* cycle (between lines 4 and 5).

Again the correctness and termination of the OS_1 algorithm are guaranteed by the correctness and termination of the BS_1 algorithm since the only difference between them is the order by which the F-boxes are explored.

6.3.2 The OS_3 Algorithm

The BS_3 algorithm may also take advantage from an ordered search in that better (wrt to a variable bound) canonical solutions may cause a faster approximation of the inner box towards the outer box.

The new algorithm, OS_3, is obtained from the original BS_3 function (see figure 6.7) by simply calling (in lines 19 and 23) function *orderedSearch* whose parameters must be previously set. The *orderedSide* parameter must be set instead of the *splitSide* parameter (lines 8 and 13). The *orderedVariable* parameter is updated to the current variable x_i before the function call in lines 19 and 23.

The correctness and termination of the OS_3 algorithm are guaranteed by the correctness and termination of the BS_3 algorithm.

6.4   The Tree Structured Approach

In all previous algorithms the search for a different variable bound does not take full advantage from the search for previous bounds. The main reason for such behaviour is that the data structures used for representing the search space, either a stack or a list of F-boxes, are optimised during the search for a particular bound and cannot be easily reused in the search for different bounds.

Alternatively, a different data structure for the representation of the search space may be used, which is simultaneously optimised with respect to each variable bound, enabling an efficient search on any of these dimensions. In this section we present a tree data structure for the representation of the complete search space which is complemented by a set of ordered lists, one for each variable bound, to keep track of the relevant F-boxes and the actions that must be executed on each of them.

The next subsections describe the data structure in more detail, and its use by an improved tree structured algorithm (*TSA*) for enforcing Global Hull-consistency.

6.4.1   The Data Structures

Besides the inner box, the F-box enclosing all known canonical solutions already used before, the basic data structures maintained by the algorithm are a binary tree and a vector of ordered lists.

Each node of the binary tree is an F-box which represents a sub-region of the search space that may contain solutions of the CCSP. By definition, each parent box is the smallest F-box enclosing its two children. Consequently, the union of all leaves of the binary tree determines the complete search space and the root of the tree defines the smallest F-box enclosing it.

The binary tree of F-boxes is maintained by the following set of basic operations: *size* returns the number of elements; *delete(B)* removes leaf B from the tree; *update(B,B₁)* changes the value of leaf B into B_1; *split(B,B₁,B₂)* turns leaf B into the parent box of two new leaves B_1 and B_2.

The vector of ordered lists has $2n$ elements (where n is the number of variables of the CCSP) each being an ordered list associated with each bound of each variable. Each element of a list is a pair (*F*-box, *Action*) where *Action* is a label representing the next action (PRUNE, SEARCH or SPLIT) to perform on the F-box. The list associated with the left (right) bound of variable x_i maintains the leafs of the binary tree in ascending (descending) order of their left (right) bounds for the x_i domain.

The ordered lists are maintained by the following set of basic operations: *size* returns the number of elements; *front* returns the first element; *delete(B)* removes the element with F-box=B; *insertOrdered(B)* adds a new element (*B*,PRUNE) in its appropriate position; *deleteOrdered(B)* removes all the elements with F-boxes following F-box=B in the ordered list; *update(B,A)* changes into A the *Action* label of the element with F-box=B; *reset(B)* changes into PRUNE the *Action* labels of all the elements in the list whose *F*-box is intersected by *B*;

The coordination between the different data structures is guaranteed by a set of procedures, illustrated in figures 6.10 and 6.11, and is based on the notion of relevance which is related with the inner box basic structure and is implemented by the functions in figure 6.9. To avoid oversized lists of parameters, it is assumed that the basic data structures are accessed through global variables, namely, B_{in}, T and L to represent respectively, the inner box, the binary tree and the vector of ordered lists.

The notion of relevance with respect to a variable bound is a key concept in the enforcing algorithm. Only regions of the search space outside the inner box (that encloses all the known canonical solutions) are relevant. Therefore before exploring a particular *F*-box, it is necessary to check whether this *F*-box includes some relevant region with respect to a variable bound, and eventually to extract it from the *F*-box. Figure 6.9 presents a function (*isRelevant*) for checking the relevance of an *F*-box with respect to a variable bound and a complementary function (*relevantSubbox*) to extract the relevant sub-box. Both functions have two arguments, an *F*-box *B* and a bound represented by an integer value *j* between 1 and $2n$ (where *n* is the number of variables).

---

function *isRelevant*(an *F*-box *B*, a bound *j*)
   (1)   **if** $B_{in}=\varnothing$ **then return** TRUE;
   (2)   **if** *isOdd*(*j*)   **then** $i \leftarrow (j+1)/2$; **return** $left(B[x_i]) < left(B_{in}[x_i])$;
   (3)   **else** $i \leftarrow j/2$;   **return** $right(B[x_i]) > right(B_{in}[x_i])$;
end function

function *relevantSubbox*(an *F*-box *B*, a bound *j*)
   (1) $B_1 \leftarrow B$;
   (2)   **if** $B_{in} \neq \varnothing$ **then**
   (3)   **if** *isOdd*(*j*)   **then** $i \leftarrow (j+1)/2$; $B_1[x_i] \leftarrow [left(B[x_i])..min(right(B[x_i]),left(B_{in}[x_i]))]$;
   (4)   **else** $i \leftarrow j/2$; $B_1[x_i] \leftarrow [max(left(B[x_i]),right(B_{in}[x_i]))..right(B[x_i])]$;
   (5)   **return** B_1;
end function

Figure 6.9 The relevance of an *F*-box with respect to a variable bound.

Function *isRelevant* returns a boolean value indicating if *B* is relevant with respect to the bound *j*. When the inner box is empty (i.e. no canonical solutions were found yet) all boxes are relevant with respect to any variable bound, and so the true value is returned (line 1). Otherwise, since bound *j* refers to the left/right bound of variable x_i, box *B* is relevant if the left/right bound of its x_i domain lies outside the inner box (lines 2/3).

Function *relevantSubbox* assumes that *B* is relevant and returns the largest sub-box for which all the values of the x_i domain are outside the inner box (smaller/larger for a left/right bound of variable x_i). When the inner box is empty, the whole box is relevant and is returned without changes (line 1 followed by line 5). Otherwise, it returns the original box *B* discarding the right/left subinterval of its x_i domain which lies inside the inner box (3/4).

Whenever a new canonical solution is found, the inner box must be enlarged to enclose it, and the ordered lists associated with each variable bound must be updated to remove irrelevant elements. Moreover, if the relevant search region of some element is narrowed, the next action to be executed over it must be a prune (propagating the domain reduction of the search region).

Figure 6.10 shows the procedure *updateInnerBox* to enclose a new canonical solution into the inner box B_{in}. Its unique argument is a canonical *F*-box B_{sol} representing the new found canonical solution.

---

procedure *updateInnerBox*(a canonical F-box B_{sol})

  (1) **if** $B_{in} = \varnothing$ **then** $B_{new} \leftarrow B_{sol}$; **else** $B_{new} \leftarrow B_{in} \uplus B_{sol}$; **end if**;

  (2) **for** $j{=}1$ **to** $2n$ **do**

  (3)   **if** $isRelevant(B_{new},j)$ **then**

  (4) $L[j].deleteOrdered(B_{new})$;

  (5) $L[j].reset(B_{new})$;

  (6)   **end if**

  (7) **end for**

  (8) $B_{in} \leftarrow B_{new}$;

end procedure

Figure 6.10 The procedure that updates the inner box to enclose new canonical solutions.

Procedure *updateInnerBox* firstly computes the new inner box B_{new}: an enlargement of the current inner box B_{in} that encloses the new canonical solution (line 1).

Subsequently, for each variable bound (the *for* cycle lines 2-7), if the inner box changed that bound (line 3), the ordered list associated with it is updated accordingly. All its elements that became irrelevant are eliminated (line 4) and the *Action* label of those with a changed relevant sub-box is reset to PRUNE (line 5). Finally the inner box is updated with the new value (line 8).

The procedures illustrated in figure 6.11 coordinate the changes, on the binary tree and on the vector of ordered lists, that result from the deletion, narrowing and branching of a leaf of the tree.

---

procedure *deleteLeaf*(an F-box B)

  (1) $T.delete(B)$;

  (2) **for** $j{=}1$ **to** $2n$ **do if** $isRelevant(B,j)$ **then** $L[j].delete(B)$;

end procedure

procedure *narrowLeaf*(an F-box B, an F-box B_1)

  (1) $T.update(B,B_1)$;

  (2) **for** $j{=}1$ **to** $2n$ **do**

  (3)   **if** $isRelevant(B,j)$ **then** $L[j].delete(B)$;

  (4)   **if** $isRelevant(B_1,j)$ **then** $L[j].insertOrdered(B_1)$;

  (5) **end for**;

end procedure

procedure *branchLeaf*(an F-box B, an F-box B_1, an F-box B_2)

  (1) $T.split(B,B_1,B_2)$;

  (2) **for** $j{=}1$ **to** $2n$ **do**

  (3)   **if** $isRelevant(B,j)$ **then** $L[j].delete(B)$;

  (4)   **if** $isRelevant(B_1,j)$ **then** $L[j].insertOrdered(B_1)$;

  (5)   **if** $isRelevant(B_2,j)$ **then** $L[j].insertOrdered(B_2)$;

  (6) **end for**;

end procedure

Figure 6.11 The procedures for deleting, narrowing and branching a leaf of the binary tree.

When a leaf is deleted from the tree, any associated element in the ordered lists must also be removed. When a leaf is narrowed, any associated element may be reordered and eventually eliminated if it becomes irrelevant. When a leaf is split into two new leaves, then its associated elements must be removed and two new elements considered for insertion in the lists. The first argument of all the procedures in figure 6.11, *deleteLeaf*, *narrowLeaf* and *branchLeaf*, is a leaf of the binary tree (an F-box). The second argument of procedures *narrowLeaf* and *branchLeaf*, and the third argument of procedure *branchLeaf* are *F*-boxes, B_1 and B_2, which are sub-boxes of *B*.

All three procedures firstly change the binary tree (line 1) and subsequently consider the ordered list associated with each variable bound and change their elements if needed (the *for* cycle).

The changes on the binary tree are respectively, to remove leaf *B* from the binary tree (*deleteLeaf*), to update the value of leaf *B* into B_1 (*narrowLeaf*) and to make leaf *B* as the parent box of the two new leaves B_1 and B_2 (*branchLeaf*).

The changes on each ordered list are the following: if *B* is a relevant box (wrt the bound associated with the ordered list) then the element of the list with *F*-box=*B* is removed (*deleteLeaf*: line 2; *narrowLeaf* and *branchLeaf*: line 3); if B_1 is a relevant box then procedures *narrowLeaf* and *branchLeaf* (line 4) insert a new element with *F*-box=B_1 in the appropriate position; if B_2 is a relevant box then procedure *branchLeaf* (line 5) inserts a new element with *F*-box=B_2 in the appropriate position.

6.4.2   The Actions

The tree structured algorithm for enforcing Global Hull-consistency alternates prune, search and split actions performed over specific sub-regions (*F*-boxes) of the current search space. The pruning of an *F*-box is achieved by enforcing a kB-Consistency criterion (either kB-Hull-consistency or kB-Box-consistency). The search action is performed with the goal of finding a canonical solution within an *F*-box (previously pruned) and may be implemented as a simple check of an initial guess or as a more complete local search procedure (see next chapter). The split of an *F*-box (previously searched) is done by splitting one of its variable domains at the mid point.

Any action is performed over some leaf of the binary tree, and in particular over a subbox which must be relevant with respect to some variable bound. Its consequences must be propagated throughout the data structures maintained by the algorithm. The resulting procedures implementing the prune, search and split actions are presented in figs. 6.12-14.

The first action, PRUNE, must be performed before any other action in order to reduce the relevant search space.

Figure 6.12 shows procedure *pruneAction*, which has three arguments. The first, B_t, is the leaf of the binary tree where the pruning takes place. The second, *B*, is a subbox of B_t on which the kB-Consistency requirement is enforced. The last argument, integer *j*, represents the bound for which the sub-box *B* is relevant. The search space discarded in the pruning is removed from the binary tree. This may imply to eliminate, narrow or branch leaves from the tree, which is achieved by procedures *deleteLeaf*, *narrowLeaf* and *branchLeaf* previously presented.

```
procedure pruneAction(an F-box B_t, an F-box B, a bound j)
  (1)   B' ← kB-consistency(P,B);
  (2)   if B' = ∅ then
  (3)       if B_t=B then deleteLeaf(B_t); else narrowLeaf(B_t,B_t\B); end if
  (4)   else
  (5)       if B'=B then L[j].update(B_t, SEARCH);
  (6)       else
  (7)           if B_t=B then narrowLeaf(B_t,B'); else branchLeaf(B_t,B',B_t\B); end if
  (8)           L[j].update(B', SEARCH);
  (9)       end if
  (10)  end if
end procedure
```

Figure 6.12 The procedure for pruning a subbox of a leaf of the binary tree.

Initially, *pruneAction* narrows the subbox B into B' through the *kB-consistency* algorithm (line 1).

If the result is the empty set (line 2) then subbox B must be completely discarded from the binary tree (line 3): if B is the whole leaf B_t, then it must be removed; if it is only a part of B_t, this leaf must be narrowed to $B_t \backslash B$ obtained by removing B from B_t.

If the pruning of sub-box B did not narrow the box (line 5) then the next action to be executed over leaf B_t wrt bound j must be a SEARCH. Thus, in the ordered list associated with bound j, the *Action* label of the element with F-box$=B_t$ is changed into SEARCH (line 5).

Otherwise, if sub-box B is narrowed into the non empty F-box B' then, either B is the whole leaf B_t, which must be narrowed into B', or B is a fraction of the leaf B_t, which must be branched into B' and the F-box remaining after removing B from B_t (line 7). In either case, since the new leaf B' was already pruned wrt to bound j, the respective *Action* label is changed into SEARCH (line 8).

The next action is the search action, which must only be performed over a space region if this region cannot be further pruned. This action precedes the split action in order to avoid unnecessary over branching of the binary tree.

Figure 6.13 shows the *searchAction* procedure with the same three arguments: a leaf of the binary tree B_t, its subbox B where the search takes place, and an integer value j representing the searched bound. It is assumed that the search for a canonical solution of the CCSP P within the F-box B is accomplished by the *searchSolution* function which may be implemented as a simple check of an initial guess or as local search procedure. Together with the CCSP P and the F-box B, the *searchSolution* function includes a third argument specifying the bound j, to allow its implementation in the context of finding an extreme canonical solution with respect to this variable bound (the details of the *searchSolution* function will be presented in the next chapter).

procedure *searchAction*(an F-box B_t, an F-box B, a bound j)

   (1) $B_{sol} \leftarrow$ *searchSolution*(P,B,j);

   (2)   **if** $B_{sol} \neq \emptyset$ **then** *updateInnerBox*(B_{sol}); **else** $L[j].update(B_t,$SPLIT$)$; **end if**

end procedure

Figure 6.13 Procedure for searching a canonical solution within a subbox of a binary tree leaf.

Initially procedure *searchAction* uses the *searchSolution* function to search a canonical solution of P within subbox B (line 1). If a new canonical solution is found then the inner box must be updated to enclose it (line 2). Otherwise, the next action to perform over leaf B_t wrt bound j must be a split action. Thus, in the ordered list associated with bound j, the *Action* label of the element with F-box$=B_t$ is changed into SPLIT (line 2).

The last action is the split action, which must only be performed over a space region if this region cannot be further pruned and the search for a canonical solution within it had failed.

Figure 6.14 presents the *splitAction* procedure. It has three arguments, a leaf of the binary tree B_t, its subbox B that is going to be split, and a bound j. It is assumed that there is a *splitBox* procedure identical to the one described in section 6.2 for branching an F-box B into two subboxes B_1 and B_2 (except that it contains an additional output parameter indicating which variable domain was split).

> **procedure** *splitAction*(an *F*-box B_t, an *F*-box B, a bound j)
> (1) **if** *isOdd*(j) **then** $i \leftarrow (j+1)/2$; *splitSide* $\leftarrow$ LEFT; **else** $i \leftarrow j/2$; *splitSide* $\leftarrow$ RIGHT; **end if**;
> (2) *splitBox*(B,B_1,B_2,x_k);
> (3) **if** $i{\neq}k$ **then** $I \leftarrow B_1[x_k]$; $B_1 \leftarrow B_t$; $B_1[x_k] \leftarrow I$;
> (4) *branchLeaf*($B_t,B_1,B_t\backslash B_1$);
> **end procedure**

Figure 6.14 The procedure to split a subbox of a binary tree leaf.

Initially the procedure *splitAction* determines which variable x_i and which side *splitSide* is associated with bound j (line 1). Then, the *splitBox* procedure is used to split the subbox B into B_1 and B_2 (line 2). If the split variable x_k is not that associated with bound j, then all the domains of box B_1, except the domain of variable x_k, are redefined to their respective values in the leaf B_t (line 3). Finally, the leaf B_t is branched into B_1 and the *F*-box remaining after removing B_1 from B_t (line 4). Note that the redefinition of B_1 in line 3 is necessary since otherwise the search space remaining after removing B_1 from B_t ($B_t\backslash B_1$) would not be necessarily an *F*-box.

6.4.3 The TSA Algorithm

The *TSA* algorithm for enforcing Global Hull-consistency takes advantage of the binary tree representation of the search space which allows the dynamic focussing on specific relevant regions without losing information previously obtained in the pruning process.

Within a round-robin fashion, the most relevant sub-region, with respect to each variable bound, is chosen for performing an adequate action. As a result, the binary tree structure will evolve updating the most relevant sub-regions and the respective adequate actions (see previous subsections). The algorithm stops when there are no more relevant space regions to analyse, that is, all the search space is contained within the inner box (or is proved that there are no canonical solutions).

Figure 6.15 shows the *TSA* algorithm. From a CCSP P and an *F*-Box B ($B{\subseteq}{<}I_{apx}(D_1),\ldots, I_{apx}(D_n){>}$), function *TSA* computes the largest Global Hull-consistent *F*-box within B or the empty set if it does not contain any canonical solutions.

> **function** *TSA*(a CCSP $P{=}({<}x_1,\ldots,x_n{>},D_1{\times}\ldots{\times}D_n,C)$, an *F*-box B)
> (1) $B_{in} \leftarrow \varnothing$; $T \leftarrow B$;
> (2) **for** $j{=}1$ **to** $2n$ **do** $L[j] \leftarrow (B,\text{PRUNE})$;
> (3) **repeat**
> (4)   *fixed-point* $\leftarrow$ TRUE;
> (5)   **for** $j{=}1$ **to** $2n$ **do**
> (6)   **if** $T.size(){=}0$ **then return** $\varnothing$;
> (7)   **if** $L[j].size(){>}0$ **then**
> (8)   *fixed-point* $\leftarrow$ FALSE;
> (9) $(B_t,Action) \leftarrow L[j].front()$;
> (10) $B \leftarrow relevantSubbox(B_t,j)$;
> (11)   **if** *Action*=PRUNE **then** *pruneAction*(B_t,B,j);
> (12)   **if** *Action*=SEARCH **then** *searchAction*(B_t,B,j);
> (13)   **if** *Action*=SPLIT **then** *splitAction*(B_t,B,j);
> (14)   **end if**
> (15)   **end for**;
> (16) **until** *fixed-point*=TRUE;
> (17) **return** B_{in};
> **end function**

Figure 6.15 The *TSA* algorithm.

At the beginning, the data structures are initialised: the inner box is empty (line 1); the binary tree contains only a single box with the original domains (line 1); each ordered list associated with each variable bound contains a single element which points to the unique leaf of the tree and defines the next adequate action to be a prune action (line 2).

The algorithm proceeds by alternating, the most relevant action with respect to each variable bound until all canonical bounds are found. This is implemented as a *for* cycle (lines 5-15) (with $2n$ iterations, one for each bound) within a *repeat* cycle (lines 3-16) which only terminates when the inner box encloses all the search space (returned at line 17), or the search space becomes empty (line 6).

Each iteration of the *for* cycle is only executed if the associated ordered list is not empty (line 7), otherwise the respective extreme bound would have already been found. If the bound was not found yet then, its most relevant leaf B_l and the respective next action *Action* are determined from the first element of the associated ordered list (line 9). Subsequently, a relevant subbox is computed by the *relevantSubbox* function (line 10) and the appropriate action is performed on it (line 11, 12 or 13).

The correctness of the algorithm is guaranteed by the correctness of the *kB-Consistency* function for pruning the search space and the properties of the ordered lists which only become empty when the respective canonical bound is found. Since the algorithm only terminates when all the lists become empty, this guarantees that all the canonical bounds have been found when the algorithm stops.

The algorithm is guaranteed to terminate since at least after 3 iterations performing different actions on the same leaf of the binary tree, either the relevant search space is reduced (either by the prune or the search action) or the leaf is split into two smaller leafs. After a finite number of these iterations the relevant search space must necessarily be reduced since the leafs eventually become canonical, and either the prune or the search action must then succeed. Since the relevant search space is finite, eventually it will be completely eliminated, terminating the algorithm.

6.5   Summary

In this chapter Global Hull-consistency was proposed as an alternative consistency criterion in continuous domains. Several different approaches were suggested for enforcing Global Hull-consistency, their enforcing algorithms were explained and its termination and correctness properties were derived. In the next chapter a local search procedure is proposed for interval constraints and its integration with the Global Hull-consistency enforcing algorithms is described.

Chapter 7

Local Search

The definition of Global Hull-consistency, demanding the existence of extreme canonical solutions with respect to each variable bound, requires enforcement strategies not only to prune the search space, proving the non-existence of canonical solutions, but also to localise canonical solutions within a box.

In the approaches of the previous chapter, whereas the pruning of the search space is achieved by a specialised partial consistency enforcement algorithm, the localisation of new canonical solutions has no specific method and is a consequence of reducing a sub-region of the search space into a canonical box that cannot be further pruned[1].

However, the enforcing algorithms that maintain an inner box enclosing all the known canonical solutions, might benefit from anticipating the localisation of new canonical solutions since the outward relevant search space would be reduced as a result of the enlargement of these inner box. Consequently, these algorithms could take advantage from the integration of a specialised approach for searching new canonical solutions.

A natural approach for searching new canonical solutions is to apply local search techniques. Local search techniques navigate through points of the search space by inspecting some local properties of the current point, and choosing a nearby point to jump to. In the CCSP context, the points of the search space are complete real valued instantiations of all its variables (degenerated F-boxes) and the navigation should be oriented towards the simultaneous satisfaction of all its constraints (the solutions of the CCSP).

Local search techniques are commonly used for solving optimisation problems, which may be seen as CSPs where the goal is to find solutions that optimise (minimise or maximise) an objective function. In general, these are based on numerical methods working on floating point arithmetic [16, 58, 101] for efficiently obtaining local optimisers, which may be embedded within an interval branch and bound approach [100, 129, 66, 78, 121, Jan92, CGM93] to guarantee global optimality.

Within the context of Global Hull-consistency enforcement, the goal is not necessarily to minimise a given objective function but rather to satisfy the set of constraints of a CCSP. Nevertheless, from some initial point chosen within an F-box, a floating point numerical method may be used for converging to another point of the F-box which is a local minimiser of some function representing how "distant" is a point from satisfying all the constraints[2].

In the next section a local search approach is proposed for integration with the Global Hull-consistency enforcement algorithms described in the previous chapter. In section 7.2 alternative local search approaches are analysed. Section 7.3 discusses which algorithms could benefit from such integration and how this could be implemented.

---

[1] An exception must be made to the *TSA* algorithm which includes the *searchSolution* function for finding new solutions.
[2] In the following this function will be denoted the distance function.

7.1   The Line Search Approach

The local search approach proposed for finding canonical solutions within an F-box is a line search approach, that is, the movement is always done along lines of the multidimensional space. From a particular current point, a convenient direction must be defined, determining a line on which the new point is searched. The chosen direction is obtained by the Newton-Raphson method for multidimensional root finding of nonlinear systems of equations. The new chosen point along this direction must lie within the original F-box and ensure a sufficient decrease of the distance function.

Figure 7.1 presents function *searchSolution* that implements the local search approach. It uses the *kB-consistency* algorithm for verifying if a canonical F-box is a canonical solution of the CCSP. Function *Newton-Raphson* (explained in the next subsection) calculates the vector that defines the line search direction and function *lineMinimisation* (presented in the subsection 7.1.2) computes a new point along that line. Function *searchSolution* has three arguments, a CCSP P, an F-box B (with $B \subseteq <I_{apx}(D_1),\ldots, I_{apx}(D_n)>$), and a bound j. The output is a canonical solution of P within B or the empty set if no solution was found.

---

function *searchSolution*(a CCSP $P=(<x_1,\ldots,x_n>,D_1\times\ldots\times D_n,C)$, an F-box B, a bound j)

(1)   **for** $i=1$ **to** n **do** $r[i] \leftarrow \lfloor center(B[i]) \rfloor$;

(2)   **if** *isOdd*(j) **then** $r[(j+1)/2] \leftarrow left(B[(j+1)/2])$; **else** $r[j/2] \leftarrow right(B[j/2])$; **end if**;

(3)   **repeat**

(4)   **for** $i=1$ **to** n **do if** $r[i] \neq right(B[i])$ **then** $B_{sol}[i] \leftarrow [r[i]..r[i]^+]$ **else** $B_{sol}[i] \leftarrow cright(B[i])$;

(5)   **if** $kB\text{-}consistency(P,B_{sol}) \neq \varnothing$ **then return** B_{sol};

(6) $r_{old} \leftarrow r$;

(7) $\delta r \leftarrow Newton\text{-}Raphson(P,B,j,r)$;

(8) $r \leftarrow lineMinimisation(P,B,r,\delta r)$;

(9)   **until** $r_{old} = r$;

(10)   **return** $\varnothing$;

end function

Figure 7.1 The local search algorithm.

Initially a starting point (a degenerate F-box) within the search box B is chosen to be the current point r. If the goal is to find the left (right) bound of variable x_i, the point is the mid point of the box (line 1) except that the ith domain is instantiated with the smallest (largest) x_i value within B (line 2). This heuristic bias the search towards the extreme bound that characterises the context in which the search is performed.

The remainder of the algorithm (lines 3 through 9) is a *repeat* cycle that implements the local navigation from point to point until a convergence point is reached (line 9) or a canonical solution is found (line 5).

At each iteration of the *repeat* cycle, the current point is firstly enlarged into a non-degenerate canonical box (line 4). If the empty set is not obtained when the *kB-consistency* algorithm is applied to this canonical box then it must be a canonical solution and is returned (line 5). Otherwise, the current point is saved (line 6) and a multidimensional vector is obtained based on the Newton-Raphson method for multidimensional root finding (line 7). Subsequently, a minimisation process obtains a new point inside the search box and within the line segment defined by the current point and the point obtained by applying the multidimensional vector to the current point (line 8).

A convergence point is reached when the next point is the same as the previous one. In this case the *repeat* cycle terminates (line 9) and the empty set is returned (line 10).

The algorithm is correct in that it either returns a canonical solution within the search box (guaranteed by the application of the *kB-consistency* algorithm) or the empty set.

The algorithm terminates since the *lineMinimisation* function ensures the minimisation of the distance function for which any solution of the CCSP is a zero, and the convergence to a local minimum is detected in line 9 terminating the *repeat* cycle.

7.1.1   *Obtaining a Multidimensional Vector - the Newton-Raphson Method*

The ultimate goal of obtaining a multidimensional vector is to apply it to the current point to find a solution of a CCSP. The idea is to reduce the problem of finding a solution of a CCSP into the problem of finding a root of a multidimensional vector function F, which can be tackled by an appropriate numerical method, such as the Newton-Raphson Method [115, 133].

If the vector function F is defined in such a way that a zero of each element F_i satisfies some constraint form the CCSP (and a non zero value implies that the constraint is not satisfied) then, any zero of F must satisfy simultaneously all the associated constraints.

In the case of equality constraints, the associated element on the vector function may be defined by the real expression of the left hand side of the constraint. For example, if the CCSP has two equality constraints $c_1 \equiv x_1^2 + x_2^2 - 1 = 0$ and $c_2 \equiv x_1 \times (x_2 - x_1) = 0$ then $F_1 \equiv x_1^2 + x_2^2 - 1$ and $F_2 \equiv x_1 \times (x_2 - x_1)$ would define a vector function F whose zeros are CCSP solutions.

If there are inequality constraints in the CCSP then they will only be included in the vector function if they are not satisfied in the current point. If an inequality constraint is already satisfied in the current point then, there is no advantage in forcing it to became zero (transforming the inequality into an equality constraint). On the contrary, if the inequality is not yet satisfied then at least the zero value must be obtained for its satisfaction. Consider the example of figure 5.1 with two inequality constraints $c_1 \equiv x_1^2 + x_2^2 - 2^2 \leq 0$ and $c_2 \equiv (x_1-1)^2 + (x_2-1)^2 - 2.5^2 \geq 0$. If the current point is $(2.0, 2.0)$ satisfying none of the constraints, then the vector function would be defined by the elements $F_1 \equiv x_1^2 + x_2^2 - 2^2$ and $F_2 \equiv (x_1-1)^2 + (x_2-1)^2 - 2.5^2$. However, at the current point $(0.0, 0.0)$ only constraint c_2 is not satisfied and consequently, the vector function would only contain the single element $F_1 \equiv (x_1-1)^2 + (x_2-1)^2 - 2.5^2$ associated with it.

Figure 7.2 shows how function F could be obtained for a particular current point. The two arguments are the CCSP P and a degenerated F-box r representing the current point.

---

function *defineFunction*(a CCSP $P = (\langle x_1, \ldots, x_n \rangle, D_1 \times \ldots \times D_n, \{c_1, \ldots, c_m\})$, a degenerated F-box r)

(1) $i = 0$;

(2)   **for** $j = 1$ **to** m **do**

(3)   **case** $c_j \equiv$

(4) $e_j = 0 : i \leftarrow i + 1; F_i \equiv e_j;$

(5) $e_j \leq 0 :$ **if** $left(E_j(r)) > 0$ **then** $i \leftarrow i + 1; F_i \equiv e_j;$ **end if**;

(6) $e_j \geq 0 :$ **if** $right(E_j(r)) < 0$ **then** $i \leftarrow i + 1; F_i \equiv e_j;$ **end if**;

(7)   **end case**;

(8)   **return** F;

end function

Figure 7.2 The definition of the vector function F.

The algorithm is a *for* cycle (lines 2 through 7), where each of the CCSP constraints is analysed for deciding whether it must be associated with a component of the vector function F.

For equality constraints (line 4), a new component of the vector function is defined by the real expression of the left hand side of the constraint expression. Otherwise, (lines 5 or 6), the new component is only added to the vector function if the constraint is not satisfied at the current point, that is, its approximate interval arithmetic evaluation (represented as $E_j(r)$) does not satisfy the constraint.

Once obtained the multidimensional function F, a multidimensional vector δr, corresponding to one step of the Newton-Raphson iterative method for multidimensional root finding, is computed. The Newton-Raphson method is known to rapidly converge to a root given a sufficiently good initial guess. Local quadratic convergence was firstly proved by Runge in 1899 and later, under more general assumptions by Kantorovich in 1948. The book by Ortega and Rheinboldt [115] is a classical reference for many of the convergence theoretical results of the Newton-Raphson method.

The method aims at computing a multidimensional real vector δr (the Newton vector) which applied to the current point r reaches a root of the multidimensional function:

$$F(r+\delta r) = 0 \qquad (1)$$

Expanding the multidimensional function F in Taylor series around the current point r, and neglecting the higher order terms, the following approximation is obtained:

$$F(r+\delta r) = F(r) + J(r)\delta r \qquad (2)$$

(where $J(r)$ is the Jacobian matrix at point r)

which, due to 1, may be rewritten as:

$$J(r)\delta r = -F(r) \qquad (3)$$

A solution of (3) in order to δr, is an estimation of the corrections to the current point r that move all functions F_i closer to zero, simultaneously. Equation (3) represents a linear system with n unknowns (the δr value with respect to each variable) and m algebraic equations (one for each component of the vector function F). Depending on the number of linear independent equations, the system may have zero, one or several solutions. Since there are no guarantees about the non-singularity of the Jacobian matrix $J(r)$, it may be impossible to solve the system by inverting it by a classical method as the Gauss-Jordan elimination or LU decomposition.

Hence, it is convenient to use a numerical technique called Singular Value Decomposition (SVD) adequate for any set of linear equations. The numerical aspects of the SVD technique are discussed in Golub and Van Loan's textbook [60] whereas practical implementational issues may be found in [45]. Efficient implementations of the SVD technique are presented in [54] (Fortran version) and [118] (C version) which are based on the original algorithm introduced by Golub and Reinsch in [59], whereas a new algorithm is presented in [62].

The basic concepts of the SVD technique are introduced next, and its potential application for finding a suitable multidimensional vector is discussed subsequently.

A singular value decomposition of a matrix $A \in \mathbb{R}^{m \times n}$ is a factorisation of the form:

$$A = U\Sigma V^{\mathrm{T}}$$

where $U \in \mathbb{R}^{m \times m}$ and $V \in \mathbb{R}^{n \times n}$ are both orthogonal matrices ($U^{\mathrm{T}}U=I$ and $V^{\mathrm{T}}V=I$), and $\Sigma \in \mathbb{R}^{m \times n}$ is a matrix where all the non-diagonal elements $\Sigma_{i,j}$ ($i \neq j$) are zero. The diagonal elements $\Sigma_{i,i}$ (denoted σ_i) are non-negative, and are called the singular values of A and the columns of U and V are the left and right singular vectors. The singular values of a matrix are uniquely defined, but not the corresponding orthogonal matrices.

Any matrix $A \in \mathbb{R}^{m \times n}$, either singular or non-singular, may be factorised into an SVD where the diagonal elements of Σ (the singular values) are in descending order ($\sigma_i \geq \sigma_j$ if $i \leq j$), the first r ($r \leq \min(m,n)$) being positive and the others zero. The value r is the rank of the matrix A (the number of independent rows or columns). The right singular vectors that are columns v_i of V whose same-numbered singular values σ_i are zero define an orthonormal basis V_0 for the null space of A, which is the subspace of $\mathbb{R}^n$ defined by all the elements x that satisfy $Ax=0$.

Any matrix $A \in \mathbb{R}^{m \times n}$, either singular or non-singular, has a unique pseudoinverse $A^\dagger \in \mathbb{R}^{n \times m}$ (a formal definition of a matrix pseudoinverse may be found in [133]), that can be obtained from its SVD by:

$$A^\dagger = V\Sigma^\dagger U^\mathrm{T}$$

where $\Sigma^\dagger \in \mathbb{R}^{n \times m}$ is a matrix where all the non-diagonal elements $\Sigma^\dagger_{i,j}$ ($i \neq j$) are zero and the diagonal elements $\Sigma^\dagger_{i,i}$ (denoted $\sigma^\dagger_i$) are zero if the respective singular value σ_i is zero, or $1/\sigma_i$ otherwise. When A is a square and non-singular matrix, its pseudoinverse coincides with its inverse, that is, $A^\dagger = A^{-1}$.

Consider the generic linear system:

$$Ax=b \quad \text{with} \quad A \in \mathbb{R}^{m \times n},\ b \in \mathbb{R}^m \text{ and } x \in \mathbb{R}^n$$

Such system may have zero, one or infinitely many solutions. A least-squares solution of the system is a vector $\tilde{x} \in \mathbb{R}^n$, that minimises the least-squares distance between Ax and b, that is:

$$\left\| A\tilde{x} - b \right\|_2 = \min_{x \in \mathfrak{R}^n} \left\| Ax - b \right\|_2$$

When there are solutions, one or many, the least-squares solutions are the real solutions of the system ($\left\| A\tilde{x} - b \right\|_2 = 0$). When there are no solutions, the least-squares solutions are the vectors, among all the possible vectors, that, although not solving the system, provide the closest approximation of the right hand side.

The least-squares solution of minimal norm is the smallest vector $\hat{x}$ which is a least-squares solution of the system:

$$\left\| \hat{x} \right\|_2 = \min_{\tilde{x}} \left\| \tilde{x} \right\|_2$$

For any linear system $Ax=b$ there is always a unique least-squares solution of minimal norm $\hat{x}$, which may be obtained by multiplying the pseudoinverse of matrix A and the right hand side vector b:

$$\hat{x} = A^\dagger b$$

From the SVD, not only the pseudoinverse $A^\dagger$ may be computed, allowing the computation of the least-squares solution of minimal norm $\hat{x}$, but also an orthonormal basis V_0 for the null space of A may be defined, allowing the assessment of other least-squares solutions. This last observation results from the fact that any vector obtained by adding a linear combination of the vectors within V_0 to $\hat{x}$ must give an equal approximation of the right hand side b and so, it must also be a least-squares solution of the system. If V_0 contains k vectors, $v_1, \ldots, v_k$, for any linear combination $\alpha_1, \ldots, \alpha_k$:

$$A(\hat{x} + \alpha_1 v_1 + \ldots + \alpha_k v_k) = A\hat{x} + \alpha_1 A v_1 + \ldots + \alpha_k A v_k = A\hat{x} + 0 + \ldots + 0 = A\hat{x}$$

Moreover, based on the factorisation of the matrix A into $U\Sigma V^\mathrm{T}$ it is easy to assess whether the least-squares solution of minimal norm $\hat{x}$ obtained is a real solution of the system, or a mere approximation. In fact, the original system may be transformed into a diagonal system:

$$Ax=b \Leftrightarrow U\Sigma V^\mathrm{T}x=b \Leftrightarrow U^\mathrm{T}U\Sigma V^\mathrm{T}x=U^\mathrm{T}b \Leftrightarrow \Sigma V^\mathrm{T}x=U^\mathrm{T}b \Leftrightarrow \Sigma z=d \quad \text{with } z=V^\mathrm{T}x \text{ and } d=U^\mathrm{T}b$$

and a solution of the diagonal system exists if and only if $d_i=0$ whenever $\sigma_i=0$ or $i>n$.

In the context of the Newton-Raphson method, the linear system $J(r)\delta r = -F(r)$ must be solved to estimate the corrections δr to the current point r that approximates F to zero. We must take into account that the quality of the approximation given by the Taylor series of equation 2 will in general decrease when larger neighbourhoods around the current point are considered.

If the system has one or more solutions, the application of the SVD technique to calculate the pseudoinverse $J(r)^{\dagger}$ of the jacobian matrix and computing the least-squares solution of minimal norm $\delta r = -J(r)^{\dagger}F(r)$, results in the smallest solution vector, corresponding to the smallest correction leading to a zero of F according to the approximation of equation 2. In this case, besides finding solutions of the system, that one requiring smaller corrections is chosen, since the quality of the approximation given by the equation is probably better.

If the system has no solutions, the least-squares solution of minimal norm $\delta r = -J(r)^{\dagger}F(r)$ computed by the SVD is the smallest vector of those that give the best possible correction leading to a zero of F according to the approximation of equation 2. Since, in this particular case, the search seems to be approaching a non-zero local minimum, it may be preferable to choose some other least-squares solution, which could be obtained by adding to δr a linear combination of the right singular vectors that define an orthonormal basis for the null space of $J(r)$. Moreover, this linear combination could be determined to maximise/minimise the extreme bound that characterises the local search context.

The complete algorithm for obtaining the multidimensional vector is implemented as function *Newton-Raphson* presented is figure 7.3. From a CCSP P, an F-box B representing the search box, a bound j representing the local search context, and a degenerate F-box r representing the current point, it computes the multidimensional vector δr based on the Newton-Raphson method. The algorithm uses function *defineFunction* for the definition of the vector function F that must be associated with a CCSP at a given point. A function *defineJacobian* computes the Jacobian matrix at a given point and a procedure *SVD* computes the SVD factorisation of a matrix. This procedure has four arguments, the first is an input argument representing the original matrix J_r, and the remaining are output arguments representing the matrices U, Σ and V such that $J_r = U\Sigma V^{\mathrm{T}}$. A function *hasSolution* is used for verifying whether the linear system has a solution based on the relevant information (Σ and d) about the equivalent diagonal system ($\Sigma z = d$). A function *defineOrthonormalBasis* is used to, based on the singular values in Σ, extract from the set of right singular vectors V those that constitute an orthonormal basis V_0 of the null space. A function *linearCombination* is used for obtaining a linear combination of these vectors in order to redirect the vector δr towards the bound j of the search box B.

---

function *Newton-Raphson*(a CCSP P, an F-box B, a bound j, a degenerated F-box r)

(1) $F \leftarrow defineFunction(P,r)$;

(2) $J_r \leftarrow defineJacobian(F,r)$;

(3) $SVD(J_r,U,\Sigma,V)$;

(4) $d \leftarrow -U^{\mathrm{T}}F(r)$;

(5) $\delta r \leftarrow V\Sigma^{\dagger}d$;

(6)   **if not** *hasSolution*(Σ,d) **then**

(7) $V_0 \leftarrow defineOrthonormalBasis(\Sigma,V)$;

(8) $\delta r \leftarrow \delta r + linearCombination(B,j,r,\delta r,V_0)$;

(9)   **end if**

(10)   **return** δr;

end function

Figure 7.3 The algorithm that computes the *Newton-Raphson* vector.

Initially the vector function $\mathbf{F}$, associated to the CCSP P at the current point r is defined (line 1). Then its Jacobian matrix J_r at that point is computed (line 2) and decomposed by the *SVD* procedure (line 3) obtaining the matrices U, Σ and V such that $J_r=U\Sigma V^{\mathrm{T}}$.

The computation of the least-squares solution of minimal norm of the linear system $J(r)\delta r=-\mathbf{F}(r)$, which is $\delta r=J(r)^{\dagger}(-\mathbf{F}(r))=V\Sigma^{\dagger}U^{\mathrm{T}}(-\mathbf{F}(r))$, is performed in two stages: firstly $d=U^{\mathrm{T}}(-\mathbf{F}(r))$ is computed (line 4) and then the remaining $\delta r=V\Sigma^{\dagger}d$ is calculated (line 5). The reason for these two steps is to separately obtain vector d, which will be used by the *hasSolution* function (line 6) to verify whether the linear system has real solutions.

If the obtained vector δr is not a solution of the linear system then an orthonormal basis V_0 of the null space is extracted from the vector within V (line 7), and other least-squares solution is obtained by adding to δr a linear combination of these vectors (line 8).

Finally the obtained multidimensional vector δr is returned (line 10).

Figure 7.4 illustrates the multidimensional vectors obtained at different points of the search space for the example presented in figure 5.1, with the two inequality constraints $c_1 \equiv x_1^2+x_2^2-2^2\leq0$ and $c_2 \equiv (x_1-1)^2+(x_2-1)^2-2.5^2\geq0$.

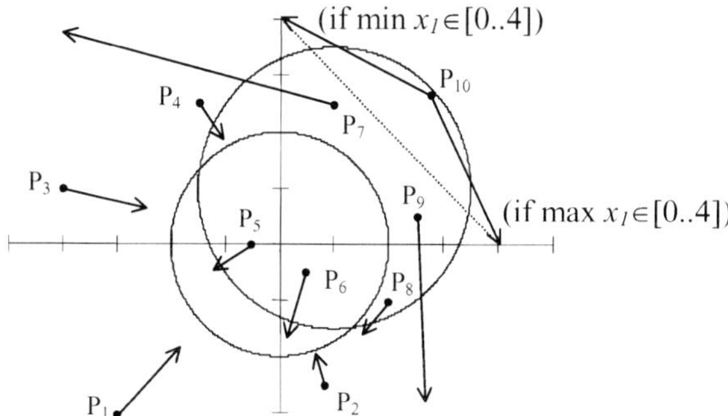

Figure 7.4 The multidimensional vectors obtained at different points of the search space.

If the current point is outside both circumferences (P_1 to P_4) then c_1 is the only constraint unsatisfied (a solution must be inside the smaller circumference) and the Newton vector points towards the centre of the smaller circumference.

Similarly, if the current point is inside both circumferences (P_5, P_6) then c_2 is the only constraint unsatisfied (a solution must be outside the larger circumference) and the Newton vector points outwards the larger circumference (with a direction opposite to the centre of the circumference).

If both constraints are unsatisfied (P_7 to P_{10}) then the Newton vector is some weighted combination of the two previous cases. In the particular case where $x_1=x_2$ (P_{10}), the resulting linear system has no solutions and the set of least-squares solutions define a straight line like the dotted line presented in the figure. Which of these points will be chosen for defining the multidimensional vector, depends upon the search box and the extreme bound which is being searched. In the example, if the search box constrain the values of the variable x_1 to be between 0 and 4 then the multidimensional vector points to (4,0) or (0,4) depending on whether the goal is to find its upper or its lower bound, respectively.

The Newton-Raphson method requires the computation of a Jacobian matrix at each of the points considered during the search. Besides implying the existence of the first derivatives at each of these points with respect to each variable, its computation may be expensive, depending on the number of variables. An alternative strategy could be based on a Secant method such as the Broyden's method [24] which works with cheaper approximations of the Jacobian matrix and still preserves similar convergence properties.

7.1.2   *Obtaining a New Point*

One problem of the Newton's iterative method is that it may fail convergence if the initial guess is not good enough (see [115]). To correct this we followed a globally convergent strategy that guarantees some progress toward the minimisation of the distance function at each iteration. This kind of strategy is often combined with the Newton-Raphson method originating the modified Newton method. The idea is that while such methods can still fail by converging into a local minimum of the distance function, they often succeed where the Newton-Raphson method alone fails convergence.

The distance function that must be minimised at each iteration of the method is defined at each point r as the Euclidian vector norm $\|F(r)\|_2$ of the vector function $F(r)$ associated with the point (see previous subsection). With the above definition, the distance function has exactly the same zeros as the vector function F. Thus, a zero of the distance function is a solution of the CCSP whereas a non zero value gives a measure of how distant a point is from satisfying simultaneously all the constraints.

If there are one or more solutions of the linear equation $J(r)x = -F(r)$ then the Newton vector δr defines a descent direction for the distance function, that is, the inner product between the gradient of the distance function and the Newton vector δr is always negative. This can be easily verified since the gradient of the distance function is:

$$\nabla \|F(r)\|_2 = \nabla \left(F(r)^T F(r) \right)^{1/2} = \frac{F(r)^T J(r)}{\|F(r)\|_2}$$

which multiplied by the Newton vector gives:

$$\nabla \|F(r)\|_2 \delta r = \frac{F(r)^T J(r) \delta r}{\|F(r)\|_2}$$

If δr is a solution of the equation then $J(r)\delta r$ must be equal to $-F(r)$ and so:

$$\nabla \|F(r)\|_2 \delta r = \frac{F(r)^T \left(-F(r) \right)}{\|F(r)\|_2} = -\frac{F(r)^T F(r)}{\|F(r)\|_2} = -\frac{\|F(r)\|_2^2}{\|F(r)\|_2} = -\|F(r)\|_2$$

which is necessary negative since the Euclidian vector norm $\|F(r)\|_2$ is always positive.

In the particular case where there are no solutions for $J(r)x = -F(r)$, vector δr is no longer guaranteed to be a descent direction for the distance function since $J(r)\delta r$ is not $-F(r)$ but rather some approximation of it (which may be represented as $-F(r)+e$ where e is a vector whose size is minimised at each least-squares solution of the system). In this case the inner product between the gradient of the distance function and the Newton vector δr is:

$$\nabla \|F(r)\|_2 \delta r = \frac{F(r)^T \left(-F(r)+e \right)}{\|F(r)\|_2} = -\frac{F(r)^T \left(F(r)-e \right)}{\|F(r)\|_2}$$

which is only negative if:

$$\sum_{i=1}^{m} \left(F_i(r)^2 - F_i(r)e_i \right) > 0$$

Since, the size of the vector e is minimised at each least-squares solution of the system the odds are that the above inequality is satisfied. Moreover, since in this particular case a suitable linear combination of the right singular vectors will be used to choose some other least-squares solution, different from the minimal norm, the above inequality could be used as an additional constraint to enforce the resulting vector to be a descent direction for the distance function.

If the Newton vector δr defines a descent direction for the distance function $\|F(r)\|_2$ then it is guaranteed that it is always possible to obtain along that direction a point closer to a zero of F (in the sense that its distance function value is smaller than the current one). This new point must lie in the segment defined by:

$$r + \lambda \delta r \quad \text{with } \lambda \in [0..1]$$

The strategy to obtain the new point consists on trying different λ values, starting with the largest possible value without exceeding the search box limits, and backtracking to smaller values until a suitable point is reached. If the current point is close enough to the solution then the Newton step has quadratic convergence. Otherwise a smaller step is taken still directed towards a solution (or a mere local minimum of the distance function), guaranteeing convergence.

Since the current point r must be within the search box B, constraining the new point to be within the search box limits is the same as imposing

$$r_i + \lambda \delta r_i \in B[x_i]$$

for each variable x_i ($1 \le i \le n$). Defining α_i associated to each x_i as:

$$\alpha_i = \begin{cases} \min\left(\left\lfloor \left| \dfrac{left(B[x_i])}{\delta r_i} \right| \right\rfloor, 1\right) & \text{if} \quad \delta r_i < 0 \\[4mm] 1 & \text{if} \quad \delta r_i = 0 \\[4mm] \min\left(\left\lfloor \left| \dfrac{right(B[x_i])}{\delta r_i} \right| \right\rfloor, 1\right) & \text{if} \quad \delta r_i > 0 \end{cases} \quad .$$

The new point is kept inside the search box, when λ does not exceed any of these α_i values:

$$\lambda \in [0.. \min_{1 \le i \le n}(\alpha_i)]$$

Instead of changing the maximum value of λ, an equivalent alternative is to keep the λ value between 0 and 1 but to adjust the Newton vector δr by multiplying it by the constant $\min_{1 \le i \le n}(\alpha_i)$.

With this adjusted Newton vector, the backtrack search consists on starting at $\lambda = 1$ and trying consecutively smaller values of λ until an acceptance criterion is achieved or a convergence point is reached.

The acceptance criterion should not only guarantee that the new point decreases the distance function, but also that it avoids a too slow convergence rate (see [46]). The latter can be achieved by requiring the average rate of decrease of the distance function $\|F(r + \lambda \delta r)\|_2 - \|F(r)\|_2$ to be at least some fraction $k_{accept} \in [0..1]$ of the initial rate decrease $\nabla \|F(r)\|_2 \delta r$.

A convergence point is reached when the value of λ nears the canonical precision, making the canonical approximations of r and $r + \lambda \delta r$ indistinct. A practical criterion is to consider that a convergence point is reached whenever the value of λ is smaller than some predefined threshold k_{stop} (close to the canonical precision).

The choice of the consecutively smaller values of λ to be considered during the backtracking search should be based on some model for the function $g(\lambda)$ defined as:

$$g(\lambda) = \|F(r + \lambda \delta r)\|_2 \qquad \text{with } g'(0) = \nabla \|F(r)\|_2 \delta r$$

If a quadratic model is used then:

$$g(\lambda) = a\lambda^2 + b\lambda + c \qquad \text{and} \qquad g'(\lambda) = 2a\lambda + b$$

From the value of g at λ_1 (a previous value of λ) the following equations are obtained:

$$g(0) = a0^2 + b0 + c \qquad\qquad g'(0) = 2a0 + b \qquad\qquad g(\lambda_1) = a\lambda_1^2 + b\lambda_1 + c$$

which define the quadratic model:

$$g(\lambda) = \frac{g(\lambda_1) - g'(0)\lambda_1 - g(0)}{\lambda_1^2} \lambda^2 + g'(0)\lambda + g(0)$$

The value of λ that minimises the above quadratic function can be easily computed by calculating the unique zero of its derivative:

$$\lambda_{\min} = -\frac{g'(0)\lambda_1^2}{2(g(\lambda_1) - g'(0)\lambda_1 - g(0))}$$

With this quadratic model, the value of λ that should be considered after λ_1 is $\lambda_{\min}$. Since, in general, there are no guarantees about the value $\lambda_{\min}$, it is convenient, in order to avoid too smaller or too larger steps, to bound the next λ value relatively to the previous value λ_1. For example, λ could be forced to be between $0.1\lambda_1$ and $0.5\lambda_1$ by choosing $\lambda_{\min}$ if it is within this range or the closest of these bounds otherwise.

If instead of a quadratic model, a cubic model or other higher order model is used for representing function g, then a similar approach would have to be considered. In this case, the complete characterisation of the model would require the knowledge of the function value at several previous values of λ (k-1 values of λ different from zero for a k order model).

The function presented in figure 7.5 implements a backtracking search along the line segment defined by the current point and the Newton vector. From a CCSP P, an F-box B representing the search box, a degenerate F-box r representing the current point and a degenerate F-box δr representing the Newton vector, it computes the next point to jump to. The algorithm uses function *defineFunction* (see previous subsection) for the definition of vector function F associated to a CCSP at a given point. It is assumed that functions *quadraticModel* and *cubicModel* compute the minimum of an univariate function G assuming a quadratic or a cubic model respectively. These functions need as input the values of G and its derivative at point zero, as well as the value of G at one or two other points, for the quadratic and the cubic models, respectively (see [118] for a practical implementation of these functions).

```
function lineMinimization(a CCSP P, an F-box B, a degenerated F-box r, a degenerated F-box δr)
(1)   F ← defineFunction(P,r); G₀ ← F(r);
(2)   αₘᵢₙ ← 1;
(3)   for i=1 to n do
(4)       if δrᵢ<0 then αₘᵢₙ ← min(αₘᵢₙ,⌊left(B[xᵢ])/δrᵢ⌋)  end if;
(5)       if δrᵢ>0 then αₘᵢₙ ← min(αₘᵢₙ,⌊right(B[xᵢ])/δrᵢ⌋) end if;
(6)   end for;
(7)   δr ←αₘᵢₙδr ;
(8)   G'₀ ←‖F(r)‖₂ δr ;
(9)   if G'₀ ≥ 0 then return r;
(10)  λ ← 1;
(11)  repeat;
(12)      F ← defineFunction(P,r+λδr); Gλ ← F(r+λδr);
(13)      if Gλ < G₀ + k_accept λ G'₀ then return r+λδr;
(14)      if λ = 1 then λ_new ← quadraticModel(G₀, G'₀ ,λ,Gλ);
(15)      else λ_new ← cubicModel(G₀, G'₀ ,λ,Gλ,λ_prev,G_λprev);
(16)      λ_prev ←λ; G_λprev ← Gλ;
(17)      λ ← max(λ_new,0.1λ_prev); λ ← min(λ,0.5λ_prev);
(18)  until λ< k_stop;
(19)  return r;
end function
```

Figure 7.5 The algorithm that computes a new point along the Newton's vector direction.

Initially vector function F, associated with the CCSP P at the current point r is defined and its value G_0 is computed (line 1). Subsequently, the Newton vector δr is adjusted in order to guarantee that the new point is within the search box B. The minimum of the α_i values associated to each variable x_i is determined (lines 2 through 6) and the Newton vector δr is updated by multiplying it by the obtained value (line 7). The slope G'_0 at the current point in the direction of the Newton vector is calculated (line 8) and, if nonnegative, the algorithm stops returning the current point r (line 9).

The backtrack search along the Newton's vector direction is implemented as a *repeat* cycle (lines 11-18). At each iteration of the cycle, a λ value is tested and if it does not meet the acceptance criterion (according to the definition given previously in the current subsection) a new λ value is calculated for the next iteration. The verification of the acceptance criterion implies the redefinition of the vector function F associated with the respective point $r+\lambda\delta r$ and the computation of its value G_λ (line 12). If the acceptance criterion is satisfied then the new point $r+\lambda\delta r$ is returned (line 13). Otherwise a new λ value is calculated, which minimises the function G according to its quadratic (line 14 - only used for the first iteration) or cubic (line 15) model. The previous values of λ and G_λ are saved (line 16) and the new λ value is guaranteed to be between 0.1 and 0.5 of the previous λ value (line 17).

The *repeat* cycle is exited either when a new point satisfying the acceptance criterion is found (line 13) or when the λ value is smaller than the predefined threshold k_{stop} (line 18). In this last case, the current point r, which is a convergence point, is returned (line 19).

The algorithm is correct since it either returns a new point with an associated value of the distance function smaller than at the current point (this is guaranteed by the acceptance criterion) or it returns the current point if no better point was found.

The algorithm terminates because at each iteration of the *repeat* cycle the λ value is at least halved and in a finite number of iterations the decrease of the λ value is such that is necessary smaller than the threshold k_{stop}.

Figure 7.6 illustrates the new points obtained by applying the *lineMinimization* algorithm on the multidimensional vectors exemplified in figure 7.4. The dashed lines represent the original Newton vectors and the crosses are the points associated with λ values that did not satisfy the acceptance criterion. It is assumed that the search box is big enough to include the complete Newton vector (except in the case where x_1 must be minimised or maximised between 0 and 4).

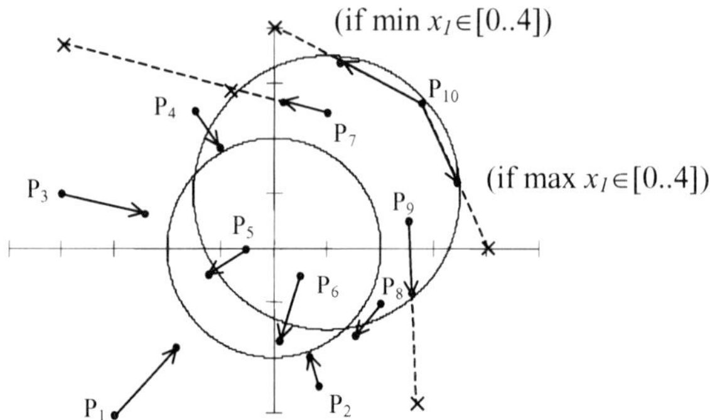

Figure 7.6 The new points obtained by the *lineMinimization* algorithm over the vectors of figure 7.4.

As shown in the figure, only for points in the first quadrant, the new point obtained is not the same obtained by applying the complete Newton vector. These points are too far from a solution of the CCSP for the quadratic approximation implicit in the Newton method to be effective, and a smaller step in Newton's direction is preferable. On the contrary, the other points are close enough of a solution so that the full Newton step is the best choice, eventually leading to quadratic convergence towards a solution.

Figure 7.7 shows the number of iterations of the complete local search algorithm (function *searchSolution*), necessary for reaching a convergence point (assuming a 4 digits precision) by starting at each of the previous points. It is assumed that the search box is big enough to include all the full Newton vectors computed during the local search.

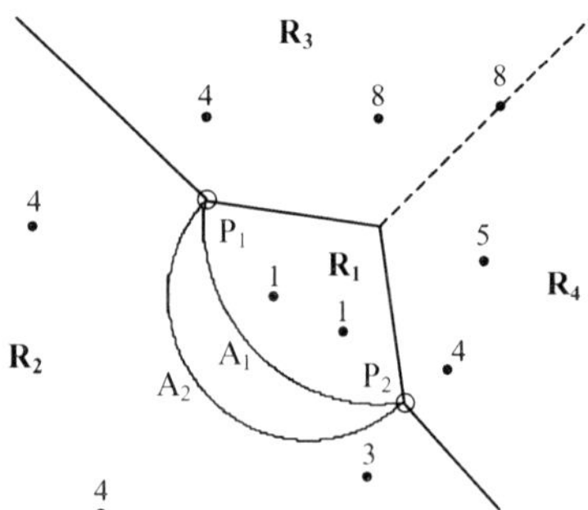

Figure 7.7 Convergence of the local search algorithm.

The number of necessary iterations is shown above the respective starting point, and is smaller when this is closer to a solution of the CCSP. Moreover, for this particular CCSP, the complete search space may be partitioned into four different regions (represented in the figure as $\mathbf{R_1}$ through $\mathbf{R_4}$) whose points share the same convergence behaviour.

From any starting point within the regions $\mathbf{R_1}$ or $\mathbf{R_2}$, the convergence point is a solution near the closest point of the arcs A_1 or A_2, respectively. If the starting point is within the regions $\mathbf{R_3}$ or $\mathbf{R_4}$, excluding the dotted line bordering these two regions, then the convergence points are solutions near points P_1 or P_2, respectively. In the particular case where the starting point is in the dotted line[3], the convergence point may be to a solution near point P_1 or P_2 depending on the context of the local search: P_1 if either the x_1 left bound or the x_2 right bound is searched; P_2 if either the x_1 right bound or the x_2 left bound is searched.

Consequently, for this particular CCSP, from any starting point and with a big enough search box, the local search algorithm converges towards the closest solution of the CCSP, being trapped in no local minimum.

The introduction of the line search minimisation along the Newton's vector direction is justified to achieve a globally convergent behaviour which could not be guaranteed by the Newton-Raphson method alone. However, since the local search is to be integrated within a branch and bound algorithm orienting and constraining the search into the most relevant regions of the search space, a globally convergent behaviour is desirable but not strictly necessary.

An alternative strategy could be to use the line minimisation algorithm exclusively for keeping the new point within the search box bounds and ignoring all the backtracking search along the Newton's vector. These simplified version would still present quadratic convergence for good initial guesses and would rapidly fail for starting points distant from the CCSP solutions.

7.2   Alternative Local Search Approaches

An alternative to the proposed local search approach based on the Newton-Raphson method for multidimensional root finding, could be based on the collapsing of the multiple dimensions into a single one by considering a nonnegative scalar function whose zeros are solutions of the CCSP. One such function could be the distance function defined in the previous subsection.

Once reduced the problem into an unconstrained multivariate minimisation problem, an efficient minimisation method [46] could be applied for searching the solutions of the CCSP. The choice of the more appropriate method should take into account the computational effort required for the calculation of the first derivatives of the scalar function and the storage required.

An efficient method not requiring the computation of derivatives is Powell's method [22], with a storage requirement of order n^2 (where n is the number of variables). Alternative methods, both requiring the computation of first derivatives, are the Conjugate Gradients methods (Fletcher-Reeves algorithm [53], Polak-Ribiere algorithm [117]) and the Quasi-Newton methods (Davison-Fletcher-Powell algorithm [52], Broyden-Fletcher-Goldfarb-Shanno algorithm [50]), with a storage requirement of order n and n^2, respectively.

The major drawback of these alternative local search strategies is that the early collapsing of the various dimensions of the vector function, whose components represent each unsatisfied constraint, implies the lack of information about each individual constraint and makes them more vulnerable to local minima. Strictly following a descent path of a

---

[3] To be precise, the singular point $(1,1)$ must also be included in this case and not within region $\mathbf{R_1}$.

scalar function such as the distance function often ends up in a local minimum which cannot be improved by local refinement.

However, since the local search is to be integrated within a branch and bound algorithm, the convergence to local minima is not restrictive, making any efficient local search strategy also competitive.

To avoid the convergence to local minima that are not solutions of the CCSP, other alternative approaches could be based on strategies used for solving constrained optimisation problems [57, 51, 17], namely Penalty methods or Lagrange-Multiplier methods. These methods reduce the constrained problems into one or a sequence of unconstrained multivariate minimisation problems by adding to the objective function terms that reflect the violation of each constraint.

The Penalty methods [49], which are easier to implement, introduce for each violated constraint a penalisation term, whose degree of penalisation must be tuned either before of during the optimisation process. This tuning is not trivial since too large penalties may result in a very irregular search space whereas too small penalties may lead convergence to unfeasible points.

The Lagrange-Multiplier methods [16], widely used for their numerical stability and accuracy, use Lagrange multipliers to combine the constraints with the original objective function obtaining a new multivariate function denoted Lagrangian function. All the minima of the Lagrangian function are minima of the original objective function which are guaranteed to satisfy the set of constraints. The price to pay for such nice properties is the increase on the number of variables that must be considered for the unconstrained minimisation of the Lagrangian function. Besides the original variables an additional variable (the respective Lagrange multiplier) is introduced for each constraint.

In our context, which does not explicitly requires the minimisation of a particular objective function, such optimisation methods could be applied, either by considering the distance function as the original objective function to minimise, or by performing an optimisation with respect to the variable bound that characterises the context of the local search.

Nevertheless, any of the above alternative approaches would still require the adoption of some strategy to bound the local search within the original search box limits. In the line search method this was accomplished by the line minimisation algorithm which avoided the larger Newton steps that escaped the search box limits. The possible drawback of such strong restriction, eventually avoiding the convergence to a solution just because the search path is not entirely contained in the search box, is largely compensated within the branch and bound context that guides the search to the most relevant subregions of the search space.

7.3 Integration of Local Search with Global Hull-Consistency Algorithms

The local search algorithm was originally conceived for providing the *TSA* algorithm with a specialised method for the localisation of new canonical solutions within a search box. In this algorithm the local search function *searchSolution* is naturally integrated as a step of the *searchAction* procedure (see figure 6.13).

The integration of the local search algorithm with the other enforcing algorithms presented in the previous chapter may be accomplished by minor changes in their generic search mechanisms (*backtrackSearch* or *orderedSearch*).

The functions *backtrackSearch* (figure 6.3) and *orderedSearch* (figure 6.8) may easily accommodate the local search algorithm by including a new argument j for representing the search context and by changing line 5 into:

(5) $B_{sol} \leftarrow$ *searchSolution*(P,B,j); **if** $B_{sol} \neq \varnothing$ **then return** B_{sol};

This approach may be adopted for the integration of the local search with the backtrack search algorithms BS_1 and BS_3 as well as with their ordered modifications OS_1 and OS_3, respectively.

A similar approach could be used for integrating local search with the BS_0 and BS_2 algorithms. However, the BS_0 algorithm does not profit from local search since no inner box is maintained, and with the above change, the BS_2 algorithm could no longer guarantee that the first canonical solution found by the generic search is the leftmost/rightmost (which is a crucial property for the correcteness of the algorithm).

In order to allow a correct integration with the BS_2 algorithm the first canonical solution found must be an extreme bound and so, the generic search should not terminate whenever a new solution is found, but rather when this solution is extreme with respect to the search context.

Figure 7.8 presents the function *backtrackSearchWithLocalSearch*, which is a modified version of the *backtrackSearch* function to integrate the local search algorithm, that may be used by any of the backtrack search algorithms (including the BS_2 algorithm). It is identical to the original *backtrackSearch* function of figure 6.3 except that it includes an additional third argument, the bound j, to represent the context of the local search, and the original lines 5-6 are replaced by the lines 5-16. It assumes that all the algorithms use a *splitStrategy* LW (largest width) or RR (round robin) except the BS_2 algorithm that uses a *splitStrategy* x_i (identifying the variable that must be firstly split).

---

function *backtrackSearchWithLocalSearch*(CCSP $P=(X,D,C)$, **inout** stack S of F-boxes, bound j)
(1)   **while** $S.size()>0$ **do**
(2) $B \leftarrow S.pop()$;
(3) $B \leftarrow$ *kB-consistency*(P,B);
(4)   **if** $B \neq \varnothing$ **then**
(5) $B_{sol} \leftarrow$ *searchSolution*(P,B,j);
(6)   **if** $B_{sol} \neq \varnothing$ **then**
(7)   **if** *splitStrategy* = LW **or** *splitStrategy* = RR **then return** B_{sol};
(8)   **if** *splitSide* = LEFT **and** *splitStrategy* = x_i **then**
(9)   **if** *left*$(B_{sol}[x_i])$= *left*$(B[x_i])$ **then return** B_{sol};
(10)   **else** $B_1 \leftarrow B$; $B_1[x_i] \leftarrow$ [*left*$(B[x_i])$..*left*$(B_{sol}[x_i])$)]; $B_2 \leftarrow B_{sol}$; **end if**;
(11)   **end if**;
(12)   **if** *splitSide* = RIGHT **and** *splitStrategy* = x_i **then**
(13)   **if** *right*$(B_{sol}[x_i])$= *right*$(B[x_i])$ **then return** B_{sol};
(14)   **else** $B_1 \leftarrow B$; $B_1[x_i] \leftarrow$ [*right*$(B_{sol}[x_i])$..*right*$(B[x_i])$)]; $B_2 \leftarrow B_{sol}$; **end if**;
(15)   **end if**;
(16)   **else** *splitBox*(B,B_1,B_2); **end if**;
(17) $S.push(B_2)$;
(18) $S.push(B_1)$;
(19)   **end if**;
(20)   **end while**;
(21)   **return** $\varnothing$;
end function

Figure 7.8 The modified generic backtrack search algorithm with local search.

The overall functioning of the *backtrackSearchWithLocalSearch* is similar to the *backtrackSearch* function except that a canonical solution is searched through a call to the local search function (line 5) and if a new solution is found (line 6), its consequences must be processed (lines 7-15). If the algorithm is not the BS_2 algorithm then the backtrack search terminates returning the new canonical solution (line 7). In the case of the BS_2

algorithm (lines 8-13) with a *splitStrategy* x_i, the backtrack search only terminates if the new solution is the leftmost/rightmost wrt x_i (line 9/13); otherwise (line 10/14), the two new F-boxes to include in the stack are the new canonical solution B_2 and the sub-box B_1 of B that remains relevant after considering this new solution (notice that B_1 is inserted at the top of the stack and so, the new canonical solution will only appear at the top if B_1 contains no solutions).

The correcteness and termination properties of the algorithm may be derived similarly to the case of the original *backtrackSearch* function as long as the *searchSolution* function is correct and terminates.

7.4   Summary

In this chapter a local search procedure based on a line search minimisation along a direction determined by the Newton-Raphson method was proposed for integration with the interval constraints framework. Its integration with Global Hull-consistency enforcement was presented for each of algorithms suggested in the previous chapter. In the next chapter preliminary results obtained with the application of Global Hull-consistency criterion are presented and compared with weaker consistency alternatives. The integration of local search within the best Global Hull enforcing algorithms is also illustrated on a simple problem.

Chapter 8

Experimental Results

In the context of decision making the trade-of between precision in the definition of the solution space and the computational efforts required to achieve it must be a major concern when solving CCSPs. In this context, our proposed approach to enforce a Global Hull-consistency may be an appropriate choice, achieving acceptable precision with relatively low computational cost. Such effort depends of course on the algorithms used to enforce such consistency. Among the set of algorithms we developed, the one that integrates constraint propagation within a tree-structured representation of the domain (algorithm TSA presented in section 6.4), has clearly shown the best performance.

In this chapter we present results obtained by imposing Global Hull-consistency in a number of problems and compare them with those obtained with (various levels of) alternative high order consistency requirements. In the first section a simple example motivates the need for strong consistency requirements such as Global Hull-consistency. Sections 8.2 and 8.3 present two practical examples illustrating the benefits of using TSA for enforcing Global Hull-consistency. In section 8.4 we compare the efficiency of TSA with the other algorithms for enforcing Global Hull-consistency presented in chapter 6, and address the potential benefit of including local search.

We implemented all Global Hull-consistency and kB-consistency (section 5.2) algorithms, based on the procedures for achieving 2B-consistency (Box-consistency) available in the OpAC library (a C++ interval constraint language [61]), and executed in a Pentium III computer at 500 MHz with 128 Mbytes memory.

8.1 A simple example

To understand the pitfalls often arisen with kB-consistency we have considered a small problem consisting of two constraints:

$$x^2 + y^2 \leq 1 \qquad \text{and} \qquad x^2 + y^2 \geq 2$$

Of course, the two constraints are unsatisfiable. Figure 8.1 illustrates the problem, which requires the solutions to be within the inner circle (first constraint) and outside the outer circle (second constraint).

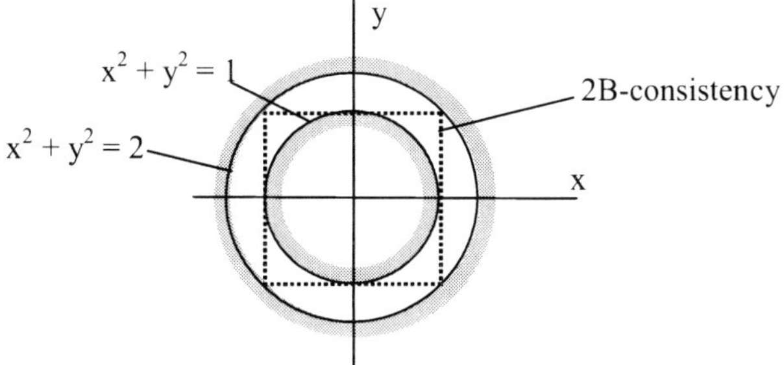

Figure 8.1 A simple unsatisfiable constraint problem.

However, 2B-consistency (Box-consistency) does not detect such inconsistency, merely pruning the initial unbounded domains of the variables to the interval [-1.001..1.001] (with 10^{-3} precision). With these domains (obtained by applying 2B-consistency to the first constraint) the second constraint does not prune the domain of any of its variables when the other is fixed to any of its bounds. Of course, x = ±1 is only compatible with y = 0 in the first constraint whereas it requires y = ± 1 in the second, but the local nature of 2B-consistency does not detect this situation. Since there are only 2 variables involved, 3B-consistency is equivalent to Global Hull-consistency and detects the inconsistency.

With an increased number of variables the insufficiency of 2B-consistency also arises in higher order consistencies. For example, in the following problem

$$x^2 + y^2 + z^2 \leq 2 \qquad \text{and} \qquad x^2 + y^2 + z^2 \geq 3$$

rather than detecting inconsistency, as Global Hull-consistency does, 3B-consistency prunes the variables to [-1.001 .. 1.001] whereas 2B-consistency performs even worse, pruning the domains to [−1.416..1.415].

In general, the difference obtained with kB-consistency and Global Hull-consistency is not so significant, but still different bounds are obtained.

For example with the slightly modified problem

$$x^2 + y^2 + z^2 \leq 2 \qquad \text{and} \qquad (x-0.5)^2 + y^2 + z^2 \geq 2.25$$

the results obtained are shown in Table 8.1. Both 2B- and 3B-consistency, although faster than Global Hull-consistency, report quite inaccurate upper bounds for variable x.

Table 8.1 Pruning domains in a trivial problem.

	2B	3B	Global Hull
x	[-1.416 .. 1.415]	[-1.415 .. 1.002]	[-1.415 .. 0.001]
y	[-1.416 .. 1.415]	[-1.415 .. 1.415]	[-1.415 .. 1.415]
z	[-1.416 .. 1.415]	[-1.415 .. 1.415]	[-1.415 .. 1.415]
t (ms)	10	50	1860

8.2   The Census Problem

The Census problem models the variation with time of a population with limited growth by means of a parametric differential equation (logistic). The equation has an analytical solution of the form:

$$x(t) = \frac{kx_0 e^{rt}}{x_0\left(e^{rt} - 1\right) + k}$$

Given a set of observations $v_0,\ldots,v_n$ at various time points $t_0,\ldots,t_n$, the goal of the problem is to adjust the parameters x_0, r and k of the equation to the observations.

We used the USA Census over the years 1790 (normalised to 0) to 1910 with a 10 year period. Table 8.2 shows the population values observed at those time points.

Table 8.2 US Population (in millions) over the years 1790 (0) to 1910 (120).

t_i	0	10	20	30	40	50	60	70	80	90	100	110	120
v_i	3.929	5.308	7.239	9.638	12.866	17.069	23.191	31.433	39.818	50.155	62.947	75.994	91.972

Figure 8.2 illustrates the problem showing its best fit solution ($x_0 = 4.024$, $r = 0.031$ and $k = 198.2$). Such solution minimises the expression: $\sum_i \left(x(t_i) - v_i\right)^2$

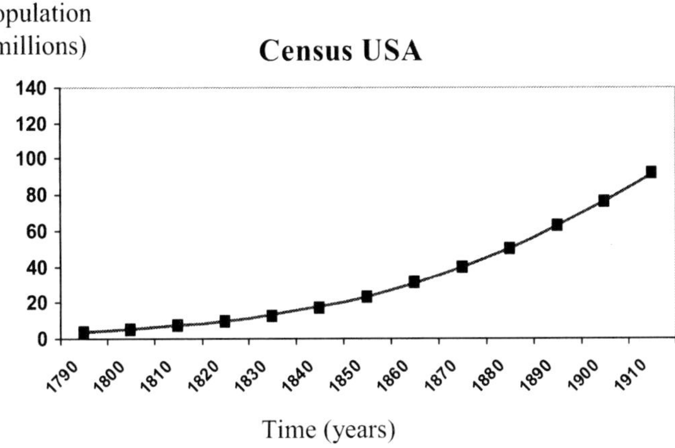

Figure 8.2 The best fit solution for the USA Census problem.

In order to adjust the parameters of the logistic equation to the observations, instead of searching for the best fit solution, we considered as acceptable a distance of up to ±1 (million) between each observed value v_i and the respective predicted value $x(t_i)$. This is achieved by imposing the following constraint[1] at each observed time point t_i:

$$x_i = \frac{kx_0 e^{0.001\,rt_i}}{x_0\left(e^{0.001\,rt_i} - 1\right) + k} \quad \text{with} \quad x_i \in [v_i - 1 .. v_i + 1]$$

In Table 8.3 we show the results of enforcing 2B-, 3B- and Global Hull-consistency on this problem (with 10^{-3} precision). The table shows the poor pruning results achieved by 2B-consistency alone, and the much better pruning achieved by Global Hull-consistency with respect to 3B-consistency.

Table 8.3 Comparing 2B-, 3B- and Global Hull-consistency in the Census problem.

	2B	3B	Global Hull
x_0	[2.929 .. 4.930]	[2.929 .. 4.862]	[3.445 .. 4.547]
k	[1.1 .. 1000]	[102.045 .. 306.098]	[166.125 .. 260.401]
r	[1.1 .. 100]	[27.474 .. 39.104]	[28.683 .. 33.714]
t(ms)	10	56 990	458 840

Although this improvement is achieved at the cost of a much longer execution time (about 8 times slower than 3B), it is important to notice that OS_3 and *TSA* algorithms for achieving Global Hull-consistency are anytime algorithms, and good results may be obtained much earlier.

Table 8.4 shows that the pruning achieved by Global Hull, at approximately 30% of the execution time spent with 3B enforcing algorithm, is already significantly better than it. For similar execution times (about 1 minute), the pruning is almost as good as the final one.

Table 8.4 Comparing anytime GH and 3B in the Census problem.

	Global Hull	Global Hull	Global Hull	3B
x_0	[3.445 .. 4.547]	[3.040 .. 4.775]	[3.142 .. 4.567]	[2.929 .. 4.862]
k	[166.125 .. 260.401]	[129.863 .. 282.040]	[148.153 .. 261.157]	[102.045 .. 306.098]
r	[28.683 .. 33.714]	[27.777 .. 36.730]	[28.646 .. 35.296]	[27.474 .. 39.104]
t (ms)	458 840	15 030	55 110	56 990

---

[1] The parameter r, with much smaller values than the other parameters, is re-scaled into the interval [1..100] by multiplying it by a factor of 0.001 (its best fit value is now 31.0).

Figure 8.3 illustrates such pruning results. The black area represents the uncertainty about the trajectory of a logistic function with the parameters ranging within the box obtained by enforcing Global Hull-consistency on this problem (the second column of table 8.4). A slightly wider uncertainty (with the extra dark-gray area) is obtained if the Global Hull-consistency enforcing algorithm is interrupted after 1 minute of execution time (the time for enforcing 3B-consistency). However, a much wider uncertainty (with the extra light-gray area) must be considered if the box is obtained by enforcing 3B-consistency (the last column of table 8.4).

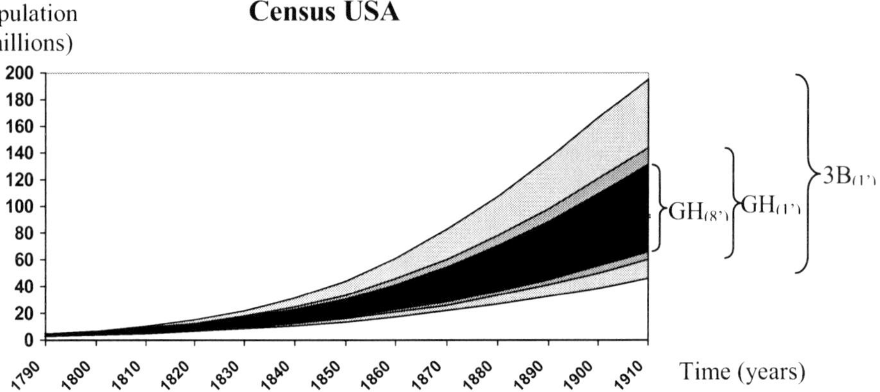

Figure 8.3 Comparing 3B with GH, and GH with similar execution time (about 1').

Also notice that, since there are only 3 variables, imposing 4B-consistency provides the same final results than Global Hull-consistency. However, as shown in the table 8.5, the algorithm is much slower. The table compares the pruning achieved during the CPU execution time. The values are given as a percentage of the area of the box obtained by Global Hull with respect to the area of the box obtained by 4B at the same execution times (both with 10^{-3} precision).

Table 8.5 Comparing anytime GH and 4B in the Census problem.

GH/4B	27.1%	43.0%	67.3%	81.6%	100%
t (min)	0.25	1	8	16	180

When the Global Hull algorithm stops, after about 8 minutes of CPU time, it obtains a domains box that is approximately 2/3 of current box of the 4B algorithm (at the same time). The equivalent pruning results, which are theoretically achievable, take in practice about 22 times more CPU time with the 4B algorithm than with the Global Hull algorithm.

Finally we compare the performance of the TSA algorithm with different precision requirements, namely, 10^{-3}, 10^{-6}, 10^{-9} and 10^{-12}. Table 8.6 shows the results where the unit of time is the execution time with 10^{-3} precision. The first row indicates the precision ε required; the second row shows the time t_0 at which the pruning results were already identical to those obtained with 10^{-3} precision; the third row presents the total execution time t_{final}; and the fourth row indicates the storage used by the algorithm in terms of the maximum number of F-boxes considered.

Table 8.6 Comparing GH with different precision requirements in the Census problem.

ε	10^{-3}	10^{-6}	10^{-9}	10^{-12}
t_0	1	1.2	1.7	2.4
t_{final}	1	2.9	5.0	7.5
F-boxes	1290	2938	4711	6314

Clearly, for this problem, there is not an explosion of execution time or storage requirements with increasing precision requirements. Both, time and storage, seem to increase linearly with the number of significant digits required. Moreover, if we consider the uncertainty obtained with 10^{-3} precision, identical results may be obtained much before the ending of the algorithm.

8.3 Protein Structure

The next problem we report is a simplification of that of finding the structure of a protein from distance constraints, e.g. obtained from Nuclear Magnetic Resonance data (see [87, 88]). The simplified problem uses Euclidean distance constraints similar to those presented in section 8.1 above, where variables x_i, y_i and z_i represent the centres of atom a_i.

In this problem, we place 6 atoms, whose centres must all be, at least, 1Å apart. For some atom pairs, the square of the distances are provided and shown in table 8.7.

Table 8.7 Square distances between pairs of atoms of the protein.

Atoms	a_1	a_2	a_3	a_4	a_5	a_6
a_1		2	4	-	4	4
a_2	2		2	4	2	-
a_3	4	2		2	4	4
a_4	-	4	2		2	2
a_5	4	2	4	2		-
a_6	4	-	4	2	-	

To solve the problem we placed 3 atoms (a_1-a_3) arbitrarily in the XY plane and the distances allow that the other 3 atoms are either above the plane (positive Z) or the corresponding quiral solution below the plane (negative Zs).

In table 8.8 we show the pruning achieved on the values of atoms a_4 to a_6 by 2B-, 3B- and Global Hull-consistency (as before, results with 4B-consistency are similar to those with Global Hull but take much longer to obtain).

Table 8.8 Comparing 2B, 3B and GH in the Protein problem.

	2B	3B	Global Hull
x_4	[-1.001 .. 1.415]	[-0.056 .. 0.049]	[-0.004 .. 0.004]
y_4	[0.585 .. 3.001]	[1.942 .. 2.047]	[1.996 .. 2.004]
z_4	[-1.415 .. 1.416]	[-1.415 .. 1.415]	[-1.415 .. 1.415]
x_5	[-0.415 .. 2.001]	[0.998 .. 1.002]	[0.999 .. 1.001]
y_5	[-0.001 .. 2.001]	[0.999 .. 1.001]	[0.999 .. 1.001]
z_5	[-1.415 .. 1.416]	[-1.415 .. 1.415]	[-1.415 .. 1.415]
x_6	[-2.001 .. 2.001]	[-1.110 .. 1.053]	[-1.008 .. -0.992]
y_6	[-0.001 .. 2.001]	[-0.894 .. 1.169]	[0.999 .. 1.001]
z_6	[-2.001 .. 2.001]	[-1.483 .. 1.483]	[-1.420 .. 1.402]
t (ms)	10	7 380	62 540

Given the uncertainty in the Z value of the atoms, neither 2B- nor 3B-consistency could prune the size of their other dimensions to the amount that Global Hull-consistency does. The most important difference between 3B- and Global Hull-consistency lies on atom a_6. Whereas the x_6 and y_6 are "fixed" (respectively at around -1 and 1) by Global Hull, 3B-consistency could not prune the value of these variables beyond the typical $[-1 .. 1]$ interval.

8.4   Local Search

In the above experiments Global Hull-consistency was enforced with the TSA algorithm without local search, a precision $\varepsilon = 10^{-3}$ and an underlying 2B-consistency procedure. To check whether the above timings were truly representative of Global Hull enforcement, we decided to test the efficiency of the different algorithms presented in chapter 6.

The results presented in Table 8.9 refer to another instance of the Protein structure (with 8 atoms), more representative of the kind of problems Global Hull is aimed at, in that they have many adjacent solutions, i.e. the final domains of all variables are relatively large.

Table 8.9 Comparing various Global Hull-consistency enforcing algorithms.

			Time (s)			Max Storage (F-boxes)		
	k	LS	$\varepsilon=10^{-1}$	$\varepsilon=10^{-2}$	$\varepsilon=10^{-3}$	$\varepsilon=10^{-1}$	$\varepsilon=10^{-2}$	$\varepsilon=10^{-3}$
OS_1	2	n	7.34	600+	600+	407	55711	63465
		y	87.52	78.48	600+	407	414	25580
	3	n	23.57	139.84	600+	27	339	1584
		y	12.80	50.39	191.41	2	15	51
BS_2	2	n	2.99	11.37	600+	30	47	60
		y	14.97	38.46	600+	14	16	24
	3	n	41.01	150.21	359.69	22	37	53
		y	19.14	75.98	234.65	3	8	25
OS_3	2	n	5.19	600+	600+	132	66410	63445
		y	23.29	20.72	600+	51	104	25037
	3	n	25.51	185.99	600+	19	695	1676
		y	13.08	45.41	162.50	2	10	32
TSA	2	n	10.16	60.96	600+	332	6786	76506
		y	44 79	47.59	600+	221	621	16764
	3	n	42.52	184.38	600+	52	207	843
		y	38.63	97.88	275.85	99	185	392

The first thing to notice is the importance of the precision used. Of course, with less precision, all algorithms are faster, since less canonical *F*-boxes are considered. More interestingly, as precision increases (smaller ε) local search becomes more important. The reason for this is that with larger canonical *F*-boxes the underlying 2B-consistency algorithm does not prune them and so accepts them as canonical solutions, making the local search for "real" solutions useless. With higher precision, canonical *F*-boxes are smaller and the 2B-consistency algorithm does not detect solutions as easily, as the pruning of most *F*-boxes does not result into canonical *F*-boxes. Hence, the advantage of local search in such situations.

Local search is also shown to improve the memory requirements, since it often finds solutions near the intended bounds of the variables in the *F*-boxes under consideration, rather than simply bisecting them (thus originating additional boxes).

Memory requirements are also much lower when the underlying enforcement procedure is 3B-consistency (rather than 2B), as the pruning achieved in any *F*-box is much more significant. Because of its better pruning, and despite its higher complexity, enforcing 3B-consistency provides in general better results than 2B-consistency, as precision increases.

Regarding the variety of Global Hull-consistency enforcing algorithms, their differences are less evident. Given the discussion above it might be important to impose some thresholds on the execution time of the algorithms, in which case the OS_1 and BS_2 have the disadvantage of not being anytime algorithms. Although OS_3 proved better than *TSA* in this problem, this behaviour is not observed consistently in other problems, and *TSA* offers the advantage of keeping a tree-based compact description of the feasible space, which is very convenient for an interactive use as envisaged in [69]. Moreover, and although not visible in the table, the anytime results of *TSA* are consistently better than those obtained with OS_3.

8.5 Summary

In this chapter our proposals were tested on simple examples such as the USA census problem and the protein structure problem. The pruning and time results obtained with the Global Hull-consistency approach (with *TSA* algorithm) were compared with those obtained by enforcing 2B-, 3B- and 4B-consistency. The integration of local search within the best Global Hull enforcing algorithms was discussed. This ends the first part of this dissertation. The next part is dedicated to handling constraints over ordinary differential equations, where the constraint methods discussed so far will be used.

Part II

INTERVAL CONSTRAINTS FOR DIFFERENTIAL EQUATIONS

Chapter 9

Ordinary Differential Equations

Differential equations are equations involving derivatives. Ordinary Differential Equations (ODEs) involve derivatives with respect to a single independent variable. The order of an ODE is determined by the highest derivative appearing in the equation.

A first-order ODE will be generically represented as:

$$\frac{dy}{dt} = f(y(t), t)$$

where, for historical reasons, the independent variable, often associated with time, is denoted as t. If it is clear from the context, the above representation may be shortened to:

$$y' = f(y, t)$$

A system of n first-order ODEs is the set of equations:

$$y_1' = f_1(y_1, ..., y_n, t)$$

$$...$$

$$y_n' = f_n(y_1, ..., y_n, t)$$

which, if there is no ambiguity with the single equation case, may be represented in vector notation as:

$$y' = f(y, t) \qquad \text{with} \quad y' = \begin{bmatrix} y_1' \\ ... \\ y_n' \end{bmatrix} \quad \text{and} \quad f(y, t) = \begin{bmatrix} f_1(y_1, ..., y_n, t) \\ ... \\ f_n(y_1, ..., y_n, t) \end{bmatrix}$$

Any explicit[1] differential equation of order m defined as:

$$y^{(m)} = f(y, y', y'', ..., y^{(m-1)}, t)$$

may be transformed into an equivalent system of m first-order ODEs:

$$z' = f(z, t) \qquad \text{with} \quad z' = \begin{bmatrix} z_1' \\ ... \\ z_{m-1}' \\ z_m' \end{bmatrix} \quad \text{and} \quad f(z, t) = \begin{bmatrix} z_2 \\ ... \\ z_m \\ f(z_1, z_2, ..., z_m, t) \end{bmatrix}$$

Given the above property we will only consider, without loss of generality, systems of first-order ODEs and will denote them as ODE systems.

Definition 9-1 (ODE system). An ODE system O is a system of n first-order ODEs defined as:

$$y' = f(y, t) \text{ with } y' = \begin{bmatrix} y_1' \\ ... \\ y_n' \end{bmatrix} \quad \text{and} \quad f(y, t) = \begin{bmatrix} f_1(y_1, ..., y_n, t) \\ ... \\ f_n(y_1, ..., y_n, t) \end{bmatrix}$$

When the independent variable t does not appear explicitly in function f, that is, $y' = f(y)$, the ODE system is called *autonomous*[2]. ❏

---

[1] A differential equation is called explicit if the highest derivative is isolated in the left-hand side.

[2] Any ODE system with n equations may be transformed into an equivalent autonomous system with $n+1$ equations by replacing any occurrence of t with a new dependent variable y_{n+1} and by adding the associated derivative equation: $y'_{n+1} = 1$.

An ODE system may be regarded as a restriction on the sequence of values that y can take over t. Informally, for any particular instantiation of y and t, it determines the evolution of y at time t associated with an increment of t. An ODE system does not determine a unique sequence of values of y associated with t; it rather characterises a family of functions whose slope must satisfy the equations for all values of t. This family of functions are the solutions of the ODE system.

Definition 9-2 (Solution of an ODE system). Consider an ODE system O as defined in 9-1. The function $s(t) = \begin{bmatrix} s_1(t) \\ ... \\ s_n(t) \end{bmatrix}$ is a solution of O wrt the interval $[t_0..t_1]$ iff:

$$\forall_{t \in [t_0..t_1]} \ \frac{ds}{d} = f(s(t),t)$$

In order to uniquely identify a particular function from the solutions of an ODE system, further information must be added. Traditionally this is accomplished either by completely specifying the value of y associated with a particular value of t (initial condition), or by partially specifying the value of y associated with different values of t (boundary conditions).

Given an ODE system and a value for y at a given t_0, the initial value problem (IVP) aims at determining the value of y for other values of t. A solution of the ODE system that satisfies the initial condition is a solution of the IVP.

Definition 9-3 (Solution of an IVP). Consider the IVP I defined by the ODE system O as expressed in 9-1 and the initial condition $y(t_0) = y_0 = \begin{bmatrix} v_{10} \\ ... \\ v_{n0} \end{bmatrix}$. Function $s(t) = \begin{bmatrix} s_1(t) \\ ... \\ s_n(t) \end{bmatrix}$ is a solution of I wrt an interval $[t_a..t_b]$ that includes t_0 iff:

$$s(t_0) = y_0 \ \wedge \ \forall_{t \in [t_a..t_b]} \ \frac{ds}{d} = f(s(t),t)$$

In general it is not possible to solve an IVP analytically, that is, the function $s(t)$ that is the solution of the IVP cannot be represented in a closed-form expression. Consequently, several approaches have been proposed for solving this kind of problems numerically.

The classical numerical approaches attempt to compute numerical approximations of the solution $s(t)$ at some discrete points of t. These methods, addressed in the next section, are usually very efficient but do not provide any guarantee on the accuracy of the approximations or even the existence of a unique solution.

Interval approaches, discussed in section 9.2, attempt to produce bounds for the solution $s(t)$ not only at some discrete points of t but also for all the continuous range of intermediate values between any two consecutive discrete points. These methods, also known as validated methods, verify the existence and uniqueness of a solution for the IVP.

The relative unpopularity of the direct application of interval methods to ODE problems (at least compared with their numerical alternatives) stems from the additional computational effort required and, in many early approaches, to the insufficient tightness provided by the enclosing bounds. To overcome such difficulties, research has been carried out to take advantage of the efficiency and soundness of constraint technique, as overviewed in section 9.3.

9.1   Numerical Approaches

Classical numerical approaches for solving the initial value problem consider a sequence of discrete points t_0, t_1, ..., t_m for which the solution is approximated. The distances $h_i=t_{i+1}-t_i$ between two consecutive points do not need to be equally spaced. At each new point t_{i+1}, the solution $s(t_{i+1})$ is approximated by a value s_{i+1} computed from the approximated values at the previous points. Whether this computation requires only the most recent value (s_i) or also other earlier values (s_{i-1}, s_{i-2}, ...) qualifies if the method is a single-step or a multistep method.

A k-step method provides a formula that approximates the solution function at the next discrete point t_{i+1} which may be generically[3] represented as:

$$s_{i+1} = \Phi(s_{i-k+1},...,s_j) \qquad \text{where } j=i \text{ or } j=i+1$$

If the approximate value s_{i+1} does not appear in the right-hand side of the equation ($j=i$), the formula is explicit and its evaluation is straightforward. Otherwise (if $j=i+1$), the formula is implicit since its evaluation may involve solving the above vector equation (a system of possibly nonlinear equations) where s_{i+1} appears in both sides.

To avoid solving the vector equation, some methods, known as predictor-corrector methods, use both an explicit formula and an implicit formula. Initially, the explicit (predictor) formula is used to obtain a first value approximation for s_{i+1}. Then the implicit (corrector) formula is used, with the s_{i+1} of the right-hand side replaced by the predicted value, to obtain a final improved value.

Although there are many variations within the different numerical methods for solving IVPs they may be classified into four general categories: Taylor series, Runge-Kutta, multistep and extrapolation methods.

Taylor series methods are single-step methods that use the Taylor series expansion of the solution function around a point, to obtain an approximation of its value at the next point. This series is computed up to a given order, requiring the evaluation of higher order derivatives of the function.

Runge-Kutta methods are single-step methods that approximate the Taylor series methods without requiring the evaluation of derivatives of the solution function beyond the first. This is accomplished at the expense of several evaluations of the first derivative of the solution function, whose expression is given by the ODE. The idea is to compute a linear combination of this function at different points to match as much as possible the Taylor series up to some higher order.

Multistep methods use a polynomial of degree k-1 to approximate either the solution function (as in backward differentiation methods) or its derivative (as in Adams methods). The coefficients of the polynomial are determined from the values of the approximated function at k different points. Implicit or explicit formulas for approximating the value of the solution function at the next point t_{i+1} may be obtained depending on whether this point is considered in the determination of the polynomial coefficients.

In contrast to the previous methods, extrapolation methods (as the Bulirsch-Stoer method [25, 133]), consider larger steps (non-infinitesimal) between consecutive discrete points of t. To compute the approximation of the solution function at the discrete point t_{i+1} from the previous one, t_i, a single-step method is used to integrate the differential equation along the interval $[t_i..t_{i+1}]$ considering an increasing sequence of finer and finer substeps. During this process an interpolating polynomial or rational function is constructed through the computed intermediate values and in the end the value at t_{i+1} is extrapolated to a zero substep size.

---

[3] For simplicity, the values of the discrete points of t are not explicitly represented as arguments of the function Φ.

A detailed overview of the above numerical methods may be found in many text books dealing with the numerical solution of initial value problems. Classical books on this subject are [72, 70, 56] and more recent overviews can be found in [63, 90, 127]. Several public domain software packages provide efficient implementations of the numerical methods, in particular for the Runge-Kutta (as the RKSUITE [21]) and the multistep methods (as the VODE [23]). For practical discussions on the software implementation see also [126, 54, 118, 5].

In the next subsection the Taylor series methods will be addressed in more detail. Subsection 9.1.2 discusses the different sources of errors and its consequences in solving an IVP numerically.

9.1.1   *Taylor Series Methods*

The simplest single-step methods are based on the Taylor series expansion of the solution function. If the solution $s(t)$ of an ODE system, as defined in 9-2, is a function which is p times continuously differentiable on the closed interval $[t_i..t_{i+1}]$ and $p+1$ times differentiable on the open interval $(t_i..t_{i+1})$, then, from the Taylor theorem, we have:

$$s(t_{i+1}) = s(t_i) + \sum_{k=1}^{p} \left(\frac{h^k}{k!} s^{(k)}(t_i) \right) + \frac{h^{p+1}}{(p+1)!} s^{(p+1)}(\xi) \quad \text{with} \quad h = t_{i+1} - t_i \quad \text{and} \quad \xi \in [t_i..t_{i+1}]$$

Since $s(t)$ is a solution of the ODE system (wrt the interval $[t_i..t_{i+1}]$), then, from definition 9-2, $\forall_{t \in [t_i..t_{i+1}]} ds/dt = f(s(t),t)$, and so:

$$\forall_{k \in [1..p+1]} s^{(k)}(t) = f^{(k-1)}(s(t),t)$$

Hence, the Taylor series expansion may be rewritten as:

$$s(t_{i+1}) = s(t_i) + \sum_{k=1}^{p} \left(\frac{h^k}{k!} f^{(k-1)}(s(t_i),t_i) \right) + \frac{h^{p+1}}{(p+1)!} f^{(p)}(s(\xi),\xi) \quad \text{w/} \quad h = t_{i+1} - t_i \text{ and } \xi \in [t_i..t_{i+1}]$$

If, additionally, $s^{(p+1)}$ is continuous throughout the closed interval $[t_i..t_{i+1}]$, it must be bounded on that interval and so, the last term of the Taylor series expansion is also bounded on $[t_i..t_{i+1}]$, being of order $O(h^{p+1})$.

Consequently, a Taylor method of order p can be obtained by neglecting this last term, providing the following formula that approximates the value s_{i+1} of the solution function at the next discrete point t_{i+1} given its approximated value s_i and at the previous point t_i:

$$s_{i+1} = s_i + \sum_{k=1}^{p} \left(\frac{h^k}{k!} f^{(k-1)}(s_i,t_i) \right) \quad \text{with} \quad h = t_{i+1} - t_i$$

The total derivatives of f may be computed recursively in terms of its partial derivatives. Since the vector function f, as defined in 9-1, is composed by n elementary functions $f_i(y_1,\ldots,y_n,t)$, the total derivatives may be obtained component wise as:

$$f_i'(y_1,\ldots,y_n,t) = \sum_{m=1}^{n} \left(\frac{\partial f_i(y_1,\ldots,y_n,t)}{\partial y_m} y_m' \right) + \frac{\partial f_i(y_1,\ldots,y_n,t)}{\partial t}$$

$$= \sum_{m=1}^{n} \left(\frac{\partial f_i(y_1,\ldots,y_n,t)}{\partial y_m} f_m(y_1,\ldots,y_n,t) \right) + \frac{\partial f_i(y_1,\ldots,y_n,t)}{\partial t}$$

$$f_i^{(k)}(y_1,\ldots,y_n,t) = \sum_{m=1}^{n} \left(\frac{\partial f_i^{(k-1)}(y_1,\ldots,y_n,t)}{\partial y_m} f_m(y_1,\ldots,y_n,t) \right) + \frac{\partial f_i^{(k-1)}(y_1,\ldots,y_n,t)}{\partial t}$$

$$\text{with} \quad f^{(k)}(y,t) = \begin{bmatrix} f_1^{(k)}(y_1,\ldots,y_n,t) \\ \ldots \\ f_n^{(k)}(y_1,\ldots,y_n,t) \end{bmatrix}.$$

In the special case of autonomous ODE systems, t does not appear explicitly in f, so neither does the last term of the above definition of the total derivatives.

Despite the existence of quite efficient methods for the automatic generation of the Taylor coefficients for autonomous ODE systems (see section 9.3), the Taylor series methods are not as competitive as other numerical methods such as the Runge-Kutta which do not require higher derivative evaluations (they do not even require the existence of such derivatives). Hence, existing software packages for solving numerically an IVP do not usually employ Taylor series methods. Nevertheless, methods based on the Taylor series (not neglecting the last term) seem to be more suitable for interval approaches, which aim at computing reliable interval bounds for the enclosure of the solution function.

9.1.2   *Errors and Step Control*

The ultimate goal of a numerical approach for solving an IVP over an interval range of t is to approximate as much as possible its solution at some discrete points placed along that interval. Usually, by starting at point t_0 (whose solution value is known: $s(t_0)=y_0$) an increasing (decreasing) sequence of discrete points is considered by adjusting the step size (the gap between two consecutive discrete points) as the calculation proceeds. The purpose of this adaptive step size policy is to keep some control over the accuracy of the approximation with minimum computational effort.

Since this effort is proportional to the total number of discrete points considered, such points should be separate from each other as much as possible. However, too wide gaps may lead to unacceptable approximation errors.

There are two different sources of errors for the numerical solution of an IVP: the discretization error (also known as truncation error) and the computational error. Whereas the first depends exclusively on the properties of the numerical method adopted, the second is due to round-off errors and errors committed in the approximated evaluation of implicit formulas.

The approximation error committed at a new point t_{i+1} is partially caused in the current step (from t_i to t_{i+1}) by the chosen method, and partially due to propagation of errors made at previous steps (from t_0 to t_i). Accordingly, there are two measures of the discretization error, the local discretization error and the global discretization error.

The local discretization error is the error committed in one step by the approximation method assuming that the previous values were exact and the absence of computational errors. If s_i and s_{i+1} are approximations of the correct solution values $s(t_i)$ and $s(t_{i+1})$, computed by the numerical method at two consecutive points, the local discretization error d_{i+1}, from t_i to t_{i+1}, is given by:

$$d_{i+1} = s_{i+1} - u(t_{i+1})$$

where u is the solution function of an IVP with the original ODE and the initial condition $u(t_i) = s_i$.

The global discretization error is the error accumulated along the whole sequence of discrete points (starting at the initial condition point t_0) again in the absence of computational errors. If s_i is the approximation of the correct solution value $s(t_i)$, computed by the numerical method along the sequence of discrete points $t_0,\ldots,\ t_i$, the global discretization error e_i accumulated so far is given by:

$$e_i = s_i - s(t_i)$$

Figure 9.1 illustrates the notions of local and global discretization errors associated with two different IVPs. The initial condition is defined at point t_0, and the approximated solution is computed at three subsequent discrete points, t_1, t_2 and t_3. The local discretization errors in each of these points are represented as d_1, d_2 and d_3, respectively.

The correct solution function is line $s(t)$ and functions $u_i(t)$ are the solutions of similar IVPs with initial conditions $y(t_i)=s_i$, respectively. The global discretization error accumulated at t_3, represented as e_3, is compared with the sum off all the previous local errors.

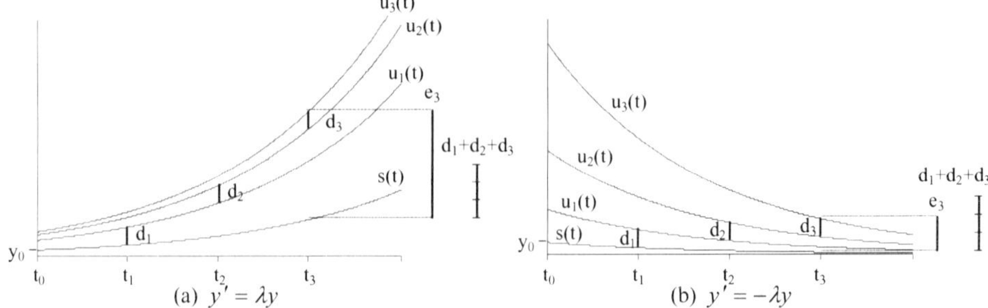

Figure 9.1 Local and global discretization errors for an unstable (a) and a stable (b) differential equation. In both equations λ is a positive real value.

The concept of stability is associated with the effect of local errors on global errors. Instability occurs when small local errors introduced during the approximation procedure are magnified into larger global error (as in Figure 9.1a). On the contrary, when local errors introduced are attenuated into a smaller global error, the problem is called stable (Figure 9.1b). Unfortunately nonlinear ODE systems are often unstable, at least in some regions.

The best that can be expected from any numerical method for solving an IVP is to maintain the inherent behaviour of the actual solution. An unstable problem cannot become stable by any numerical method. However, instability may be introduced by the numerical method in originally stable regions of the actual solution. Avoiding such instabilities may require the introduction of additional restrictions limiting the length of the step sizes.

Roughly speaking, the global error of an approximation corresponds to a sum of local errors weighted by factors associated with the stability of the equation and, in particular, with the numerical method adopted. In practice, the control of the approximation accuracy is achieved by controlling the local error up to a specified tolerance and keeping an estimation of the global error within acceptable bounds. These estimations, which may be more or less sophisticated, are all based on assumptions about the global behaviour of the numerical method for solving the IVP, but do not provide any guarantee on the actual accuracy of the approximation.

Computational errors, which are of a more random nature, increase the difficulty in defining correct global error estimates, specially for unstable problems. Moreover, this effect imposes practical limitations on the reduction of the step size. In general, when the step size is decreased, the global discretization error is decreased by the same factor raised to some power p (where p is defined as the order of the method), thus reducing the global error. However, more steps are required for covering the same interval of t. If the step size becomes too small, the accumulation of computational errors eventually exceed the reductions achieved on the local discretization errors. Consequently, smaller step sizes eventually become useless to reduce the global error, limiting in practice the maximum precision attainable for the computed approximation of the actual solution.

9.2   Interval Approaches

The first interval approaches for solving initial value problems had their sources in the interval arithmetic framework introduced by Moore [99]. They use interval arithmetic to calculate each approximation step explicitly, keeping the error term within appropriate interval bounds. Discretization and computational errors are thus encapsulated within bounds of uncertainty around the true solution function. In addition to providing guaranteed enclosures of the actual solution function, interval methods also verify the existence of a unique solution for the IVP.

Most interval methods for solving IVPs [99, 89, 48, 95, 123, 132, 103] are explicit methods based on Taylor series since this provides a simple form for the error term which can be bound as long as some enclosure of the actual solution function is provided. Moreover, the Taylor series coefficients can be efficiently computed through automated differentiation techniques, and both the step size and the order of the method may be easily modified during the approximation process. Changes on the step size do not require recomputing these coefficients, and to increase or decrease the order of the method it is sufficient to add or delete Taylor series terms.

In the mentioned approaches, each step between two consecutive points t_i and t_{i+1} generally consists of two phases. The first validates the existence of a unique solution and calculates an *a priori* enclosure of it between the two points. In the second phase, a tighter enclosure of the solution function at point t_{i+1} is obtained through interval arithmetic over a chosen numerical approximation step, with the error term bound as a result of the enclosure of the previous phase.

The next subsection presents the main ideas of the Interval Taylor Series (ITS) methods and how they take advantage from the automatic generation of the Taylor coefficients. Subsections 9.2.2 and 9.2.3 summarise the principal techniques used by ITS methods during, respectively, the first and the second phases of the enclosing steps.

It is worth mentioning that, alternatives to the explicit interval Taylor series approaches were recently proposed. An implicit interval method based on the Taylor expansion was proposed by Rihm in [124]. An interval Hermite-Obreschkoff method which outperforms the explicit interval Taylor series methods was proposed by Nedialkov in [103, 104] for improving the quality of the enclosures obtained at the second phase. In [19] Berz and Makino present a method, known as Taylor model, based on Taylor series expansions on both time and the initial conditions for solving IVPs, and the framework was extended later for dealing with implicit ODEs and differential algebraic equations [76]. Work on Runge-Kutta interval methods has been carried out by Hartmann and Petras, and presented at ICIAM'99 [71] and SCAN'2000 [116].

Some widely used software was developed based on the above interval methods. Lohner's AWA program [94, 95], written in Pascal-XSC [85], and Stauning's ADIODES package [132], written in C++, were the first public domain implementations of the explicit interval Taylor series methods. Nedialkov's VNODE [103, 105, 107], also written in C++, is a more recent package that includes an implementation of the interval Hermite-Obreschkoff method. Berz's COSY INFINITY [18] is a software package designed for beam physics applications, in a Pascal like language, which provides an efficient implementation of the Taylor model approach.

9.2.1 Interval Taylor Series Methods

Similarly to the traditional numerical Taylor Series methods (see subsection 9.1.1), ITS methods, are based on the Taylor series expansion of the solution function $s(t)$ around point t_i:

$$s(t_{i+1}) = s(t_i) + \sum_{k=1}^{p}\left(\frac{h^k}{k!}s^{(k)}(t_i)\right) + \frac{h^{p+1}}{(p+1)!}s^{(p+1)}(\xi) \quad \text{with} \quad h = t_{i+1} - t_i \text{ and } \xi \in [t_i..t_{i+1}]$$

However, instead of neglecting the error term, ITS methods use interval arithmetic to obtain reliable enclosures not only for the error term but also for every term of the series, allowing the computation of a reliable enclosure of the solution function at the point t_{i+1}.

Usually, and without loss of generality, the ITS methods assume that the ODE system is autonomous and rewrite the above equation into:

$$s(t_{i+1}) = s(t_i) + \sum_{k=1}^{p} h^k f^{[k]}(s(t_i)) + h^{p+1} f^{[p+1]}(s(\xi)) \quad \text{with} \quad h = t_{i+1} - t_i \text{ and } \xi \in [t_i..t_{i+1}]$$

where $f^{[k]}(s(t_i))$ denotes the kth Taylor coefficient of function s at the point t_i:

$$f^{[k]}(s(t_i)) = (s(t_i))_k = \frac{1}{k!} s^{(k)}(t_i)$$

For any k (>0) differentiable function $g(t)$ the relation between its kth Taylor coefficient and the $(k$-$1)$th Taylor coefficient of its derivative may be expressed as:

$$(g(t_i))_k = \frac{1}{k!} g^{(k)}(t_i) = \frac{1}{k}\left(\frac{1}{(k-1)!}\frac{d^{k-1}}{dt^{k-1}}\right)\frac{d}{dt} g(t_i) = \frac{1}{k}\left(\frac{1}{(k-1)!}\frac{d^{k-1}}{dt^{k-1}}\right) g'(t_i) = \frac{1}{k}(g'(t_i))_{k-1}$$

Since function s is the solution of an ODE system $y' = f(y)$, each kth Taylor coefficient of s at point t_i may be computed from the $(k$-$1)$thTaylor coefficient of its derivative expressed by the ODE:

$$f^{[k]}(s(t_i)) = \frac{1}{k}(f(s(t_i)))_{k-1}$$

Vector function f, as defined in 9-1, is composed by n elementary functions $f_i(y_1,...,y_n)$, and its k-Taylor coefficient may be obtained component wise:

$$(f(y))_k = \begin{bmatrix} (f_1(y_1,...,y_n))_k \\ ... \\ (f_n(y_1,...,y_n))_k \end{bmatrix}$$

In [99], Moore proposed a simple procedure for the automatic generation of Taylor coefficients of a given function. The method, applicable to functions expressed as in definition 3.2-1, allows the reliable computation of the Taylor coefficients up to a desired order. An efficient implementation of this method may be found at the public domain software package TADIFF [7] (implemented in C++).

The procedure defines recursive rules associated with each of the basic operators (and elementary functions) and use, as base of recursion, the fact that $(f(y))_0 = f(y)$. For example, the rules associated with the basic arithmetic operators ($+$, $-$, $\times$ and $/$) are:

$$(g(y)+h(y))_k = (g(y))_k + (h(y))_k \qquad (g(y)-h(y))_k = (g(y))_k - (h(y))_k$$

$$(g(y)\times h(y))_k = \sum_{r=0}^{k}(g(y))_r (h(y))_{k-r} \qquad \left(\frac{g(y)}{h(y)}\right)_k = \frac{1}{h(y)}\left((g(y))_r - \sum_{r=1}^{k}(g(y))_r \left(\frac{g(y)}{h(y)}\right)_{k-r}\right)$$

From this set of rules it is always possible to compute the k-Taylor coefficient (with $k>0$) of a function through decomposition. Note that when the function cannot be further decomposed (is expressed as a variable or a constant) its kth Taylor coefficient can be obtained from the $(k$-$1)$th Taylor coefficient of its derivative. The derivative of constant functions is 0 and so is the corresponding kth Taylor coefficient. Otherwise, if the function is represented as a variable, and in the case of an autonomous ODE system $y' = f(y)$, the variable must be an y_i and its kth Taylor coefficient may be computed from:

$$(y_i)_k = \frac{1}{k}(y_i')_{k-1} = \frac{1}{k}(f_i(y_1,...,y_n))_{k-1}$$

An efficient method using the above rules for computing the Taylor coefficients up to an order p firstly considers the Taylor coefficients of order $k=0$ for any variable and constant appearing in the expressions, then computes the kth Taylor coefficients for their compositions (accordingly to the expressions) and increments the order $k=k+1$, repeating the process until $k=p$.

The above computations may be performed either in real arithmetic (with finite precision), with real values representing the variables y_i and the constants appearing in the expressions, or in interval arithmetic where interval enclosures for these values are used instead. Whereas in the former an approximation of the Taylor coefficients is obtained, in the later reliable enclosures are computed.

With reliable enclosures for the Taylor coefficients, interval extensions (as defined in 3.2.1-1) of the Taylor series expansion of ODE solution functions may be computed. This is extensively used in ITS methods not only for enclosing the value, at point t_{i+1}, of a single solution function $s(t)$ with initial condition $s(t_i)=s_i$, but also to enclose such value for the set of solution functions whose values at the point t_i are within interval S_i.

9.2.2   *Validation and Enclosure of Solutions Between two Discrete Points*

Usually the validation and enclosure of solutions of an ODE system between two discrete points t_i and t_{i+1} is based on the Banach fixed-point theorem and the application of the Picard-Lindelöf operator (see [108, 131] for details).

The following theorem (proved in [48, 95]) may be used for defining a first order enclosure method based on the (first-order) interval Picard operator.

Theorem 9.2.2-1 (Interval Picard Operator). Let O be the autonomous ODE system of n equations defined by $y' = f(y)$. Let f be continuous and with first order partial derivatives over $t \in [t_i..t_{i+1}]$. Let S_i and S be n-ary real boxes with $S_i \subseteq S$. Let F be an interval extension of f. The interval Picard operator Φ is a vector interval function:

$$\Phi(S) = S_i + [0..h]F(S) \quad \text{with} \quad h = t_{i+1} - t_i$$

If $\Phi(S) \subseteq S$ then for every $s_i \in S_i$, the IVP defined by O and the initial condition $y(t_i)=s_i$ has a unique solution s and $\forall_{t \in [t_i..t_{i+1}]} s(t) \in \Phi(S)$ ❏

Based on the interval Picard operator, algorithms to obtain an enclosure for the set of solution functions whose values at t_i are within the box S_i may be described as follows. Firstly, a desired step size h is chosen together with an initial guess S^0 for the enclosure (with $S_i \subseteq S^0$). Then the interval Picard operator is applied to obtain the box $S = \Phi(S^0)$. If $S=\Phi(S^0) \subseteq S^0$ then, by theorem 9.2.2-1, S is an enclosure for the set of solution functions between t_i and t_i+h. Otherwise, two different strategies may be recursively applied: either the initial guess S^0 is inflated to enclose more solutions of the ODE for the same step size; or the step size is reduced to satisfy $\Phi(S^0) \subseteq S^0$ (note that for a small enough step h this property can always be satisfied). The final result of such algorithms is a box $S_{[i..i+1]}$ and a step size h (not necessarily the initially one) for which the box is an enclosure of the set of solution functions whose values at t_i are within the box S_i.

Several ITS proposals [89, 48, 95, 132] rely on the use of a first order enclosure method for the validation and enclosure of ODE solutions at its first phase. The major drawback of these approaches is that the step size restriction imposed by the (first-order) interval Picard operator is often much more severe than the limitations imposed in the second phase, based on higher order Taylor series expansions.

Alternative higher order enclosure methods were also proposed for this first phase, allowing larger step sizes more compatible with the second phase algorithms. A polynomial enclosure method was proposed by Lohner in [96] and several high-order Taylor series enclosure methods [31, 103, 109] were proposed after the original work of Moore [99]. The main advantage of the latter proposals is the possible reuse of the Taylor coefficients computations in both the first and the second phases of the ITS method.

9.2.3   *Computation of a Tight Enclosure of Solutions at a Discrete Point*

Once obtained an enclosure box $S_{[i..i+1]}$ of the set of solutions between two points, t_i and t_{i+1}, a straightforward ITS method for computing a tight enclosure at t_{i+1} is directly based on:

$$S_{i+1} = S_i + \sum_{k=1}^{p} h^k F^{[k]}(S_i) + h^{p+1} F^{[p+1]}(S_{[i..i+1]}) \quad \text{with} \quad h = t_{i+1} - t_i$$

where S_i and S_{i+1} are enclosing boxes at points t_i and t_{i+1} respectively, and $F^{[k]}(B)$ is a reliable enclosure (computed as described in subsection 9.2.1) of the kth Taylor coefficient of the solution function at any point within the box B ($\forall_{y \in B} f^{[k]}(y) \in F^{[k]}(B)$).

The above method usually leads to large overestimations of the enclosing box at point t_{i+1}. Since this box is computed from S_i (the enclosing box at point t_i) enlarged as a result of the addition of the other Taylor terms, from t_i to t_{i+1} the size of the enclosing box cannot decrease as it would be expected if, for example, the ODE system is stable within that region (see figure 9.1b).

A better approach is to use a Mean Value interval extension of the Taylor series with respect to the box S_i. This form, presented in subsection 3.2.1 of chapter 3 for scalar interval functions, can be easily extended component wise for vector interval functions. In this case, a method known as the ITS direct method is obtained:

$$S_{i+1} = c + \sum_{k=1}^{p} h^k F^{[k]}(c) + h^{p+1} F^{[p+1]}(S_{[i..i+1]}) + \left[I + \sum_{k=1}^{p} h^k J(f^{[k]}, S_i) \right] \times (S_i - c) \text{ with } h = t_{i+1} - t_i$$

where c is the mid point of box S_i and $J(f^{[k]}, S_i)$ is the Jacobian of $f^{[k]}$ evaluated at box S_i. The Jacobian may be obtained by automatic differentiation of the Taylor coefficient [6, 7].

The above form allows the decrease in size of consecutive enclosing boxes and provides a quadratic approximation (see subsection 3.2.1), quite advantageous when the boxes are small. However, the overestimation of enclosing boxes at the consecutive points may accumulate as the integration proceeds (a phenomenon known as the wrapping effect) and lead to unreasonable results.

In [48] and [103] it is shown that the overestimation made in one step of the ITS direct method for the last two terms of the expression which contains interval arguments, that is,

$$h^{p+1} F^{[p+1]}(S_{[i..i+1]}) \quad \text{and} \quad \left[I + \sum_{k=1}^{p} h^k J(f^{[k]}, S_i) \right] \times (S_i - c), \text{ is } O(h^{p+2}) \quad \text{and} \quad O(h \times width(S_i)^2),$$

respectively.

A consequence of this analysis is that whereas the error induced by the first term may be controlled by adjusting the step size and the order of the method, on the contrary, the error induced by the last term is highly dependent on the initial box S_i, which may be oversized due to previous errors. This dependency on the size of S_i may be insignificant for initial conditions represented as points or small boxes but may become quite significant with the accumulation of errors during the integration process.

In interval methods for solving IVPs, the accumulation of errors at each integration step is magnified by the effect of always enclosing (wrapping) the set of solutions at each discrete point within boxes regardless of its correct shape. In practice this means that, at each step, besides the overestimation due to the interval arithmetic evaluation of some interval extension (such as the above Mean Value form) of the real set of solutions, additional overestimation is introduced because entire boxes must be considered for representing the domains.

Several strategies have been proposed for reducing the overestimation and, in particular, for handling the wrapping effect [99, 89, 48, 95, 123]. They choose an interval extension of the Taylor series suitable for the interval arithmetic evaluation (for example minimizing the multiple occurrences of the interval variables) complemented with an effective enclosing method for representing the intermediate values of the set of solutions. The most successful enclosing methods (the Lohner's QR-Factorization method [94, 95] and its simplifications [106]) are based on changes of the coordinate system at each step of the integration process, aiming at reducing as much as possible the overestimation of the representation of domains by means of boxes.

9.3 Constraint Approaches

The application of the interval constraints framework for validated ODE solving was firstly suggested by Older [112] and Hickey [73]. Both approaches are extensions of the constraint programming language CLP(BNR) [113, 15] for handling constraints expressed as ODE systems. They represent an ODE system by a constraint network determined by a set of discrete points of t and the restrictions between the respective enclosing boxes derived from a chosen approximation step (as in the interval approaches).

In these early approaches, the goal was not to improve efficiency (compared to the interval approaches) but rather to extend the generality of a constraint approach. Once an ODE system is translated into a constraint network, the solving mechanism of a constraint language such as CLP(BNR) may efficiently propagate any restriction imposed on any of its variables (see chapter 4). Hence, not only can IVPs be solved, but also any additional information (e.g. final or boundary conditions) may propagate throughout the constraint network, reducing the enclosures, with the guarantee that no possible ODE solution function is lost.

The major drawbacks of both approaches, which limit its practical application on real world problems, are the huge number of constraints that must be maintained in the constraint network and the required specification of additional a priori information about the solutions of the ODE system.

More recently, the research group of Jansen, Deville and Van Hentenryck developed a new alternative constraint satisfaction approach to ODEs [47, 79, 80, 81]. Their goal is to extend the interval approaches with constraint propagation techniques to enhance the quality of the enclosure bounds for the solution of the IVP and to improve the efficiency of the computations. Therefore, their work focuses on overcoming the main difficulties presented by the interval approaches for IVPs and not on the full integration of ODE systems into the interval constraint framework.

The next subsections summarise the main characteristics of each of these constraint approaches.

9.3.1 Older's Constraint Approach

The constraint approach proposed by Older in [112] allows the definition of an ODE[4] as an interval arithmetic expression $F(S,T)$ which is extensively used for the generation of interval constraints (processed by the CLP(BNR) language) relating an enclosure S for the set of solution values at a discrete point t (or an interval T of values of t) with an enclosure of its derivative.

---

[4] For simplicity a single differential equation is considered, as in the original work. However, this framework can be easily extended component wise for ODE systems.

In order to solve an ODE problem, in addition to the specification of the ODE, it is also required the definition of the initial and final integration points, t_0 and t_f respectively, the number d of recursive subdivisions that will be considered between these points, and an *a priori* enclosure $S_{[0..f]}$ for the set of solution functions between these points.

Between the initial and final integration points, a sequence of $2^d - 1$ equally spaced intermediate points are considered. Two interval variables, S_i and F_i, representing enclosures of solution values and of its derivatives, are associated with each discrete point t_i and the respective interval constraints c_i are generated restraining their possible value combinations according to the ODE specification:

$$c_i \equiv F_i = F(S_i, t_i)$$

Moreover, between each two consecutive discrete points t_i and t_{i+1}, two additional constraints, $c_{i,i+1}$ and $c_{i..i+1}$, are generated based on second order Taylor series expansions around these points. With $h = t_{i+1} - t_i$ the second order Taylor series expansions of a solution function of the ODE around t_i and t_{i+1} are:

$$s(t_{i+1}) = s(t_i) + hf(s(t_i), t_i) + \frac{h^2}{2} f'(s(\xi_1), \xi_1) \qquad \text{with } \xi_1 \in [t_i..t_{i+1}]$$

$$s(t_i) = s(t_{i+1}) - hf(s(t_{i+1}), t_{i+1}) + \frac{h^2}{2} f'(s(\xi_2), \xi_2) \qquad \text{with } \xi_2 \in [t_i..t_{i+1}]$$

Taking the difference between the two above equations, it follows:

$$s(t_{i+1}) - s(t_i) = \frac{1}{2} h\big(f(s(t_i), t_i) + f(s(t_{i+1}), t_{i+1})\big) + \frac{h^2}{4}\big(f'(s(\xi_1), \xi_1) - f'(s(\xi_2), \xi_2)\big)$$

which is an equation relating the values of a solution function and of its derivative at the two discrete points t_i and t_{i+1}.

A reliable enclosure $R_{[i..i+1]}$ of the difference $\big(f'(s(\xi_1), \xi_1) - f'(s(\xi_2), \xi_2)\big)$ appearing in the last term of the equation may be obtained from the interval evaluation of the derivative of f with the argument $s(\xi)$ replaced by the a priori enclosure $S_{[0..f]}$ and ξ replaced by the interval $[t_i..t_{i+1}]$. If the result of such evaluation is an interval $F'_{[i..i+1]}$ then the magnitude of the difference, which is between two real values belonging to this interval, cannot exceed its width. Consequently a reliable enclosure is given by the interval $R_{[i..i+1]} = [-width(F'_{[i..i+1]})..width(F'_{[i..i+1]})]$.

Reliable enclosures for $s(t_i)$, $s(t_{i+1})$, $f(s(t_i), t_i)$ and $f(s(t_{i+1}), t_{i+1})$ are represented by the interval variables S_i, S_{i+1}, F_i and F_{i+1} that were initially associated with the points t_i and t_{i+1}.

Given these reliable enclosures for each term of the above equation, a first constraint $c_{i,i+1}$ is directly obtained:

$$c_{i,i+1} \equiv S_{i+1} - S_i = \frac{1}{2} h(F_i + F_{i+1}) + \frac{h^2}{4}\big(R_{[i..i+1]}\big)$$

Moreover, generalizing the equation for any intermediate point of t between t_i and t_{i+1}, and representing the enclosure for $\forall_{t \in [t_i..t_{i+1}]} s(t)$ by a new interval variable $S_{[i..i+1]}$ initialised with the a priori enclosure $S_{[0..f]}$, a second constraint $c_{i..i+1}$ is obtained:

$$c_{i..i+1} \equiv S_{[i..i+1]} - S_i = \frac{1}{2} [0..h]\big(F_i + F(S_{[i..i+1]}, [t_i..t_{i+1}])\big) + \frac{[0..h]^2}{4}\big(R_{[i..i+1]}\big)$$

With this approach the problem of integrating an ODE is transformed into a CCSP with:

    (i) $3 \times 2^d + 2$ variables, S_i $(0 \le i \le 2^d)$, F_i $(0 \le i \le 2^d)$ and $S_{[i..i+1]}$ $(0 \le i < 2^d)$;

    (ii) $3 \times 2^d + 1$ constraints, c_i $(0 \le i \le 2^d)$, $c_{i,i+1}$ $(0 \le i < 2^d)$ and $c_{i..i+1}$ $(0 \le i < 2^d)$.

There are several drawbacks with this constraint approach. Firstly, the number d of subdivisions must be specified in advance and the interpolation points must be homogeneously distributed along the whole interval of integration. Thus, the approach has no error control mechanism and is insensitive wrt the stability of the differential equation (see subsection 9.1.2).

Secondly, given the lack of an error control mechanism, the number d of subdivisions needs often to be very large, leading to a constraint network with a huge number of variables and constraints, which is difficult to handle, even for a specialised constraint programming language such as CLP(BNR).

Thirdly, the constraints between any two consecutive discrete points contain an interval constant $R_{[i..i+1]}$ which is only computed once and is not updated by propagation, even when the range of the enclosure $S_{[i..i+1]}$ for the set of solution functions between these points is changed.

Finally, the approach is highly dependent on the specification of a tight enough a priori enclosure $S_{[0..f]}$ for the set of solution functions between the initial and the final integration points. This is a main problem since there is usually no clue for the specification of such enclosure which must hold along the whole interval of integration. Moreover, if the values of the solution functions vary considerably along this interval then there is no single tight enclosure for the complete interval, limiting the practical applicability of such approach.

9.3.2 Hickey's Constraint Approach

Independently from Older's work, Hickey in [73] proposed a somewhat similar constraint approach for solving constrained ODE problems. It shares the same ideas of associating interval variables for representing the enclosures of solutions and their derivatives at discrete points of t (or between two consecutive points) and generating constraints to bound its possible values. However, instead of considering only enclosures for the solution function and its first derivative, it considers enclosures for all the derivatives of the solution function up to a given order and uses constraints based on Taylor series expansions whose remainder term does not exceed that order.

From the specification of the ODE[5], together with a sequence of integration points (between t_0 and t_f) and an integer p (representing the maximum order of the Taylor expansions used for generating constraints), the following variables are considered (for which a priori enclosures may be specified):

(i) $S_i, S_i', S_i'',\ldots, S_i^{(p)}$ associated with each discrete point t_i ($0 \le i \le f$) and representing

the enclosures for $s(t_i)$, $f(s(t_i),t_i)$, $f'(s(t_i),t_i)$, $\ldots$, $f^{(p-1)}(s(t_i),t_i)$ respectively;

(ii) $S_{[i..i+1]}, S'_{[i..i+1]}, S''_{[i..i+1]},\ldots, S^{(p+1)}_{[i..i+1]}$ associated with each interval $[t_i..t_{i+1}]$ ($0 \le i < f$) and

representing the enclosures for $\forall_{t \in [t_i..t_{i+1}]} s(t)$, $\forall_{t \in [t_i..t_{i+1}]} f(s(t),t)$,

$\forall_{t \in [t_i..t_{i+1}]} f'(s(t),t)$,$\ldots$, $\forall_{t \in [t_i..t_{i+1}]} f^{(p)}(s(t),t)$ respectively;

Expressions for each of the derivatives associated with the above variables are computed through automated differentiation and the corresponding constraints are generated, relating its value at each point (t_i) or interval ($[t_i..t_{i+1}]$).

Between each two consecutive discrete points t_i and t_{i+1}, several additional constraints are generated based on Taylor series expansions around these points. For any pair of integers k and m such that $0 \le k \le m \le p$, the following four constraints are generated[6] (assuming that the solution function s is $p+1$ times continuous differentiable and $h=t_{i+1}-t_i$):

---

[5] Again only a single differential equation is considered. The generalization for ODE systems is straightforward.

[6] In practice, these constraint are generated in a different form, more suitable for the interval arithmetic evaluation.

$$c_{i,i+1}^{k,m} \equiv S_{i+1}^{(k)} = \sum_{j=0}^{m-k}\left(\frac{h^j}{j!}S_i^{(k+j)}\right) + \frac{h^{m-k+1}}{(m-k+1)!}R_{[i..i+1]}^{(m+1)}$$

$$c_{i+1,i}^{k,m} \equiv S_i^{(k)} = \sum_{j=0}^{m-k}\left(\frac{(-h)^j}{j!}S_{i+1}^{(k+j)}\right) + \frac{(-h)^{m-k+1}}{(m-k+1)!}R_{[i..i+1]}^{(m+1)}$$

$$c_{i,[i,i+1]}^{k,m} \equiv S_{[i..i+1]}^{(k)} = \sum_{j=0}^{m-k}\left(\frac{[0..h]^j}{j!}S_i^{(k+j)}\right) + \frac{[0..h]^{m-k+1}}{(m-k+1)!}R_{[i..i+1]}^{(m+1)}$$

$$c_{i+1,[i,i+1]}^{k,m} \equiv S_{[i..i+1]}^{(k)} = \sum_{j=0}^{m-k}\left(\frac{[-h..0]^j}{j!}S_{i+1}^{(k+j)}\right) + \frac{[-h..0]^{m-k+1}}{(m-k+1)!}R_{[i..i+1]}^{(m+1)}$$

where the $R_{[i,i+1]}^{(m+1)}$ are different interval variables representing the enclosure $S_{[i,i+1]}^{(m+1)}$. These new variables are introduced to avoid the narrowing of the remainder term as a result of propagation of the Taylor series expansion. This narrowing can only be achieved through the last two constraints and is propagated to the remainder terms via an additional unidirectional CLP(BNR) constraint defined as:

$$c_i^m \equiv R_{[i,i+1]}^{(m+1)} \text{ is } S_{[i,i+1]}^{(m+1)}$$

Note that the constraint $c_{i,i+1}^{0,p}$ is the natural interval extension of the Taylor series expansion of order p around t_i, which is the base of the ITS methods (see subsection 9.2.1). Note also that the ranges of the variables $S_{[i..i+1]}^{(p+1)}$ are not constrained, implying the specification of their a priori enclosures.

This approach presents the same general drawbacks of the previous approach. It lacks an error control mechanism, generates a huge constraint network and requires the specification of the integration points and tight *a priori* enclosures for the values that each variable can take along the whole interval of integration.

9.3.3   *Jansen, Deville and Van Hentenryck's Constraint Approach*

The approach proposed by Janssen et al [47, 79, 80, 81] is a more direct extension of interval approaches for solving IVPs. It does not translate the IVP into a CCSP through a constraint network, but rather follows the traditional methods. It performs the integration step by step, from the initial to the final integration point, using reliable interval methods improved with constraint propagation techniques.

Each integration step is quite similar to the two phases process of the interval methods. Firstly, the validation and enclosure of solutions between the two discrete points is achieved by a first order enclosure method based on the interval Picard operator (cf. 9.2.2). Afterwards, a tight enclosure of the solution at the next discrete point is computed. The novelty of the approach is the subdivision of this second phase into a predictor process, for computing an initial enclosure, and a pruning (corrector) process, for narrowing this enclosure, both based on constraint techniques.

The predictor process is based on a traditional ITS method for the generation of an interval extension of the Taylor series (see subsection 9.2.3). In particular, [47] suggests the ITS direct method complemented with a co-ordinate transformation strategy, based on Lohner's QR-factorisation method [94], to reduce the wrapping effect.

As a result, some (vector) interval expression $E(S_i, S_{[i..i+1]}, t_i, t_{i+1})$ is obtained for the computation of an enclosure S_{i+1} at the next point t_{i+1} from the enclosure at the previous point t_i and the enclosure $S_{[i..i+1]}$ between the two points obtained in the first phase.

A major difficulty of the interval methods is the overestimation obtained from the direct interval evaluation of such expression. In this approach, the overestimation is reduced by the application of constraint techniques for the decomposed evaluation of the interval expression (see theorem 3.2.1-4), and several such techniques are proposed [47].

A first technique is based on the piecewise interval extension of the solution function, defined as the smallest enclosure for the expression $E(S_i,S_{[i..i+1]},t_i,t_{i+1})$ when evaluated at each element of S_i:

$$S_{i+1} = B_{hull}(\{E(B_{hull}(s_i),S_{[i..i+1]},t_i,t_{i+1}) \mid s_i \in S_i\})$$

where $B_{hull}(S)$ is the smallest F-box enclosing S.

To compute such enclosure, $2n$ unconstrained optimisation problems are generated for finding the minimum/maximum possible value of each component of S_{i+1}. For the practical implementation of these optimisation problems, an appropriate constraint programming language such as Numerica [135] is used.

Since the above multidimensional optimisation problems may be computationally expensive, alternative coarser approximations of the piecewise interval extension were proposed, using projections into a smaller number of variables.

A box-piecewise interval extension is defined as the intersection on every component j of the piecewise interval extensions obtained when all the other components are replaced by the respective interval constants:

$$S_{i+1} = \bigcap_{1 \le j \le n} B_{hull}(\{E(<I_1,\ldots,I_{j-1},I_{apx}(r_j),I_{j+1},\ldots, I_n>,S_{[i..i+1]},t_i,t_{i+1}) \mid r_j \in I_j\}) \text{ w/ } S_i = <I_1,\ldots,I_n>$$

Despite providing a less accurate enclosure, this is computationally less expensive, only requiring the solution of unidimensional unconstrained optimisation problems.

Other alternative techniques lie half-way between the two extreme alternatives above. If instead of using projections into a single component, any k components are considered then, box(k)-piecewise interval extensions may be obtained from the solution of k-dimensional optimisation problems.

For the pruning process, two alternative techniques were proposed: a one-step method which uses the forward step backwards [47] and a multistep method that uses Hermite interpolation polynomials [79, 80, 81].

The one-step method generates a constraint c_j from each component j of the vector interval expression E, which is the same used in the predictor process but applied backwards (from t_{i+1} to t_i), and enforces box-consistency wrt the enclosure S_{i+1}:

$$c_j \equiv I_j = E_j(S_{i+1},S_{[i..i+1]},t_{i+1},t_i) \qquad \text{with } S_i = <I_1,\ldots, I_n>$$

The multistep method aims at narrowing the predicted enclosure S_{i+1} for the solution functions at time t_{i+1} from the knowledge about reliable enclosures at k previous discrete points. It is based on the fact that if a real function at $t_{i-k+1},\ldots,t_i,t_{i+1}$ passes through $S_{i-k+1},\ldots,S_i,S_{i+1}$, with the derivatives $s'_{i-k+1},\ldots,s'_i,s'_{i+1}$, then there is a unique polynomial of degree $2k+1$, the Hermite polynomial, which simultaneously interpolates both the real function and its derivative at these points (see [133] for details). Moreover, it is possible to bind the error of the Hermite polynomial for the approximation of the real function at some $t \in [t_{i-k+1}..t_{i+1}]$ and so, it is possible to derive an interval extension of the original real function based on the Hermite polynomial. Consequently, denominating $HP(S_{i-k+1},\ldots,S_i,S_{i+1},t)$ the Hermite interval polynomial obtained from enclosures at $k+1$ discrete points (and their respective derivative satisfying the ODE) the following relation holds for any solution function s of the ODE:

$$\forall t \in [t_{i-k+1}..t_{i+1}] \; s(t_{i-k+1}) \in S_{i-k+1} \wedge \ldots \wedge s(t_{i+1}) \in S_{i+1} \Rightarrow s(t) \in HP(S_{i-k+1},\ldots,S_i,S_{i+1},t)$$

Additionally, through differentiation on HP, an interval extension DHP of the derivative of any solution function s may be obtained:

$$\forall_{t\in[t_{i-k+1}..t_{i+1}]} \; s(t_{i-k+1})\in S_{i-k+1} \wedge \ldots \wedge s(t_{i+1})\in S_{i+1} \Rightarrow \frac{ds(t)}{dt} \in DHP(S_{i-k+1},\ldots,S_i,S_{i+1},t)$$

The multistep pruning methods use the interval expressions *HP* and *DHP* together with the interval expression *F* defining the ODE to generate a set of constraints based on:

$$DHP(S_{i-k+1},\ldots,S_i,S_{i+1},t) = F(HP(S_{i-k+1},\ldots,S_i,S_{i+1},t),t)$$

which can be rewritten in a suitable Mean Value form (see [79]). By choosing a particular value for $t\in[t_{i-k+1}..t_{i+1}]$, and enforcing box-consistency (or other type of consistency) on the enclosure S_{i+1}, this box may be effectively narrowed.

The method is improved in [80] by considering several successive multistep constraints together for pruning simultaneously several enclosing boxes. The best choice of the particular value for $t\in[t_{i-k+1}..t_{i+1}]$ is independent from the ODE and may be precomputed before the integration process starts [81]. Experimental results on several benchmarks were presented confirming the advantages of the constraint approach compared with the best interval approaches.

9.4   Summary

In this chapter ordinary differential equations and initial value problems were introduced. Classical numerical approaches for solving IVPs were overviewed, and sources of errors and its consequences were discussed. Interval approaches for solving IVPs were reviewed and, in particular, Interval Taylor Series methods were fully described. The existing approaches that apply interval constraints for ODE solving were surveyed. In the next chapter our proposal of Constraint Satisfaction Differential Problems for handling differential equations is presented.

Chapter 10

Constraint Satisfaction Differential Problems

In this work, we propose an interval constraint approach for dealing with differential equations, by considering each ODE system together with related additional information as a special kind of CSP. We will refer such approach as a Constraint Satisfaction Differential Problem (CSDP).

Whereas in a CCSP the values of all the variables are real numbers and their domains are sets of real numbers represented by real intervals, in a CSDP there is a special variable whose values are functions and whose domain is a set of functions. Such special variable is named the solution variable and represents the functions that are solutions of the ODE system and satisfy all the additional restrictions.

The other variables of the CSDP are all real valued variables, and will be denoted as restriction variables, which represent each of the required restrictions. Solving the CSDP may be seen as a correct narrowing procedure for reducing the domains of the restriction variables without loosing any possible solution.

The full integration of a CSDP within a CCSP (as defined in 2.2-2) is accomplished by sharing common variables (the restriction variables of the CSDP) and by considering the CSDP as a special constraint restraining the possible value combinations of those variables. For pruning the domains of its variables, this constraint has an associated narrowing function derived from the procedure for solving the CSDP.

This chapter characterises a CSDP. The next section identifies its variables and restrictions. Section 10.2 addresses its integration within a larger CCSP framework. Section 10.3 discusses some modelling issues. The procedure for solving a CSDP is presented in the next chapter.

10.1 CSDPs are CSPs

A CSDP is a CSP with a special variable (the solution variable x_{ODE}), a special constraint (the ODE constraint c_{ODE}) and other constraints and variables for representing additional required restrictions. Before presenting the definition of a CSDP, the special variable and constraint must be characterised together with the type of constraints allowed for the representation of additional restrictions.

Let $S_{[t_0..t_1]}$ be the set of all n-ary vector functions s from the real interval $[t_0..t_1]$ to $\mathbb{R}^n$:

$$S_{[t_0..t_1]} = \{\, s \mid s : [t_0..t_1] \subset \mathbb{R} \to \mathbb{R}^n \,\}$$

The association of an n-ary ODE system $y' = f(y,t)$ (as defined in 9-1) with a real interval $[t_0..t_1]$ may be seen as a restriction on the set of functions $S_{[t_0..t_1]}$ to a subset defined by those functions that are solutions of the ODE system with respect to $[t_0..t_1]$ (as defined in 9-2). This is represented in the CSP framework (see definitions 2-1 and 2-2) by considering the solution variable x_{ODE} with the initial domain $D_{ODE} = S_{[t_0..t_1]}$ and the ODE constraint $c_{ODE} = (\langle x_{ODE} \rangle, \rho_{ODE})$ where

$$\rho_{ODE} = \{\, \langle s \rangle \in D_{ODE} \mid \forall_{t \in [t_0..t_1]} \; \frac{ds}{dt} = f(s(t),t) \,\}$$

The specification of additional information, such as initial conditions, boundary conditions, or other more complex restrictions on the ODE solutions is represented in the CSP framework by a finite set of binary constraints, denoted ODE restrictions. Each of these ODE restrictions $c_r=(<x_{ODE},x_i>,\rho_r)$ defines a relation ρ_r between the solution variable x_{ODE} and some other interval variable x_i of the CSDP. The relation ρ_r must be defined through a function $r : S_{[t_0..t_1]} \to \mathbb{R}$ in the following way:

$$\rho_r = \{<s,v>\in<D_{ODE},D_i> \mid v = r(s)\}$$

In the following subsection we will define several such functions to account for value, maximum, minimum, time and area restrictions. First we present a formal definition of a CSDP that summarises the above concepts.

Definition 10.1-1 (CSDP). Let $y' = f(y,t)$ be an n-ary ODE system as defined in 9-1. Let $[t_0..t_1]$ be a real interval and $S_{[t_0..t_1]}$ be the set of all n-ary vector functions from the real interval $[t_0..t_1]$ to $\mathbb{R}^n$. A CSDP is a CSP $P=(X,D,C)$ (see definition 2-2) where:

$$X =<x_{ODE},x_1,\ldots,x_m>$$
$$D =<D_{ODE},D_1,\ldots,D_m> \quad \text{with } D_{ODE}=S_{[t_0..t_1]} \text{ and } D_i \,(1\leq i\leq m) \text{ real intervals}$$
$$C =\{c_{ODE}\} \cup C_r$$

and:

$$c_{ODE}=(<x_{ODE}>, \rho_{ODE}) \quad \text{with}$$

$$\rho_{ODE} = \{<s>\in<D_{ODE}> \mid \forall_{t\in[t_0..t_1]} \frac{ds}{dt} = f(s(t),t)\}$$

$$\forall_{c_r\in C_r} c_r=(<x_{ODE},x_i>,\rho_r) \text{ w/ } 1\leq i\leq m, \rho_r=\{<s,v>\in<D_{ODE},D_i>|v = r(s)\} \text{ and } r : S_{[t_0..t_1]} \to \mathbb{R}$$

x_{ODE} is called the solution variable, c_{ODE} is called the ODE constraint, each variable x_i $(1\leq i\leq m)$ is called a restriction variable and each constraint $c_r\in C_r$ is an ODE restriction.   ❑

The following subsections describe different typical ODE restrictions which can be combined together within the same CSDP for the specification of many common ODE problems. Several examples of CSDPs will be presented for illustration purposes. They are all based on one of the two following ODE systems and their respective CSDPs.

Example P1:

The first example is the unary ODE system $y'(t) = -y(t)$ defined for $t\in[0.0..4.0]$. This is represented by CSDP $P1=(<x_{ODE}>,<D_{ODE}>,\{c_{ODE}\})$ where:

$$D_{ODE} = S_{[0.0..4.0]} = \{s \mid s : [0.0..4.0] \to \mathbb{R}\}$$
$$c_{ODE}=(<x_{ODE}>, \rho_{ODE}) \quad \text{with } \rho_{ODE} = \{<s>\in D_{ODE} \mid \forall_{t\in[0.0..4.0]} s'(t) = -s(t)\}$$

Without further constraints, CSDP $P1$ has an infinite number of solutions which may be represented analytically as $s(t) = ke^{-t}$ for any real number k. Figure 10.1 shows CSDP $P1$ together with some of its solutions.

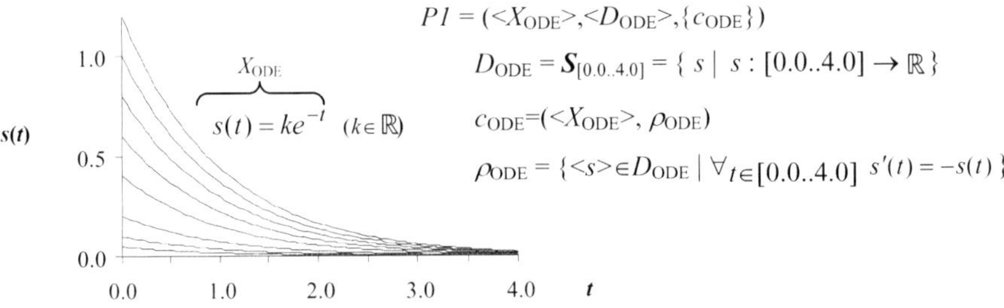

Figure 10.1 CSDP $P1$, representing the ODE system $y'(t) = -y(t)$ for $t\in[0.0..4.0]$.

<u>Example P2:</u>

The second ODE system is the following binary system[1] defined for $t \in [0.0..6.0]$:

$$y_1'(t) = -0.7 y_1(t)$$

$$y_2'(t) = 0.7 y_1(t) - \frac{\ln(2)}{5} y_2(t)$$

This is represented by CSDP $P2 = (\langle x_{ODE}\rangle, \langle D_{ODE}\rangle, \{c_{ODE}\})$ where:

$D_{ODE} = S_{[0.0..6.0]} = \{ s \mid s : [0.0..6.0] \to 3^2 \}$

$c_{ODE} = (\langle x_{ODE}\rangle, \rho_{ODE})$

$$\rho_{ODE} = \{ \langle s \rangle \in D_{ODE} \mid \forall_{t \in [0.0..6.0]} \left(s_1'(t) = -0.7 s_1(t) \ \wedge \ s_2'(t) = 0.7 s_1(t) - \frac{\ln(2)}{5} s_2(t) \right) \}$$

Again, without further constraints, CSDP $P2$ has an infinite number of solutions. These solutions may be represented analytically as:

$$s(t) = \begin{cases} s_1(t) \\ s_2(t) \end{cases} = \begin{cases} (b/a - 1)k_1 e^{-at} \\ k_1 e^{-at} + k_2 e^{-bt} \end{cases}$$

where $a = 0.7$, $b = \dfrac{\ln(2)}{5}$, and k_1, k_2 are real numbers.

Figure 10.2 shows CSDP $P2$ together with some of its solutions. For keeping the illustration in two dimensions, each component (s_1 and s_2) of each solution (s) is represented in a different graphic sharing the same time axis (in the graphics corresponding solutions share the same line type).

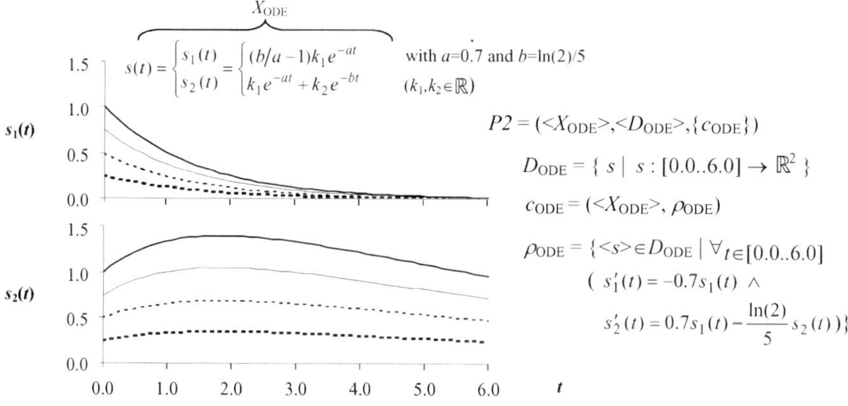

Figure 10.2 CSDP $P2$, representing a binary ODE system for $t \in [0.0..6.0]$.

In the following subsections, the short notation "P with C and D" is used to designate a CSDP similar to P but with the additional constraints (and respective variables) appearing in C. The initial domains of the new restriction variables are specified in D.

10.1.1　*Value Restrictions*

An initial condition, which together with an ODE system defines an IVP (see definition 9-3), specifies the value that a solution of the ODE system must have at a particular point of t. Hence, if s is such solution and t_p such time point, the initial condition specifies a value $s_j(t_p)$ for each component s_j ($1 \leq j \leq n$) of s at point $t = t_p$. A boundary condition is similar to an initial condition, except that only a proper subset of the components of s are specified.

---

[1] See section 12.2 for a "physical" justification of the ln(2)/5 constant.

In a CSDP, initial and boundary conditions are represented by a set of constraints denoted Value restrictions. Each of these constraints is an ODE restriction that relates an ODE solution with the value of one of its components at a particular point of t.

Definition 10.1.1-1 (Value restrictions). Let CSDP $P=(X,D,C)$ be defined as in 10.1-1. Let j be an integer $(1 \leq j \leq n)$ representing a component of the n-ary system, $t_p \in [t_0..t_1]$ a real value representing a point of t, $c_r \in C_r$ an ODE restriction and x_i its restriction variable.

(i) c_r is a Value restriction wrt j and t_p, denoted $Value_{j,t_p}(x_i)$, iff: $r(s) = s_j(t_p)$ ❏

Consider the IVP determined by the ODE of CSDP $P1$ (see figure 10.1) and the initial condition $y(0.0) = 1.0$. This IVP is represented by a CSDP similar to $P1$ but with an extra restriction variable x_1 with the interval domain $D_1=[1.0]$ and an extra value restriction $Value_{1,0.0}(x_1)$, which associates this variable with the value of the 1st (and unique) component of the solution at $t=0.0$. Additionally, if we are interested in the value of the solution of the above IVP at $t=4.0$, then a new restriction variable x_2 and a new value restriction $Value_{1,4.0}(x_2)$, associating that variable with such value, are required. Since there is no a priori information about this value, the initial domain D_2 of the new restriction variable is the interval $[-\infty..+\infty]$. The result is CSDP $P1a$ as follows:

$$P1a = P1 \text{ with } \{Value_{1,0.0}(x_1), Value_{1,4.0}(x_2)\} \text{ and } \{D_1=[1.0], D_2=[-\infty..+\infty]\}$$

CSDP $P1a$ has a single solution $<e^{-t}, 1.0, e^{-4.0}>$, shown in figure 10.3. Therefore, solving problem $P1a$ completely would narrow D_2 from $[-\infty..+\infty]$ to $[e^{-4.0}]$ which is the required value at $t=4.0$.

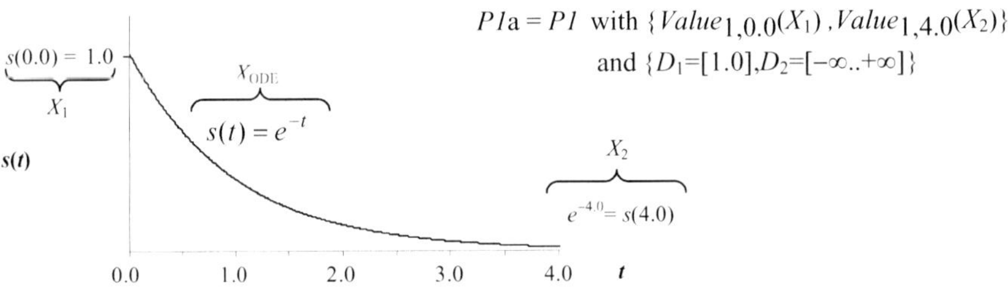

Figure 10.3 CSDP $P1a$, representing the IVP: $y(0.0) = 1.0$ and $y'(t) = -y(t)$ for $t \in [0.0..4.0]$.

A real interval I (or a real box B for multidimensional ODE systems) may be used instead for the specification of the initial condition at some point t_p of t. In this case a solution of the ODE must have at time t_p a value within interval I (or within box B). This can be easily accommodated in the CSDP framework by considering the interval I as the initial domain of the respective variable restriction (or, in the multidimensional case, by initializing each variable restriction with the correspondent component of B).

Considering the previous example with $y(0.0) = [0.5..1.0]$ (instead of $y(0.0) = 1.0$), the solutions of this problem may be expressed analytically as $s(t) = ke^{-t}$ for any $k \in [0.5..1.0]$. CSDP $P1b$, illustrated in figure 10.4, represents the problem.

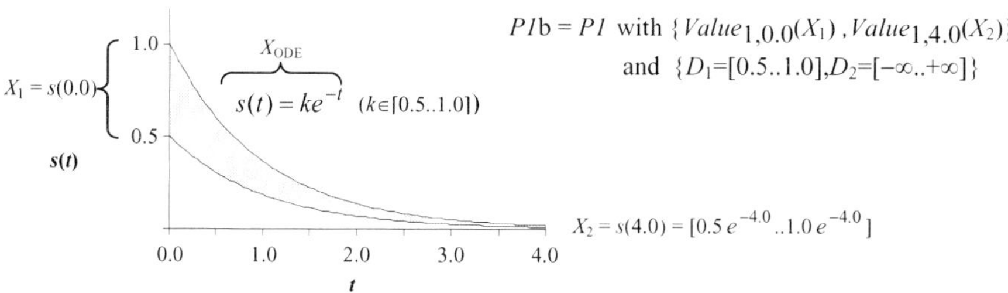

Figure 10.4 CSDP $P1$b, representing the IVP: $y(0.0) = [0.5..1.0]$ and $y'(t) = -y(t)$ for $t \in [0.0..4.0]$.

The grey area in the figure represents the set of all solutions of $P1$b. Solving this CSDP, would narrow D_2 from $[-\infty..+\infty]$ to $[0.5\,e^{-4.0}..1.0\,e^{-4.0}]$, the smallest interval containing all the real values of these solutions at time t=4.0.

A boundary value problem may be derived from CSDP $P2$ (see figure 10.2) by adding boundary restrictions specifying the value of the solutions components at different points of time. The specification of real values for y_2 at any two different time points t_{p1} and t_{p2} completely determines the values of k_1 and k_2 of the analytical solution, restricting the infinite set of solutions of $P2$ to a single solution.

For example, let us consider the boundary conditions $y_2(0.0)$=0.75 and $y_2(6.0)$=1.0 for the ODE system of CSDP $P2$, for which we are interested in solution values at time t=3.0. This new CSDP, $P2$a, is derived from $P2$ by adding one value restriction for each boundary condition and two more value restrictions (for obtaining the solution value of each component at time t=3.0). Figure 10.5 illustrates the problem and shows its single solution.

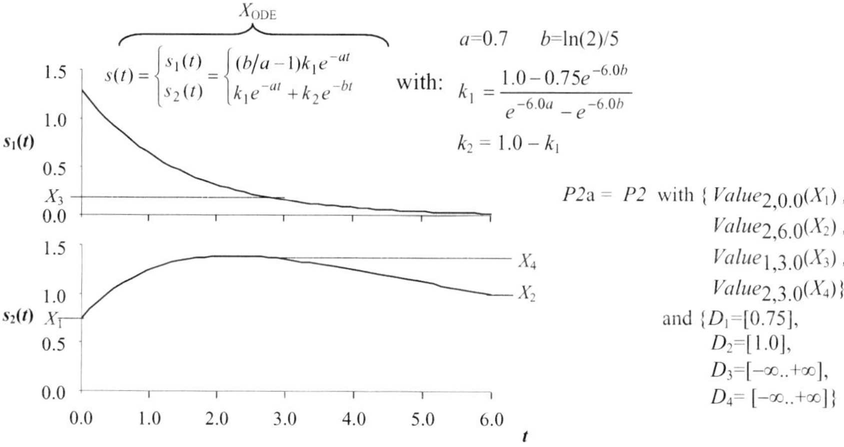

Figure 10.5 CSDP $P2$a, representing a boundary value problem.

Similarly to the previous examples, instead of real values, interval values may be used for the specification of the boundary conditions.

10.1.2   *Maximum and Minimum Restrictions*

Traditionally, the numerical problems dealing with ODEs are initial and boundary value problems, which, as we have seen in the previous subsection, can be easily represented in the CSDP framework through an adequate set of Value restrictions.

However, thinking of an ODE solution as a continuous vector function, and in particular thinking of each of its components as a continuous real function, several other conditions of interest may be imposed.

Important properties of a continuous function are its maximum and minimum values. In the CSDP framework, a Maximum restriction is an ODE restriction that relates an ODE solution component with its maximum value within an interval of time[2]. A Minimum restriction is similar.

Definition 10.1.2-1 (Maximum and Minimum restrictions). Let CSDP $P=(X,D,C)$ be defined as in 10.1-1. Let j be an integer ($1{\leq}j{\leq}n$) representing a component of the n-ary system, $[t_{p0}..t_{p1}]{\subseteq}[t_0..t_1]$ be a real interval, $c_r{\in}C_r$ be an ODE restriction and x_i its restriction variable.

(i) c_r is a Maximum restriction wrt j and $[t_{p0}..t_{p1}]$ (denoted $Maximum_{j,[t_{p0}..t_{p1}]}(x_i)$) iff:

$$r(s) = s_j(t_p) \quad \text{with } t_p{\in}[t_{p0}..t_{p1}] \text{ and } \forall_{t{\in}[t_{p0}..t_{p1}]} s_j(t) \leq s_j(t_p)$$

(ii) c_r is a Minimum restriction wrt j and $[t_{p0}..t_{p1}]$ (denoted $Minimum_{j,[t_{p0}..t_{p1}]}(x_i)$) iff:

$$r(s) = s_j(t_p) \quad \text{with } t_p{\in}[t_{p0}..t_{p1}] \text{ and } \forall_{t{\in}[t_{p0}..t_{p1}]} s_j(t) \geq s_j(t_p)$$ ❑

Consider CSDP $P2$ (figure 10.2) with boundary condition $s_1(0.0)=1.25$ specifying the value of the first solution component at $t=0.0$ and an additional restriction requiring the maximum value of the second solution component between $t=1.0$ and $t=3.0$ to lie within interval $[1.1..1.3]$. Moreover, let us assume that we are interested in the value of the second solution component at $t=6.0$.

Figure 10.6 illustrates the problem, CSDP $P2b$, showing its solutions (represented in the figure by the grey area).

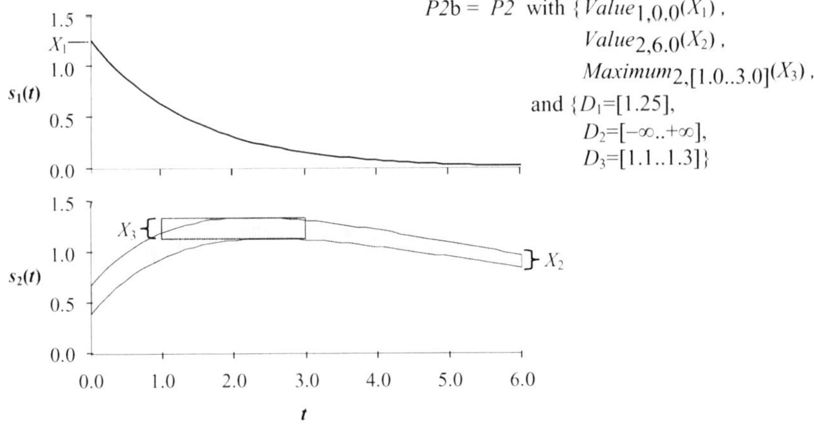

Figure 10.6 CSDP $P2b$, representing a problem with a Maximum restriction.

By solving CSDP $P2b$ exactly, domain D_2 would be narrowed from $[-\infty..+\infty]$ to the smallest interval containing the second component values at time $t=6.0$ of all possible solutions.

---

[2] In finite domains a similar type of constraint, the global commulative constraint, may be imposed for requiring that the usage of some resource cannot exceed some threshold during some interval of time.

10.1.3 Time and Area Restrictions

Other important property of a continuous real function, particularly useful for modelling many biophysical problems, regards the length of time in which its value remains above/below some predefined threshold. In this case, we are no longer interested in some particular value of the function, but rather to compute the time during which its value exceeds (or not) the threshold.

A Time restriction is an ODE restriction that captures such property from an ODE solution component given an interval of time T (specifying where to consider the time) and a threshold k. Such restriction is defined through a real function r which is a unity integral over a region determined by those points within T that satisfy the goal.

Definition 10.1.3-1 (Time restrictions). Let CSDP $P=(X,D,C)$ be defined as in 10.1-1. Let j be an integer ($1 \leq j \leq n$) representing a component of the n-ary system, $[t_{p0}..t_{p1}] \subseteq [t_0..t_1]$ a real interval, $\diamond \in \{\leq,\geq\}$, k a real value, $c_r \in C_r$ be an ODE restriction and x_i its restriction variable.

(i) c_r is a Time restriction wrt j, $[t_{p0}..t_{p1}]$ and $\diamond k$ (denoted $Time_{j,[t_{p0}..t_{p1}],\diamond k}(x_i)$) iff:

$$r(s) = \int_R dt \quad \text{with } R = \{\; t \in [t_{p0}..t_{p1}] \mid s_j(t) \diamond k \;\} \qquad \Box$$

The previous definition for the real function r may be generalised to other integrand functions to represent other properties of an ODE solution component. In particular, the area above (or under) the specified threshold may be obtained if the integrand measures the distance between the function value and this threshold value. This leads to the following definition of the Area restrictions.

Definition 10.1.3-2 (Area restrictions). Let CSDP $P=(X,D,C)$ be defined as in 10.1-1. Let j be an integer ($1 \leq j \leq n$) representing a component of the n-ary system, $[t_{p0}..t_{p1}] \subseteq [t_0..t_1]$ a real interval, $\diamond \in \{\leq,\geq\}$, k a real value, $c_r \in C_r$ be an ODE restriction and x_i its restriction variable.

(i) c_r is an Area restriction wrt j, $[t_{p0}..t_{p1}]$ and $\diamond k$ (denoted $Area_{j,[t_{p0}..t_{p1}],\diamond k}(x_i)$) iff:

$$r(s) = \int_R |s_j(t) - k|\,dt \qquad \text{with } R = \{\; t \in [t_{p0}..t_{p1}] \mid s_j(t) \diamond k \;\} \qquad \Box$$

The following is an example of a CSDP that adds Time and Area restrictions to CSDP $P2b$ (figure 10.6). It requires that at least half of the time (between 0.0 and 6.0) the second solution component has a value no less than 1.1. Moreover, we are interested in the area of the solutions (the second component) above that threshold. This is represented by CSDP $P2c$ illustrated in figure 10.7 (only the graphic of the second component $s_2(t)$ is shown in the figure; the first component is as in figure 10.6). This CSDP includes Time and Area restrictions, which are associated with new restriction variables x_4 and x_5, respectively. For imposing the required Time restriction, x_4 is initialised to an interval whose left bound is half of the total time (the right bound is the total time). Variable x_5, initially unbounded, represents the value of the required areas of all possible CSDP solutions.

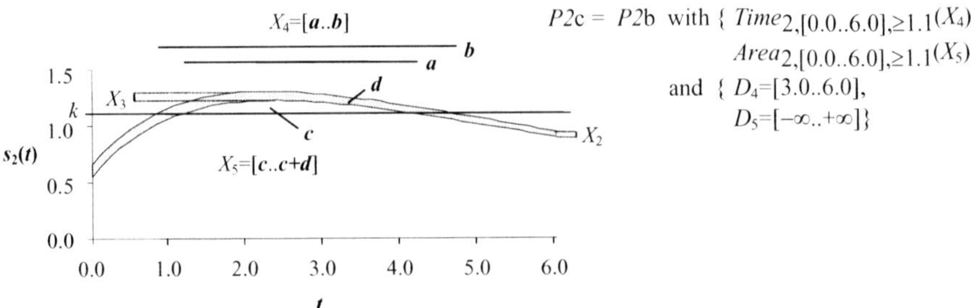

Figure 10.7 CSDP *P2c*, representing a problem with Time and Area restrictions.

The set of solutions of *P2c* is restricted by the Time restriction to the grey area represented in figure 10.7 (smaller than that in figure 10.6), since solving *P2c* exactly, narrows the final domains of x_2 and x_3 with respect to problem *P2b*. Figure 10.7 also shows the final domains of the Time variable x_4 (whose right bound was reduced from 6.0 to **b**) and the Area variable x_5 (where **c** is the area between the lower curve and the straight line k; and **d** is the area above that line and between the two curves).

10.1.4   *First and Last Value Restrictions*

The final important properties of real functions that we will represent as ODE restrictions are related with particular points of time for which the value of the real function satisfies some criterion.

First (Last) Value restrictions relates an ODE solution component s_j with the first (last) point of time t_p (within some interval $t_p \in [t_{p0}..t_{p1}]$) such that the criterion is satisfied for $s_j(t_p)$. If the criterion cannot be satisfied by any of those points the restriction fails. The possible criteria are the $s_j(t_p)$ to be less or equal, or greater or equal, than some predefined threshold k.

The real function r which defines such restriction is itself defined in two complementary and exclusive parts. If there is no point within the interval $[t_{p0}..t_{p1}]$ satisfying the criterion then it returns a value outside this interval $(+\infty/-\infty)$, which will make the constraint unsatisfiable. If there is one or more points within the interval $[t_{p0}..t_{p1}]$ satisfying the criterion then one of them must be the first (last), which is returned. In this case, for any precedent (subsequent) point the criterion cannot be satisfied.

Definition 10.1.4-1 (First and Last Value restrictions). Let CSDP $P=(X,D,C)$ be defined as in 10.1-1. Let j be an integer $(1 \leq j \leq n)$ representing a component of the n-ary system, $[t_{p0}..t_{p1}] \subseteq [t_0..t_1]$ a real interval, $\diamond \in \{\leq, \geq\}$, k a real, $c_r \in C_r$ be an ODE restriction and x_i its restriction variable.

(i) c_r is a First Value restriction wrt j, $[t_{p0}..t_{p1}]$ and $\diamond k$ (denoted *firstValue*$_{j,[t_{p0}..t_{p1}],\diamond k}(x_i)$)

iff:
$$r(s) = \begin{cases} +\infty & \text{if } \forall_{t \in [t_{p0}..t_{p1}]} \neg(s_j(t) \diamond k) \\ t_p & \text{if } t_p \in [t_{p0}..t_{p1}], s_j(t_p) \diamond k \text{ and } \forall_{t \in [t_{p0}..t_{p1}]} (t < t_p \Rightarrow \neg(s_j(t) \diamond k)) \end{cases}$$

(ii) c_r is a Last Value restriction wrt j, $[t_{p0}..t_{p1}]$ and $\diamond k$ (denoted *lastValue*$_{j,[t_{p0}..t_{p1}],\diamond k}(x_i)$)

iff:
$$r(s) = \begin{cases} -\infty & \text{if } \forall_{t \in [t_{p0}..t_{p1}]} \neg(s_j(t) \diamond k) \\ t_p & \text{if } t_p \in [t_{p0}..t_{p1}], s_j(t_p) \diamond k \text{ and } \forall_{t \in [t_{p0}..t_{p1}]} (t > t_p \Rightarrow \neg(s_j(t) \diamond k)) \end{cases} \qquad \square$$

Consider the example in figure 10.4 (CSDP *P1*b) with an additional First Value restriction requiring that, within time interval [0.0..2.0], the first solution value not exceeding 0.25 occurs sometime between 1.0 and 2.0. This is represented in CSDP *P1*d of figure 10.8 where, associated with the First Value restriction, a new restriction variable x_3 was introduced with domain D_3=[1.0..2.0].

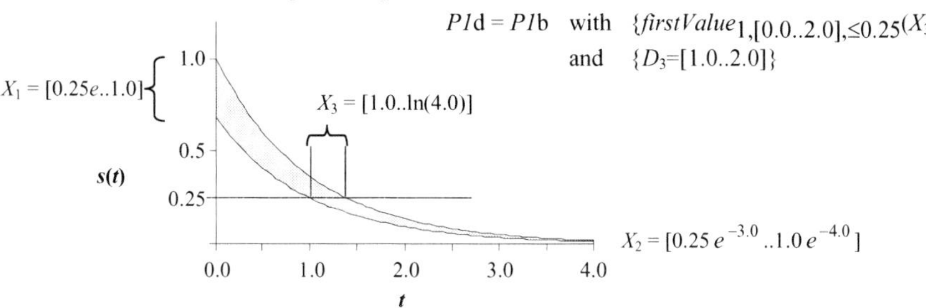

Figure 10.8 CSDP *P1*d, representing a problem with a First Value restriction.

The grey area, representing the set of possible solutions, is reduced, comparatively to CSDP *P1*b, which implies the narrowing of x_1 and x_2 variable domains. Moreover, the final right bound of variable x_3 is also reduced since the Value restrictions prevent any solution from having its first value not exceeding 0.25 after time t=ln(4.0).

10.1.5   First and Last Maximum and Minimum Restrictions

A natural extension of the First (Last) Value restrictions is to consider the special case where the threshold k is the maximum or minimum value of the real function. In this case the defining function r no longer needs to be defined in parts since within any interval $[t_{p0}..t_{p1}]$ there must exist a point that maximises (minimises) the real function.

Definition 10.1.5-1 (First and Last Maximum and Minimum restrictions). Let CSDP $P=(X,D,C)$ be defined as in 10.1-1. Let j be an integer ($1{\leq}j{\leq}n$) representing a component of the n-ary system and $[t_{p0}..t_{p1}]{\subseteq}[t_0..t_1]$ an interval. Let $c_r{\in}C_r$ be an ODE restriction and x_i its restriction variable.

(i) c_r is a First Maximum restriction wrt j and $[t_{p0}..t_{p1}]$ ($firstMaximum_{j,[t_{p0}..t_{p1}]}(x_i)$) iff:

$$r(s) = t_p \quad \text{with } t_p{\in}[t_{p0}..t_{p1}] \text{ and } \forall_{t\in[t_{p0}..t_{p1}]} [(s_j(t) \leq s_j(t_p)) \wedge (t < t_p \Rightarrow s_j(t) < s_j(t_p))]$$

(ii) c_r is a First Minimum restriction wrt j and $[t_{p0}..t_{p1}]$ ($firstMinimum_{j,[t_{p0}..t_{p1}]}(x_i)$) iff:

$$r(s) = t_p \quad \text{with } t_p{\in}[t_{p0}..t_{p1}] \text{ and } \forall_{t\in[t_{p0}..t_{p1}]} [(s_j(t) \geq s_j(t_p)) \wedge (t < t_p \Rightarrow s_j(t) > s_j(t_p))]$$

(iii) c_r is a Last Maximum restriction wrt j and $[t_{p0}..t_{p1}]$ ($lastMaximum_{j,[t_{p0}..t_{p1}]}(x_i)$) iff:

$$r(s) = t_p \quad \text{with } t_p{\in}[t_{p0}..t_{p1}] \text{ and } \forall_{t\in[t_{p0}..t_{p1}]} [(s_j(t) \leq s_j(t_p)) \wedge (t > t_p \Rightarrow s_j(t) < s_j(t_p))]$$

(iv) c_r is a Last Minimum restriction wrt j and $[t_{p0}..t_{p1}]$ ($lastMinimum_{j,[t_{p0}..t_{p1}]}(x_i)$) iff:

$$r(s) = t_p \quad \text{with } t_p{\in}[t_{p0}..t_{p1}] \text{ and } \forall_{t\in[t_{p0}..t_{p1}]} [(s_j(t) \geq s_j(t_p)) \wedge (t > t_p \Rightarrow s_j(t) > s_j(t_p))] \quad \square$$

10.2   Integration of a CSDP within an Extended CCSP

In the previous section, a CSDP is defined as a CSP specialised in the specification of constraints restraining the set of possible solutions of an ODE system, and consequently, the possible values of the restriction variables associated with properties of such solutions. Each of these properties is characterised by a function r which associates a real value to each possible solution of the ODE system. Hence, the restriction variables are real valued variables and their domains are represented as real intervals.

In section 2.2, a CCSP is defined as a CSP specialised in the specification of constraints as numerical equalities or inequalities, restraining the set of possible values of real variables with initial domains ranging over real intervals.

Since the restriction variables are similar to the variables of a CCSP, the CCSP framework may be extended to allow the specification of constraints as CSDPs, sharing some CCSP variables. A CSDP may be seen as a constraint (see definition 2-1), where the constraint scope is the set of its restriction variables and the constraint relation is the set of their possible combination values from the whole set of solutions (as defined in 2-5) of the CSDP. The following definition formalises an extended CCSP, including constraints specified as CSDPs.

Definition 10.2-1 (Extended Continuous Constraint Satisfaction Problem). An extended CCSP is a CSP $P=(X,D,C=C_{CCSP} \cup C_{CSDP})$ where each domain is an interval of $\mathbb{R}$ and each constraint relation is defined either as a numerical equality or inequality, or as a constraint satisfaction differential problem:

    i) $D=<D_1,\ldots,D_n>$ where D_i is a real interval $(1 \leq i \leq n)$

    ii) $\forall_{c \in C_{CCSP}}$ c is defined as $e_c \diamond 0$ where e_c is a real expression and $\diamond \in \{\leq,=,\geq\}$

    iii) $\forall_{c \in C_{CSDP}}$ $c=(s,\rho)$ is defined as a CSDP $(<x_{ODE},x_1',\ldots,x_m'>,D',C')$

where $s=<x_1',\ldots,x_m'> \subseteq X$, $D'=D[s]$ and $\rho = \{d[s] \mid d \in D' \wedge \forall_{(s_c,\rho_c) \in C'} d[s_c] \in \rho_c \}$ ❑

The next chapter presents a procedure for handling a CSDP, aiming at pruning the domains of its restriction variables, implemented as a function *solveCSDP*. From an *F*-box representing the domains of the restriction variables, it returns a smaller *F*-box where some value combinations that can be proved to be inconsistent with the CSDP are discarded.

Hence, when a CSDP is used for the definition of an extended CCSP constraint, a function NF_{CSDP} may be associated to it to discard the same value combinations of its scope variables that would be discarded by the *solveCSDP* function. As long as the *solveCSDP* function is correct, not eliminating any possible CSDP solution, and contracting, returning a smaller *F*-box, the associated function NF_{CSDP} is a narrowing function for the CCSP according to definition 4.1-1.

The following definition formalises a narrowing function NF_{CSDP} associated with a constraint defined by a CSDP. The *solveCSDP* function is used for narrowing the domains of a subset of the CCSP variables, namely those belonging to the scope of the constraint, that is, the restriction variables of the CSDP. The other variable domains are unchanged.

Definition 10.2-2 (CSDP Narrowing Functions). Let $P=(X,D,C=C_{CCSP} \cup C_{CSDP})$ be an extended CCSP as defined in 10.2-1. Each constraint $c=(s,\rho) \in C_{CSDP}$ defined as a CSDP (X',D',C') with a solving procedure *solveCSDP*, has associated the following narrowing function:

    (i) $NF_{CSDP}(A) = A'$ (with $A' \subseteq A \subseteq D$)

where $B = solveCSDP(A[s])$ (with $B \subseteq A[s] \subseteq D'[s]$)

    $\forall x_j \in X ((x_j \in s \Rightarrow A'[x_j]=B[x_j]) \wedge (x_j \notin s \Rightarrow A'[x_j]=A[x_j]))$ ❑

These additional narrowing functions associated with the CSDP constraints, together with the narrowing functions associated with the numerical constraints, completely characterise the set of narrowing functions of an extended CCSP. This set may be used by a constraint propagation algorithm (such as the *prune* function of figure 4.1) for pruning the domains of the extended CCSP variables.

As discussed in chapter 5, the set of narrowing functions together with a constraint propagation algorithm characterise a local property denoted local consistency. Moreover, higher order consistency requirements may be imposed through an algorithm (such as the *kB-consistency* function in figure 5.2) interleaving search techniques with the above constraint propagation algorithm. A generic definition of kB-Consistency (definition 5.2-1) was given for including all the consistency types commonly required in CCSPs (when $k=2$ it designates local consistency). This definition applies equally well to the case of extended CCSPs as long as local consistency is regarded, not as pure hull-consistency or pure box-consistency, but rather as the local property enforced by the constraint propagation algorithm with the set of narrowing functions associated to the constraints of the extended CCSP.

The definition of Global Hull-Consistency (definition 6-1) can also be applied to extended CCSPs if the definition of a canonical solution (definition 2.2.4-3) is redefined for this context. Moreover, any algorithm (such as those from chapter 6), for enforcing Global Hull-Consistency can be directly applied in this context except that the local search procedure must be able to deal with CSDP constraints. The next two subsections address respectively the redefinition of canonical solutions and the adaptation of the local search procedure for extended CCSPs.

10.2.1 Canonical Solutions for Extended CCSPs

In an extended CCSP, since a new type of constraints defined as CSDPs is allowed, in addition to defining how a canonical F-box may satisfy a numerical constraint, it is also necessary to define how it satisfies a CSDP constraint. In the formal definition below, a canonical box satisfies a constraint if it cannot be proved that the box does not contain any real valued combination, for the variables of the constraint scope, satisfying the constraint relation. In the case of a CSDP constraint, it is satisfied by a canonical box if it cannot be proved that the box does not contain solutions of the CSDP, that is, if the empty set is not obtained when the narrowing function associated to the CSDP constraint is applied to the box.

Definition 10.2.1-1 (Canonical Solution of an extended CCSP). Let $P=(X,D,C)$ be an extended CCSP as defined in 10.2-1. Let $B \subseteq D$ be a canonical F-box and $e_c(B)$ denote the F-interval obtained by the interval evaluation of expression e_c with argument B.

 B is a canonical solution of P iff :

  i) $\forall_{c \in C}\, c$ is defined as $e_c \diamond 0 \Rightarrow \exists_{r \in e_c(B)}\ r \diamond 0$

  ii) $\forall_{c \in C}\, c$ is defined as a CSDP $\Rightarrow NF_{\mathrm{CSDP}}(B) \neq \varnothing$

where e_c is a real expression, $\diamond \in \{\leq,=,\geq\}$ and NF_{CSDP} is the CSDP narrowing function defined in 10.2-2   ❏

There are some important consequences of the above definition regarding the extent of domain pruning that may be achieved by enforcing Global Hull-Consistency with different precision requirements.

In a CCSP with numerical constraints alone, due to the properties of the interval arithmetic evaluation (in particular theorem 3.2.1-4), it is expected that smaller intervals are obtained when evaluating an interval expression (such as e_c in the definition of the constraint c) with smaller arguments. This implies that, considering a larger finite set of F-numbers (see definition 2.2.1-2), and consequently smaller canonical F-boxes, the evaluation of the constraint expressions with canonical arguments becomes more precise, decreasing the likelihood of its wrong classification as canonical solutions. In the limit, with infinite precision arithmetic, a canonical solution is equivalent to a real valued solution. Because by definition, the pruning of domains achieved by enforcing Global Hull-Consistency is bound by canonical solutions, better results may be obtained with smaller canonical F-boxes. In the limit, with infinite precision arithmetic, the smallest box enclosing all real solutions would be obtained.

However, this desirable effect is not guaranteed in the case of extended CCSPs. In fact, even with infinite precision arithmetic, a complete real valued instantiation of its variables could be considered a canonical solution without satisfying some CSDP constraint.

One reason for such phenomenon derives from the approximate nature of any procedure based on ODE interval approaches (presented in section 9.2) for solving the CSDP. Since the CSDP narrowing function is based on this solving procedure, it is impossible to discard non solution instantiations which are enclosed within the safe bounds computed by such procedure.

Another reason for the phenomenon derives from the limitations of the particular procedure used for solving the CSDP. For example, a solving procedure (like that presented in the next chapter) for computing a safe enclosure for the ODE solutions of a CSDP, requires the initial trajectory to be bound at least at one time point of the ODE trajectory (otherwise the interval step method cannot be applied at any point, preventing the reduction of the trajectory uncertainty – see next chapter, third section). As a consequence, if the above requirement is not satisfied with a real (or interval) valued instantiation of the CSDP variables (even if such instantiation theoretically implies the elimination of the trajectory uncertainty) the solving procedure is unable to prune any trajectory enclosures.

In order to minimise the wrong labelling of canonical boxes as canonical solutions of an extended CCSP, each CSDP constraint should satisfy the solving requirement specified in definition 10.2.1-3. Before presenting such definition, the concept of a CSDP solving relaxation is introduced.

Definition 10.2.1-2 (CSDP Solving Relaxation). Let $c=(s,\rho)$ be a constraint defined as a CSDP $P=(<x_{ODE},x_1,\ldots,x_m>,D,C)$.
CSDP $P'=(X'=<x_{ODE},x_1^{(1)},x_1^{(2)},\ldots,x_1^{(k_1)},\ldots,x_m^{(1)},x_m^{(2)},\ldots,x_m^{(k_m)}>,D',C')$ is the solving relaxation of P iff it is obtained from P by:

(i) Renaming each variable x_i into $x_i^{(1)},x_i^{(2)},\ldots,x_i^{(k_i)}$ where k_i is the number of ODE restrictions in C containing x_i in its scope.

(ii) Redefining C to C' according to the renaming introduced in (i).

(iii) Unbounding the restriction variables: $\forall_{x_i} \forall_{k \in [1..k_i]} D'[x_i^{(k)}]=[-\infty..+\infty]$.

In P' there is one and only one ODE restriction $c_i^{(k)} \in C'$ for each restriction variable $x_i^{(k)}$.

The set s' of all its restriction variables contains $N = \displaystyle\sum_{i=1}^{m} k_i$ elements.   ❑

Definition 10.2.1-3 (CSDP Solving Requirement). Let constraint $c=(s,\rho)$ be defined as a CSDP $P=(X,D,C)$ and $P'=(X',D',C')$ be its solving relaxation.
Constraint c satisfies the CSDP solving requirement iff there is a minimal subset s_b of restriction variables s' such that for every F-box $B \subseteq D'$ where $\forall x \in X'$ $((x \in s_b \Rightarrow B[x]$ is degenerate)$\wedge(x \notin s_b \Rightarrow B[x]=D'[x]))$:

(i) CSDP (X',B,C') has a single solution

(ii) the result of applying the solving procedure to CSDP (X',B,C') is a bound F-box:
$$B'=solveCSDP(B[s']) \Rightarrow \forall x \in s' ((left(B'[x]) \neq -\infty) \wedge (right(B'[x]) \neq +\infty))$$
The set of variables s_b is denoted the solving base and the set s_e of remaining restriction variables is denoted the evaluated set ($s_b \cup s_e=s'$ and $s_b \cap s_e=\varnothing$).   ❑

The first requirement guarantees that with infinite precision arithmetic it is theoretically decidable whether a canonical box satisfies the CSDP constraint. The second requirement guarantees that the solving procedure may contribute for such decision returning a safe approximation enclosure for every variable domain.

For the solving procedure of the next chapter, the CSDP solving requirement could be easily achieved by imposing that at least one point t_p has Value restrictions $Value_{j,t_p}(x_i)$ for all the trajectory components. The set of restriction variables would define the solving base of its solving relaxation. Note that such imposition does not imply the specification of initial bounds for the domains of these variables, and if some CSDP does not satisfy the solving requirement, an equivalent CSDP could be considered by introducing extra Value restrictions with new unbounded restriction variables.

If the CSDP solving requirement is satisfied by every CSDP constraint belonging to an extended CCSP, the improvement of the pruning results with precision is similar to that discussed for CCSPs (differing only in the limit: even with infinite precision arithmetic, the smallest box enclosing all real solutions would never be obtained).

10.2.2 Local Search for Extended CCSPs

As explained in chapter 7, the local search procedures navigate throughout points of the search space, inspecting some local properties of the current point to select a nearby point to jump to. In the CCSP context, such navigation is oriented towards the simultaneous satisfaction of all its constraints.

The search approach presented in section 7.1 is based on the definition of a vector function $F(r)$. When evaluated in a particular point r of the search space (a degenerate F-box) it returns a real value $F_j(r)$ for at each component j, which represents the amount by which some CCSP constraint is violated at that point. The goal is thus to reach a point that zeroes all function components simultaneously. From the Jacobian matrix $J(r)$ of such function at that point (which summarises the effects on the function of small local changes at the current point) a procedure was devised to compute a new better point.

The integration of a CSDP constraint implies the addition of extra components on the vector function $F(r)$ and the Jacobian matrix $J(r)$. These extra components may be obtained as long as the CSDP solving requirement is satisfied.

As in the case of a non extended CCSP, the values of the vector function $F(r)$ must be computed from the evaluation of each constraint at the current point r. However, in the case of a CSDP constraint, such evaluation no longer returns a single value, but rather a set of values (grouped in a box) representing the deviations from each ODE restriction associated with a variable from the evaluation set (see definition 10.2.1-3). The input of this evaluation is the set of values from the solving base variables at the current point. The following is a formal definition for the evaluation of a CSDP constraint at some point r.

> **Definition 10.2.2-1 (CSDP Evaluation).** Consider an extended CCSP (X,D,C) and a degenerate F-box $r \subseteq D$. Let $c \in C$ be a constraint defined as a CSDP P and its solving relaxation $P'=(X',D',C')$ with the solving base s_b and the evaluated set s_e (the set of its restriction variables is $s'=s_b \cup s_e$). Let $B \subseteq D'[s']$ be an F-box where $\forall x_i^{(k)} \in s'$ $((x_i^{(k)} \in s_b \Rightarrow B[x_i^{(k)}]=r[x_i]) \land (x_i^{(k)} \notin s_b \Rightarrow B[x_i^{(k)}]=D'[x_i^{(k)}]))$. Let $B'=solveCSDP(B)$.
> The evaluation of constraint c at point r, denoted $evaluation_{CSDP}(r)$, is the degenerate F-box $E \subseteq D'[s_e]$ where: $\forall x_i^{(k)} \in s_e\ E[x_i^{(k)}] = \lfloor center(B'[x_i^{(k)}]-r[x_i]) \rfloor$. ❑

For each CSDP constraint, the values of the new components of the vector function F at the current point r are given by its evaluation box at that point. Moreover, their derivatives with respect to any CCSP variable x_i may be approximately computed by re-evaluating the constraint at a nearby point where only the x_i domain is increased by a small amount h. The next definition characterises the new components of the vector function F and its Jacobian J, associated with each CSDP constraint.

Definition 10.2.2-2 ($F(r)$ and $J(r)$ values associated with a CSDP constraint). Consider an extended CCSP ($X=<x_1,\ldots,x_n>,D,C$) and a degenerate F-box $r \subseteq D$. Consider a constraint $c \in C$ defined as a CSDP P and its solving relaxation P' with the evaluated set s_e.

Let $f_{x_j}(r)$ denote the value of some variable x_j of s_e obtained from the evaluation of c at point r:

$$f_{x_j}(r) = evaluation_{\mathrm{CSDP}}(r)[x_j] \text{ for every } x_j \in s_e.$$

Let $h>0$ be a small real value and r_{x_i+h} be an n-ary degenerate F-box where:

$$\forall_{x \in X}((x=x_i \Rightarrow r_{x_i+h}[x]=\lfloor r[x]+h \rfloor) \wedge (x \neq x_i \Rightarrow r_{x_i+h}[x]=r[x])).$$

For each variable x_i of s_e, a new component j of the vector function F is associated and defined as:

$$F_j(r) = f_{x_j}(r)$$

The derivatives of such function at point r with respect to each variable x_i of X (which define the new line j of the Jacobian $J(r)$) are approximated by:

$$J_{ji}(r) = \frac{d}{dx_i} f_{x_j}(r) = \left\lfloor \frac{f_{x_j}(r_{x_i+h}) - f_{x_j}(r)}{h} \right\rfloor \qquad \square$$

In practice, besides the evaluation at the current point r, only k more CSDP evaluations (where k is the number of solving base variables s_b) are required for computing all the new components of F and J associated with the CSDP (the evaluation at point r_{x_i+h} is only necessary if $x_i \in s_b$). Otherwise, either x_i does not belong to the scope of the CSDP constraint (i.e. $f_{x_j}(r_{x_i+h}) = f_{x_j}(r)$ and $J_{ji}(r) = 0$) or it belongs to its evaluated set s_e, in which case $f_{x_j}(r_{x_i+h}) = f_{x_j}(r) - h$ and $J_{ji}(r) = -1$.

With the introduction of components derived from each CSDP constraint in the vector function F and its Jacobian matrix J, the local search methods described in chapter 7 may be applied to extended CCSPs.

10.3   Modelling with Extended CCSPs

This section discusses modelling capabilities of the extended CCSPs framework that take advantage of the integration of CSDP constraints and may be used for solving many real world problems.

In the next subsection, the parametric specification of ODE systems is addressed together with possible application to problems with fitting constraints. Subsection 10.3.2 discusses the representation of properties which are usually associated with the ODE systems but are naturally expressed as interval values. Subsection 10.3.3 discusses the possibilities derived from the combination of several ODE solution components and its application to the representation of periodic and distance constraints.

10.3.1 Modelling Parametric ODEs

In the definition of an ODE system (definition 9-1) it was assumed that function f depends exclusively on $y(t)$ and t. However, many important ODE problems are based on a parametric specification of the function f, which may also depend on several parameters ranging within predefined interval bounds. The semantics of such specification is the characterisation of the derivative of y as a family of functions, each one corresponding to a particular instantiation of these parameters.

$$\text{Let} \quad y' = f(y,p,t) \quad \text{with} \quad y' = \begin{bmatrix} y'_1 \\ ... \\ y'_n \end{bmatrix} \text{and } f(y,p,t) = \begin{bmatrix} f_1(y_1,...,y_n,p_1,...,p_k,t) \\ ... \\ f_n(y_1,...,y_n,p_1,...,p_k,t) \end{bmatrix} \text{be a}$$

parametric n-ary ODE system with k parameters $p_1 \in I_1,..., p_k \in I_k$. The function $s(t) = \begin{bmatrix} s_1(t) \\ ... \\ s_n(t) \end{bmatrix}$

is one of its solutions with respect to the interval $[t_0..t_1]$ iff:

$$\exists_{p_1 \in I_1},...,\exists_{p_k \in I_k} (p=[p_1,...,p_k] \wedge \forall_{t \in [t_0..t_1]} \frac{ds}{dt} = f(s(t),p,t)).$$

Since along the whole trajectory, between t_0 and t_1, the values of each parameter remain constant (its time derivative is zero), the above parametric system may be equivalently represented by a non parametric ODE system with $n+k$ equations:

$$z' = f(z,t) \quad \text{with} \quad z' = \begin{bmatrix} z'_1 \\ ... \\ z'_n \\ p'_1 \\ ... \\ p'_k \end{bmatrix} \text{and} \quad f(z,t) = \begin{bmatrix} f_1(z_1,...,z_n,p_1,...,p_n,t) \\ ... \\ f_n(z_1,...,z_n,p_1,...,p_n,t) \\ 0 \\ ... \\ 0 \end{bmatrix}$$

together with the boundary restrictions $p_1(t_p) \in I_1 \wedge ... \wedge p_k(t_p) \in I_k$ for some time point $t_p \in [t_0..t_1]$.

Any k-parametric n-ary ODE system (such as the $y' = f(y,p,t)$ above), may thus be modelled by a CSDP with a non parametric ODE system of $n+k$ equations (such as shown above with $z' = f(z,t)$) and k additional Value restrictions $Value_{n+j,t_p}(x_j)$ with initial domains $D_j = I_j$ $(1 \leq j \leq k)$.

An equivalent alternative to such representation would change the ODE constraint definition (see definition 10.1-1) to allow a direct representation of the parametric ODE system. This would imply the extension of its scope for the inclusion of k additional real valued variables $xp_1,...,xp_k$ and the redefinition of its constraint relation:

$$c_{ODE} = (<x_{ODE},xp_1,...,xp_k>, \rho_{ODE})$$

$$\rho_{ODE} = \{ <s,p_1,...,p_k> \in <D_{ODE},Dp_1,...,Dp_k> \mid p=[p_1,...,p_k] \wedge \forall_{t \in [t_0..t_1]} \frac{ds}{dt} = f(s(t),p,t) \}.$$

The domains of the new variables $xp_1,...,xp_k$ would be initialized by the respective parameter ranges: $Dp_1=I_1,...,Dp_k=I_k$. Such representation has the advantage of avoiding the introduction of new dimensions on the ODE system, but the disadvantage of introducing new variables into the ODE constraint scope and consequently on any narrowing function associated to it.

For simplicity, it is assumed in the next chapter that a CSDP is defined as in 10.1-1, and so, the first alternative is adopted for modelling parametric ODE problems.

Consider, for example, the parametric ODE fitting problem defined by the k-parametric unary system $y' = f(y,p,t)$ (with $t \in [t_0..t_l]$ and $p=[p_1,...,p_k] \in I_1 \times ... \times I_k$) and the set of m observed values $y(t_{pi})=v_{t_{pi}}$ (with $t_{p1}=t_0$, $t_{pm}=t_l$ and $\forall_{1 \leq i \leq m} t_{pi} \in [t_0..t_l]$).
The CSDP:

$(X=<x_{ODE},x_1,...,x_k,x_{p1},...,x_{pm}>,D=<D_{ODE},D_1,...,D_k,D_{p1},...,D_{pm}>,C=\{c_{ODE},c_1,...,c_k,c_{p1},...,c_{pm}\})$

where:

$D_{ODE} = \{ s \mid s : [t_0..t_l] \to 3^{k+1} \}$ and $c_{ODE}=(<x_{ODE}>, \rho_{ODE})$

$\rho_{ODE} = \{<s> \in D_{ODE} \mid \forall_{t \in [t_0..t_l]} \; s_1'(t) = f(s_1(t), s_2(t),..., s_k(t), t) \wedge s_2'(t) = 0 \wedge ... \wedge s_{k+1}'(t) = 0 \}$

$\forall_{1 \leq i \leq k} D_i=I_i$ and $c_i = Value_{i+1,t_0}(x_i)$ $\forall_{1 \leq i \leq m} D_{pi}=[-\infty..+\infty]$ and $c_{pi} = Value_{1,t_{pi}}(x_{pi})$

may be used for the definition of a CSDP constraint $c=(<x_1,...,x_k,x_{p1},...,x_{pm}>,\rho)$ relating the predicted values x_{pi} at each time point t_{pi} with some real valued instantiation of the k parameters (represented by the real variables $x_1,...,x_k$) of the system $y' = f(y,p,t)$.

An extended CCSP with such CSDP constraint together with other numerical constraints including the $x_{p1},...,x_{pm}$ variables and the respective observed values $v_{p1},...,v_{pm}$ could be used for modelling a variety of fitting problems.

For instance, several important statistical quantities, such as the total sum of squares (SS_{Tot}), the residual sum of squares (SS_{Res}), the regression sum of squares (SS_{Reg}) and the coefficient of determination (R^2) could be easily encapsulated into the respective real valued variables through the addition of the numerical constraints:

$$x_{Tot} = \sum_{i=1}^{m} \left(v_{pi} - \bar{v}\right)^2 \quad x_{Res} = \sum_{i=1}^{m} \left(v_{pi} - x_{pi}\right)^2 \quad x_{Reg} = \sum_{i=1}^{m} \left(x_{pi} - \bar{v}\right)^2 \quad x_{R^2} = \frac{x_{Reg}}{x_{Tot}} \quad \text{with } \bar{v} = \sum_{i=1}^{m} v_{pi}$$

With such constraints, fitting problems requiring these quantities to range within predefined bounds can be modelled by the appropriate specification of the initial domains of these variables. Enforcing some consistency requirement on the resulting extended CCSP, the initial parameter domains are pruned by eliminating some values for which the requirements cannot be satisfied.

Other less usual constraints, such as particular time point requirements, could also be easily added, with the guarantee that no excluded parameter value can satisfy all the constraints. Moreover, a best fit constrained problem could be modelled by such extended CCSP with a solving procedure for searching solutions that optimise some predefined criterion.

10.3.2   Representing Interval Valued Properties

As stated previously, each ODE restriction of a CSDP associates a restriction variable to some real valued property of some component of the ODE solutions. However, since the components of the ODE solutions are real functions defined along an interval of time, some of their properties are interval valued properties that cannot be expressed as single real values.

For example, the set of all the function values within an interval of time is an interval valued property of a real function (ranging along that interval of time) that cannot be expressed by a single real value. Such property, $Value_{j,T}(x)$, would be a variant of the Value restriction corresponding to the union of this real valued property along the whole interval of time T:

$$x = \bigcup_{t \in T} x_t \qquad \text{where } \forall_{t \in T} Value_{j,t}(x_t)$$

Since by definition, the restriction variable of a CSDP and all variables of a CCSP (and an extended CCSP) must be real valued variables, interval valued properties cannot be modelled by a single variable.

For integrating interval valued properties into a CSDP (and consequently into an extended CCSP), their intervals must be represented by pairs of real valued variables identifying its upper and lower bounds. Thus, any such property must be modelled by a pair of ODE restrictions associating one real variable to its maximum possible value and other real variable to its minimum possible value.

For example, the set of all the function values within an interval of time T can be modelled by a Maximum restriction for identifying its upper bound and a Minimum restriction for identifying its lower bound:

$$Maximum_{j,T}(x_{maxT}) \quad \text{and} \quad Minimum_{j,T}(x_{minT})$$

In an extended CCSP with a CSDP constraint that represents an interval valued property with two extreme real variables x_{maxT} and x_{minT}, other important information can be modelled by the addition of further numerical constraints including those variables.

For instance, a particular real value within the interval associated with the property could be modelled by the real variable x_{inT} through the additional pair of numerical constraints:

$$x_{minT} \leq x_{inT} \quad \text{and} \quad x_{inT} \leq x_{maxT}$$

The maximum distance (amplitude) between the two extremes of the property interval could be modelled by the real variable x_{ampT} through the additional pair of numerical constraints:

$$x_{ampT} = x_{maxT} - x_{minT} \quad \text{and} \quad x_{ampT} \geq 0$$

The center of the property interval could be modelled by the real variable x_{medT} through the additional numerical constraint:

$$x_{medT} = \frac{x_{maxT} + x_{minT}}{2}$$

Similarly to the case of Maximum and Minimum restrictions, First and Last restrictions (either Value, Maximum or Minimum) may be combined together for defining the smallest interval containing all the time points that satisfy the respective condition.

10.3.3 Combining ODE Solution Components

An important modelling issue of the CSDP framework is the possibility of combining several equations of an ODE system into a new equation that is added to the system as a new component. This may be used for modelling properties that do not depend exclusively on a single component of the original system but rather on some composition of a subset of its components.

Consider, for example, some vector function $y(t)$. The function is periodic if there is a positive constant k such that $y(t+k)=y(t)$ for all t. The smallest value of k that satisfies the previous condition is called the period of the function.

Suppose that the vector function is represented as an homogeneous n-ary ODE system $y' = f(y)$ (within an interval of time $[t_0..t_1]$ where $t_0=0$) with the initial value condition $y(0)=y_0$. In order to associate a real variable to the value of the period of such function it is necessary to identify the first point of time t_p (greater than 0) such that $y(t_p)=y_0$.

This is similar to the definition of a First Value restriction (see subsection 10.1.4), except that in this case the condition $y(t_p)=y_0$ does not refer to a single component but rather requires that the equality must hold simultaneously for all the components of the n-ary system.

However, the n-ary system may be transformed into an equivalent system with an extra component whose value at any time t represents the square of the distance between the vector $y(t_0) = y_0$ and $y(t)$.

This extra component defined by $y_{n+1}(t) = \sum_i^n (y_i(t) - y_i(0))^2$ is added as the $n+1$ component:

$$y'_{n+1}(t) = 2 \sum_i^n \left(y'_i(t)(y_i(t) - y_i(0)) \right) = 2 \sum_i^n \left(f_i(t)(y_i(t) - y_i(0)) \right)$$

together with the initial value $y_{n+1}(0) = 0$. With this new component the period of $y(t)$ could be associated with the real valued variable x_π by the ODE restriction[3]: $firstValue_{n+1,[t_0+\varepsilon..t_1], \le 0}(x_\pi)$

With the same kind of technique it is possible to model the distance at any time point between the trajectories of two different ODEs which may be used for imposing proximity requirements that will eventually lead to some adjustments on their parameters.

10.4   Summary

In this chapter the CSDP framework was characterised. The different types of restrictions supported by the framework were defined and illustrated with simple modelling examples. Continuous CSPs were extended for the inclusion of a new kind of constraint defined as a CSDP. The integration of CSDP constraints with the Global Hull-consistency criterion and with local search procedures was discussed. The next chapter presents a solving procedure for pruning the domains of the CSDP variables taking its restrictions into account.

---

[3] ε is some small positive value necessary for avoiding that the first time value satisfying the condition is at $t=0$.

Chapter 11

Solving a CSDP

The solving procedure for CSDPs must maintain a safe enclosure for the whole set of possible ODE solutions based on the interval approaches described in chapter 9 (section 9.2). This enclosure, used for the representation of a subset of the domain D_{ODE} of the solution variable x_{ODE}, will be called the ODE trajectory.

The solving algorithm is based on the improvement of the quality of such ODE trajectory (the reduction of the enclosing uncertainty), combined with the enforcement of the ODE restrictions through a constraint propagation algorithm (similar to the one presented in chapter 4, figure 4.1). A set of narrowing functions (see definition 4.1-1) associated with the ODE restrictions and the ODE constraint are the basis of such algorithm.

For each ODE restriction a pair of narrowing functions are defined: one reduces the domain of the restriction variable according to the ODE trajectory and the other decreases the uncertainty of the ODE trajectory according to the domain of the restriction variable.

Two narrowing functions are additionally included for reducing the uncertainty of the ODE trajectory: one propagates any domain narrowing along the trajectory, from a time point to a neighbouring point; the other links two consecutive time points through the application of an interval step method. Moreover, for each ODE restriction, a narrowing function may also be associated with the ODE constraint for improving the ODE trajectory, aiming at reducing the uncertainty of its restriction variable domain.

The next section describes the ODE trajectory, its implementation and functions for accessing and changing its contents. The narrowing functions associated with each type of ODE restrictions are presented in section 11.2. Section 11.3 presents the narrowing functions for reducing the uncertainty of the ODE trajectory. Section 11.4 describes how the previous set of narrowing functions is integrated in the constraint propagation algorithm for narrowing the domains of the CSDP variables.

11.1 The ODE Trajectory

An ODE trajectory TR is implemented as a tuple of 4 ordered lists $TR=<TP,TG,TF,TB>$.

A first list, TP, defines a sequence of k *trajectory time points* t_p along the interval of time $[t_0..t_1]$ (associated with the CSDP, cf. definition 10.1-1) together with the corresponding n-ary boxes, representing enclosures for the ODE solution values at those points. The first and last time points of such list are t_0 and t_1 respectively.

A second list, TG, defines the sequence of $k-1$ *trajectory time gaps* (between each pair of consecutive time points, t_{pi} and t_{pi+1}, of the previous list) and the associated n-ary boxes representing enclosures for the ODE solution values between those points.

The third and fourth lists, TF and TB, are auxiliary lists, representing at each trajectory time point t_p the enclosure for the ODE solution value when the interval step method was lastly applied from t_p to the next or to the previous point, respectively. This forward and backward information are used exclusively in the definition of appropriate narrowing functions for the ODE trajectory through the successive application of the interval step method over different pairs of consecutive time points (see section 11.3).

The boxes associated with the elements of these lists are represented respectively as $TP(t_p)$, $TG([t_{pi}..t_{pi+1}])$, $TF(t_p)$ and $TB(t_p)$. The intervals associated with the component j ($1 \leq j \leq n$) of the previous boxes are represented respectively as $TP_j(t_p)$, $TG_j([t_{pi}..t_{pi+1}])$, $TF_j(t_p)$ and $TB_j(t_p)$.

Figure 11.1 shows an example of an ODE trajectory (the forward and backward information is omitted) representing a safe enclosing of the set of possible ODE solutions of CSDP *P2b* (see figure 10.6). The set of possible ODE solutions is illustrated by a single line for its first component s_1 and by the grey area for its second component s_2. The ODE trajectory is defined through a sequence of seven time points and the time gaps in between. For each component, the intervals associated to each time point and time gap are represented, respectively, as a vertical line and a dashed rectangle.

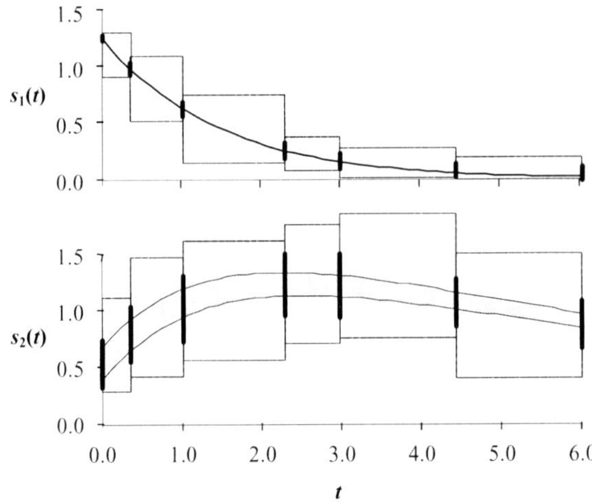

Figure 11.1 An ODE trajectory enclosing the ODE solutions of the CSDP *P2b*.

The ODE trajectory of figure 11.1 represents a subset of the D_{ODE}, defined in figure 10.2, containing all functions whose components are continuous functions enclosed by the rectangles and crossing all the vertical lines. This definition includes any possible ODE solution and so this ODE trajectory is a safe enclosing for the set of ODE solutions for CSDP *P2b*.

In general, an ODE trajectory $TR=<TP,TG,TF,TB>$ may be viewed as a finite representation of an infinite set of triples $<s,s_f,s_b>$ of functions from D_{ODE} satisfying the enclosures associated with the respective time points and gaps[1]:

   (i)   For every trajectory time point t_p: $s(t_p) \in TP(t_p)$, $s_f(t_p) \in TF(t_p)$ and $s_b(t_p) \in TB(t_p)$.

   (ii) For every trajectory time gap $[t_{pi}..t_{pi+1}]$: $\forall_{t \in [t_{pi}..t_{pi+1}]} s(t) \in TG([t_{pi}..t_{pi+1}])$.

During the solving process, the current ODE trajectory $TR=<TP,TG,TF,TB>$ is modified by the narrowing functions associated with the constraints of the CSDP. Each individual change of the trajectory is either the narrowing of some box (associated with a time point or gap) or the addition of a new time point (and subsequent reformulation of the ordered lists).

The assignments $TP(t_p) \leftarrow S$ and $TP_j(t_p) \leftarrow I$ will be used to denote respectively, the association of a box S to the time point t_p of TP and the association of an interval I to the component j of the time point t_p of TP (similar denotations will be used for assignments of

---

[1] This view is consistent with the narrowing functions definition 4.1-1, if the effect on the x_{ODE} domain of each narrowing function application is the elimination of some such triples without discarding any possible solution function s.

the lists *TG*, *TF* and *TB*). "*TR* with *C*" will be used as a short notation for an ODE trajectory *TR'* obtained from *TR* after performing all the assignments specified in *C*.

Procedures *insert*($TP(t_p)=S$) and *delete*($TP(t_p)$), respectively, introduce a new time point t_p in the ordered list of points of *TP* (initially associated with the box *S*), and remove one such time point from the list (similar denotation will be used for the lists *TG*, *TF* and *TB*).

Some of the narrowing functions are defined through the values of the ODE trajectory. The data of an ODE trajectory $TR=<TP,TG,TF,TB>$ may be accessed through the following set of auxiliary functions (where t_{p0} and t_{p1}, with $t_{p0} \le t_{p1}$, are two trajectory time points and *j* is one of its components; to simplify the notation, the *F*-interval is omitted if it denotes the entire sequence of points $[t_0..t_1]$):

(i) Functions $timePoints_{[t_{p0}..t_{p1}]}(TR)$, $leftPoints_{j,[t_{p0}..t_{p1}]}(TR)$, $rightPoints_{j,[t_{p0}..t_{p1}]}(TR)$, return a list of *F*-numbers representing for each point $t_p \in [t_{p0}..t_{p1}]$ of *TR*, its time value t_p, the left bound of its *j*-th component, $left(TP_j(t_p))$, and the right bound of its *j*-th component, $right(TP_j(t_p))$, respectively.

(ii) Function $timeGaps_{[t_{p0}..t_{p1}]}(TR)$ returns a list of *F*-intervals representing each gap $[t_{pi}..t_{pi+1}] \subseteq [t_{p0}..t_{p1}]$ of *TR*.

(iii) Functions $leftGaps_{j,[t_{p0}..t_{p1}]}(TR)$ and $rightGaps_{j,[t_{p0}..t_{p1}]}(TR)$, return a list of *F*-numbers representing for each gap $[t_{pi}..t_{pi+1}] \subseteq [t_{p0}..t_{p1}]$ of *TR*, the left bound of its *j*-th component $left(TG_j([t_{pi}..t_{pi+1}]))$, and the right bound of its *j*-th component, $right(TG_j([t_{pi}..t_{pi+1}]))$.

As usual, *min*(*L*) and *max*(*L*) return the minimum and maximum values from a list *L* of *F*-numbers.

At the beginning of the solving process, the ODE trajectory *TR* is initialised with function *initialiseTrajectory*() that:

(i) Introduces a time point for the t_p of each Value restriction and for the t_{p0} and t_{p1} of each of the other ODE restrictions (as defined in the previous chapter).

(ii) For the First and Last Value restrictions (subsection 10.1.4) and for the First and Last Maximum and Minimum restrictions (subsection 10.1.5), the initial left and right bounds of the associated restriction variable are also considered as initial time points.

(iii) Each of the time points and respective time gaps is associated to an *n*-ary box with all its components unbounded.

Considering again the example of CSDP *P2*b, whose ODE restrictions are $Value_{1,0.0}(x_1)$, $Value_{2,6.0}(x_2)$ and $Maximum_{2,[1.0..3.0]}(x_3)$, function *initialiseTrajectory*() returns, an ODE trajectory $TR=<TP,TG,TF,TB>$ with:

(i) the time points 0.0, 1.0, 3.0 and 6.0 associated to the boxes:
$$TP(0.0) = TP(1.0) = TP(3.0) = TP(6.0) = <[-\infty..+\infty],[-\infty..+\infty]>$$
$$TF(0.0) = TF(1.0) = TF(3.0) = TF(6.0) = <[-\infty..+\infty],[-\infty..+\infty]>$$
$$TB(0.0) = TB(1.0) = TB(3.0) = TB(6.0) = <[-\infty..+\infty],[-\infty..+\infty]>$$

(ii) the time gaps [0.0..1.0], [1.0..3.0] and [3.0..6.0] associated to the boxes:
$$TG([0.0..1.0])=TG([1.0..3.0])=TG([3.0..6.0])=<[-\infty..+\infty],[-\infty..+\infty]>$$

11.2 Narrowing Functions for Enforcing the ODE Restrictions

According to definition 4.1-1, the narrowing functions associated with the ODE restrictions must satisfy the contractness and correctness properties.

The contractness property is easily guaranteed by preventing the enlargement of any interval domain, either from a restriction variable or from a component of any box of the ODE trajectory. The insertion of a new time point cannot enlarge the subset of functions represented by the ODE trajectory since it represents an additional restriction to such functions (assuming, of course, that when inserting a point $t_p \in [t_{pi}..t_{pi+1}]$ the boxes associated to the two new gaps $TG([t_{pi}..t_p])$ and $TG([t_p..t_{pi+1}])$ are included in the box associated with the whole gap $TG([t_{pi}..t_{pi+1}])$).

For a narrowing function that reduces the domain of an ODE restriction variable, the correctness property may be achieved by identifying, within the ODE trajectory, the functions that maximise and minimise the values of such variable and guaranteeing that its new domain includes those values.

When a narrowing function reduces the trajectory uncertainty from the domain I of a restriction variable, this reduction is achieved through the narrowing of one or more boxes of the ODE trajectory (namely the interval component j associated with the ODE restriction). Correctness is guaranteed if considering in isolation each narrowed interval (without any other interval reductions), there are no discarded functions with a value (of the restriction variable) within I.

The definition of narrowing functions in the following subsections are based on the above properties.

11.2.1 Value Narrowing Functions

From definition 10.1.1-1, the ODE restriction $Value_{j,t_p}(x_i)$ relates a function with the value of its j component at time t_p. If $TR=<TP,TG,TF,TB>$ is an ODE trajectory representing a set of possible functions, then those that maximise/minimise the value of its j component at time t_p must have this value equal to $right(TP_j(t_p))/left(TP_j(t_p))$[2] (values for the restriction variable x_i outside the interval $TP_j(t_p)$ cannot be associated with any function represented in the ODE trajectory TR). On the other hand, if the values for the restriction variable x_i lie within F-interval I_i, any function whose j's component value at time t_p is outside I_i may be safely discarded from the ODE trajectory.

The above reasoning justifies the following formal definition for the narrowing functions associated with a Value restriction. The definition (and all the following subsequent narrowing function definitions) represents any domain element A of a narrowing function NF, ($A \subseteq D$ for a CSDP (X,D,C)), as a tuple where only the relevant variable domains are shown. All the other domains are kept unchanged by NF.

Definition 11.2.1-1 (Value Narrowing Functions). Let CSDP $P=(X,D,C)$ be defined as in 10.1-1. Let $TR=<TP,TG,TF,TB>$ be the ODE trajectory representing the domain of x_{ODE} and I_i the domain of x_i. The ODE Value restriction $c \equiv Value_{j,t_p}(x_i) \in C$ (defined in 10.1.1-1) has associated the following pair of narrowing functions:

    (i) $NF_1(<TR,...,I_i,...>) = <TR,...,TP_j(t_p) \cap I_i,...>$

    (ii) $NF_2(<TR,...,I_i,...>) =< TR$ with $\{TP_j(t_p) \leftarrow TP_j(t_p) \cap I_i\},...,I_i,...>$ ❑

If an enclosure $TP_j(t_p)$ of some component j of the trajectory TR becomes empty at some point t_p then TR can no longer represent any function and the domains box that includes TR (and represents the Cartesian product of its components) also becomes empty.

The previous definition could be exemplified with narrowing functions for the Value restrictions of CSDP $P2b$. For instance, for the ODE trajectory enclosure of figure 11.1, narrowing function NF_1 associated with the restriction $Value_{2,6.0}(x_2)$ would narrow an

---

[2] Note that, due to the initialisation procedure, t_p must be a time point of the ODE trajectory. Similarly, any point used in the definition of any narrowing function is, due to the initialisation procedure, within the time points of the ODE trajectory.

initially unbounded domain of x_2 to the interval $TP_2(6.0)$ (represented as the vertical line of the second component graphic at time 6.0). Conversely, if an interval I_2 represents the domain of x_2 then, narrowing function NF_2 could narrow the associated interval $TP_2(6.0)$ of the ODE trajectory into $TP_2(6.0) \cap I_2$. If, by any of the above narrowing operations, the empty set is obtained then the ODE restriction cannot be satisfied and the CSDP has no solutions.

11.2.2 Maximum and Minimum Narrowing Functions

According to definition 10.1.2-1, the ODE restriction $Maximum_{j,[t_{p0}..t_{p1}]}(x_i)$ relates a function with the maximum value of its j component within time interval $[t_{p0}..t_{p1}]$. If $TR=<TP,TG,TF,TB>$ is an ODE trajectory representing a set of possible functions then their maximum value within $[t_{p0}..t_{p1}]$ cannot exceed the maximum of the right bounds of their j component enclosures (possibly corresponding to several time points and gaps) within that interval. Moreover, since for every time point $t_p \in [t_{p0}..t_{p1}]$ of TR all functions must have its j component value within the interval $TP_j(t_p)$, they must have some value not less than the maximum of the left bounds of these intervals. On the other hand, if the values for the restriction variable x_i lie within the F-interval I_i, then any function whose j component has values higher than $right(I_i)$ within the time interval $[t_{p0}..t_{p1}]$ may be safely discarded from the ODE trajectory.

A similar reasoning, but with respect to the minimum value, may be used for deriving the narrowing functions associated with the ODE restriction $Minimum_{j,[t_{p0}..t_{p1}]}(x_i)$. Both are formally defined as follows.

Definition 11.2.2-1 (Maximum and Minimum Narrowing Functions). Let CSDP $P=(X,D,C)$ be defined as in 10.1-1. Let $TR=<TP,TG,TF,TB>$ be an ODE trajectory representing the domain of x_{ODE}, I_i the domain of x_i and $c \in C$ a Maximum or Minimum restriction (as defined in 10.1.2-1).

If $c \equiv Maximum_{j,[t_{p0}..t_{p1}]}(x_i)$, let:

$a=max(leftPoints_{j,[t_{p0}..t_{p1}]}(TR))$, $b=max(rightGaps_{j,[t_{p0}..t_{p1}]}(TR))$ and $I=[-\infty..right(I_i)]$.

If $c \equiv Minimum_{j,[t_{p0}..t_{p1}]}(x_i)$, let:

$a=min(leftGaps_{j,[t_{p0}..t_{p1}]}(TR))$, $b=min(rightPoints_{j,[t_{p0}..t_{p1}]}(TR))$ and $I=[left(I_i)..+\infty]$.

The ODE restriction c has associated the following pair of narrowing functions:

(i) $NF_1(<TR,\ldots,I_i,\ldots>) = <TR,\ldots,[a..b] \cap I_i,\ldots>$

(ii) $NF_2(<TR,\ldots,I_i,\ldots>) =< TR',\ldots,I_i,\ldots>$

where $TR' = TR$ with

$\{ \forall t_p \in timePoints_{[t_{p0}..t_{p1}]}(TR)\ TP_j(t_p) \leftarrow TP_j(t_p) \cap I,$

$\forall [t_{pi}..t_{pi+1}] \in timeGaps_{[t_{p0}..t_{p1}]}(TR)\ TG_j([t_{pi}..t_{pi+1}]) \leftarrow TG_j([t_{pi}..t_{pi+1}]) \cap I \}$ ❏

Figure 11.2 illustrates the above definition for the narrowing functions associated with restriction $Maximum_{2,[1.0..3.0]}(x_3)$ to CSDP $P2b$ for the ODE trajectory TR represented in figure 11.1.

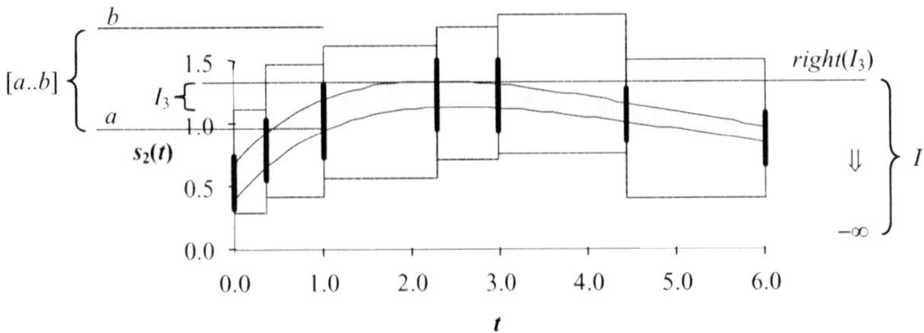

Figure 11.2 Narrowing functions associated with a Maximum restriction.

The figure shows the values of a and b for the definition of the narrowing function NF_1. It is clear that the maximum value of any trajectory function within the time interval [1.0..3.0] cannot be outside the dashed box, upper bounded by $b=max(rightGaps_{2,[1.0..3.0]}(TR))$ and lower bounded by $a=max(leftPoints_{2,[1.0..3.0]}(TR))$. On the other hand, if the interval I_3=[1.1..1.3] represents the domain of x_3 then, any value of the second solution component represented by TR cannot exceed 1.3 (within the time interval [1.0..3.0]) and so the NF_2 narrowing function may discard the region of the ODE trajectory (represented in the figure as rectangles and vertical lines within the dashed box) above this value, that is, outside $[-\infty..1.3]$.

11.2.3   Time and Area Narrowing Functions

To simplify the specification of the narrowing functions associated with the Time and Area restrictions, two auxiliary functions $Time_{\diamond k}(I)$ and $Area_{\diamond k}(I)$ are defined (where $\diamond \in \{\leq,\geq\}$ and k is a real number) for representing respectively, the values of time and area that will be considered at some time gap. The argument I of these functions represents the possible solution values along the time gap. The argument $\diamond k$ specifies a filtering condition. We are only interested in solution values within the interval $I_{\diamond k}$ which is $I_{\geq k}=[k..+\infty]$ or $I_{\leq k}=[-\infty..k]$.

The definitions of the $Time_{\diamond k}(I)$ and $Area_{\diamond k}(I)$ functions, below, consider 3 different situations:

$$Time_{\diamond k}(I) = \begin{cases} [1] & \text{if } I \subseteq I_{\diamond k} \\ [0] & \text{if } I \cap I_{\diamond k}=\varnothing \\ [0] \uplus [1] & \text{otherwise} \end{cases} \qquad Area_{\diamond k}(I) = \begin{cases} |I - [k]| & \text{if } I \subseteq I_{\diamond k} \\ [0] & \text{if } I \cap I_{\diamond k}=\varnothing \\ [0] \uplus |I - [k]| & \text{otherwise} \end{cases}$$

If all the solution values satisfy the condition ($I \subseteq I_{\diamond k}$), function $Time_{\diamond k}(I)$ returns the degenerate interval [1] specifying that the whole width of the time gap must be considered. Function $Area_{\diamond k}(I)$ returns an interval, bounded by the maximum and minimum distances of the solution values to threshold k.

If none of the solution values satisfy the condition ($I \cap I_{\diamond k}=\varnothing$), both functions return the degenerate interval [0] specifying that there is nothing to be considered.

If some solution values satisfy the condition and others do not ($I \not\subseteq I_{\diamond k}) \wedge (I \cap I_{\diamond k}\neq\varnothing$) then, both functions must take into account that there are possible solution values that must be considered, as in the first case, and there are possible solution values that should not be considered, as in the second case. This is represented by returning the union hull of the intervals returned in the previous cases.

Figure 11.3 illustrates function $Area_{\geq k}(I)$ when applied to each of the above possible cases. The width of the rectangle represents the time gap and the height the possible solution values I.

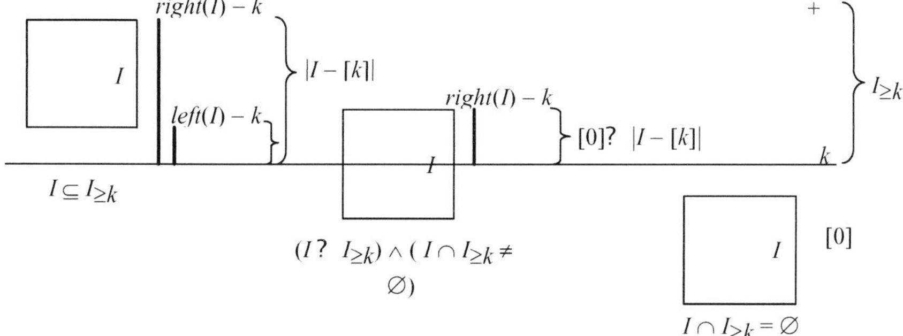

Figure 11.3 The three possible cases for the definition of the $Area_{\geq k}(I)$ function.

The auxiliary functions $Time_{\diamond k}(I)$ and $Area_{\diamond k}(I)$ are used for the specification of the narrowing function associated with the Time and Area restrictions. Note that for these restrictions no narrowing function can be defined to reduce the uncertainty of an ODE trajectory from the domain of the restriction variable. Both Time and Area values are compound from the function values for every time point within an interval of time and the existence of high or low peaks for a short period of time cannot be prevented. These do not affect significantly the overall area/time to be considered, but avoid a safe narrowing of any interval enclosing.

Definition 11.2.3-1 (Time and Area Narrowing Functions). Let CSDP $P=(X,D,C)$ be defined as in 10.1-1. Let $TR=\langle TP,TG,TF,TB\rangle$ be the ODE trajectory representing the domain of x_{ODE}, I_i the domain of x_i, and $c\in C$ a Time or Area restriction (as defined in 10.1.3-1 and 10.1.3-2, respectively).

If $c \equiv Time_{j,[t_{p0}..t_{p1}],\diamond k}(x_i)$, let $F(X) \equiv Time_{\diamond k}(X)$.

If $c \equiv Area_{j,[t_{p0}..t_{p1}],\diamond k}(x_i)$, let $F(X) \equiv Area_{\diamond k}(X)$.

The ODE restriction c has associated the following narrowing function:

(i) $NF_1(\langle TR,...,I_i,...\rangle) = \langle TR,..., \Big(\sum I_{apx}((t_{pi+1}-t_{pi})\times F(TG_j([t_{pi}..t_{pi+1}]))) \Big) \cap I_i,...\rangle$

$$[t_{pi}..t_{pi+1}] \in timeGaps_{[t_{p0}..t_{p1}]}(TR)$$ □

Figure 11.4 illustrates the Time and Area narrowing functions for CSDP $P2c$ (see figure 10.8), assuming that TR is the ODE trajectory represented in figure 11.1 (only the enclosing bounds associated with the time gaps are represented in figure 11.4).

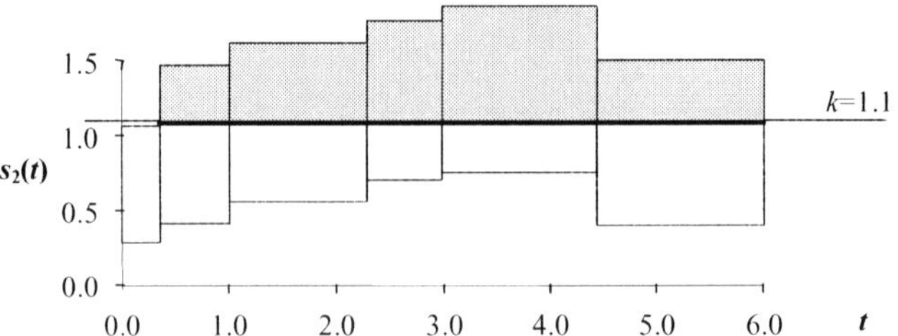

Figure 11.4 Time and Area narrowing functions for CSDP *P2*c.

From the narrowing function associated with restriction $Time_{2,[0.0..6.0],\geq1.1}(x_4)$, x_4 is guaranteed to be between 0 and the length of the horizontal solid line. From the narrowing function associated with restriction $Area_{2,[0.0..6.0],\geq1.1}(x_5)$, x_5 will lie between 0 and the value of the dashed area.

11.2.4 First and Last Value Narrowing Functions

The specification of narrowing functions associated with the First and Last Value restrictions is based on the auxiliary function $timeEnclosure_{j,[t_{p0}..t_{p1}]}$ where j is an integer representing a solution component and $[t_{p0}..t_{p1}]$ is an *F*-interval representing a time interval.

The function is defined in the pseudocode of figure 11.5. In addition to the solution component j and the interval of time $[t_{p0}..t_{p1}]$, it has three arguments: an ODE trajectory *TR*, an *F*-interval *I* and a label *type* $\in$ {*first*, *last*}. If the label *type* is set to *first/last* then it returns the smallest interval of time $[a..b]$ which may be obtained by narrowing $[t_{p0}..t_{p1}]$ without discarding any value of *t* which may be the first/last of any function *f* represented by the component *j* of *TR* such that $f(t) \in I$.

---

function $timeEnclosure_{j,[t_{p0}..t_{p1}]}$(an ODE trajectory *TR*=<TP,TG,TF,TB>, an *F*-interval *I*,
         a label *type* $\in$ {*first*, *last*})

(1) $a \leftarrow +\infty$; $b \leftarrow -\infty$;

(2) $L_{Gaps} \leftarrow timeGaps_{[t_{p0}..t_{p1}]}(TR)$;

(3)   **repeat**

(4) $[t_{pi}..t_{pi+1}] \leftarrow L_{Gaps}.pop\_front()$;

(5)   **if** $TG_i([t_{pi}..t_{pi+1}]) \cap I \neq \varnothing$ **then** $a \leftarrow t_{pi}$; **end if**;

(6)   **until** $a \neq +\infty$ **or** $L_{Gaps}.size()=0$;

(7)   **if** $a=+\infty$ **then return** $\varnothing$;

(8) $L_{Gaps} \leftarrow timeGaps_{[a..t_{p1}]}(TR)$;

(9)   **repeat**

(10) $[t_{pi}..t_{pi+1}] \leftarrow L_{Gaps}.pop\_back()$;

(11)   **if** $TG_i([t_{pi}..t_{pi+1}]) \cap I \neq \varnothing$ **then** $b \leftarrow t_{pi+1}$; **end if**;

(12)   **until** $b \neq -\infty$;

(13) $L_{Points} \leftarrow timePoints_{[a..b]}(TR)$;

(14)   **repeat**

(15)   **if** *type*=*first* **then** $t_p \leftarrow L_{Points}.pop\_front()$; **else** $t_p \leftarrow L_{Points}.pop\_back()$;

(16)   **if** $TP_i(t_p) \subseteq I$ **then**

(17)   **if** *type*=*first* **then return** $[a..t_p]$; **else return** $[t_p..b]$; **end if**;

(18)   **end if**;

(19)   **until** $L_{Points}.size()=0$;

(20)   **return** $[a..b]$;

end function

Figure 11.5 The definition of the *timeEnclosure* function.

The correctness of such function is achieved by firstly discarding all the extreme left (lines 2-6) and right (lines 8-12) time intervals $[t_{pi}..t_{pi+1}]$ where the values of all the associated functions are guaranteedly outside I (for which $TG_j([t_{pi}..t_{pi+1}]) \cap I = \varnothing$). Secondly, the obtained interval $[a..b]$ is further narrowed. If the label *type* is set to *first* (*last*) this interval can be safely narrowed to $[a..t_p]$ ($[t_p..b]$) (lines 13-19) if somewhere in the sequence of time ($t_p \in [a..b]$) it is guaranteed that the value of any represented function is within I (if $TP_j(t_p) \subseteq I$). In this case, and given the continuity of any function f represented by component j of TR, the value t which is the first (last) time such that $f(t) \in I$ must be within such interval, that is, $t \in [a..t_p]$ ($t \in [t_p..b]$).

On the other hand, the knowledge that the value of the restriction variable x_i, of a First/Last Value restriction must be within the real interval I_i, may be used to safely discard from the ODE trajectory TR any function with values satisfying the condition $\diamond k$ before/after that interval of time. If the first/last value of I_i does not coincide with first/last value of the relevant time interval $[t_{p0}..t_{p1}]$ then, by continuity, any function at that extreme point of I_i either equals k or does not satisfy the condition $\diamond k$.

Definition 11.2.4-1 formalises the pair of narrowing functions associated with First and Last Value restrictions.

Definition 11.2.4-1 (First and Last Value Narrowing Functions). Consider CSDP $P=(X,D,C)$ as defined in 10.1-1. Let $TR=<TP,TG,TF,TB>$ be an ODE trajectory representing the domain of x_{ODE}, I_i the domain of x_i and $c \in C$ a First or Last Value restriction (as defined in 10.1.4-1). Let I be an interval with all the real values satisfying $\diamond k$, and $\bar{I}$ its complement: $I=[k..+\infty]$ and $\bar{I}=[-\infty..k]$ if $\geq k$; or $I=[-\infty..k]$ and $\bar{I}=[k..+\infty]$ if $\leq k$.

If $c \equiv firstValue_{j,[t_{p0}..t_{p1}],\diamond k}(x_i)$, let $type=first$, $a=left(I_i)$ and $T_{out}=[t_{p0}..a]$.

If $c \equiv lastValue_{j,[t_{p0}..t_{p1}],\diamond k}(x_i)$, let $type=last$, $a=right(I_i)$ and $T_{out}=[a..t_{p1}]$.

The ODE restriction c has associated the following pair of narrowing functions:

(i) $NF_1(<TR,\ldots,I_i,\ldots>) = <TR,\ldots, timeEnclosure_{j,I_i}(TR,I,type),\ldots>$

(ii)
$$NF_2(<TR,\ldots,I_i,\ldots>) = \begin{cases} \varnothing & \text{if } \exists_{t_p \neq a \in timePoints_{T_{out}}(TR)}\ TP_j(t_p) \subseteq I \ \vee \\ & \qquad (width(T_{out})>0 \ \wedge \ TP_j(a) \cap \bar{I} = \varnothing) \\ <TR',\ldots,I_i,\ldots> & \text{otherwise} \end{cases}$$

where $TR' = TR$ w/ $\{\ \forall_{t_p \neq a\ \in timePoints_{T_{out}}(TR)}\ TP_j(t_p) \leftarrow TP_j(t_p) \cap \bar{I}$,

$\qquad TP_j(a) \leftarrow TP_j(a) \cap \bar{I}$ (if $width(T_{out})>0$),

$\qquad \forall_{[t_{pi}..t_{pi+1}] \subset timeGaps_{T_{out}}(TR)}\ TG_j([t_{pi}..t_{pi+1}]) \leftarrow TG_j([t_{pi}..t_{pi+1}]) \cap \bar{I} \ \}\ \square$

Figure 11.6 illustrates the narrowing functions associated with the First Value restriction $firstValue_{1,[0.0..2.0],\leq 0.25}(x_3)$ of CSDP *P1*d (see figure 10.9) and the ODE trajectory TR represented in the figure by the vertical solid lines and the dashed rectangles.

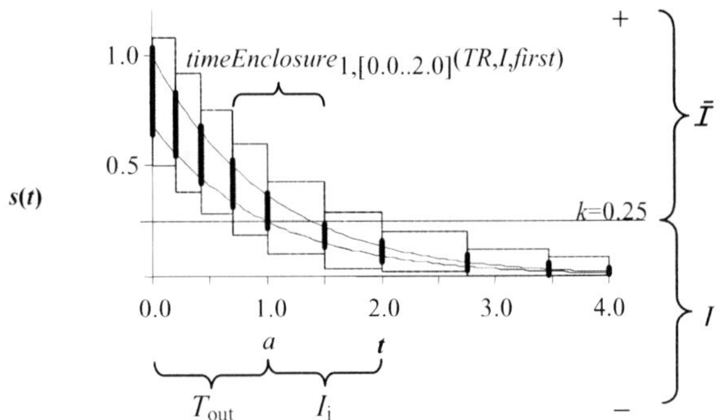

Figure 11.6 First Value narrowing functions for CSDP *P1*d.

Function $timeEnclosure_{1,[0.0..2.0]}(TR,I,first)$ first excludes from interval [0.0..2.0] the first three interval gaps for which all the associated rectangles are guaranteedly above 0.25, and then it excludes all times after t=1.5 since at this point the condition $s(t) \le 0.25$ is already satisfied. On the other hand, if the domain variable x_3 is I_i = [1.0..2.0] then the ODE trajectory is narrowed, namely the rectangle associated with time gap [0.7..1.0] and the vertical line of t=1.0 are reduced by discarding any region outside $\bar{I}$.

11.2.5   *First and Last Maximum and Minimum Narrowing Functions*

The definition of narrowing functions associated with the First and Last Maximum and Minimum restrictions is similar to the previous definition. The main difference is that in this case there is no previous knowledge about a value k determining the condition $\diamond k$ that must be satisfied. This value, the maximum or minimum value of any possible function of component j of TR, lies within an interval which can be computed similarly to definition 11.2.2-1.

According to this definition, and regarding time interval $[t_{p0}..t_{p1}]$, the maximum of any function represented by component j of TR must be between $k_0=max(leftPoints_{j,[t_{p0}..t_{p1}]}(TR))$ and $max(rightGaps_{j,[t_{p0}..t_{p1}]}(TR))$. However, if $I_i \subseteq [t_{p0}..t_{p1}]$ is the domain of x_i that must contain the maximum of any such function, then this maximum cannot exceed $k_1=max(rightGaps_{j,I_i}(TR))$. Consequently, the maximum of any function must be between k_0 and k_1. This allows the definition of the interval $I_0=[k_0..+\infty]$ to contain any possible maximum value, the interval $I=[k_1..+\infty]$ to be greater or equal than any function value within I_i and its complement $\bar{I}=[-\infty..k_1]$ to contain any possible value of any function within T_{out} (where T_{out}, defined as in definition 11.2.4-1, represents the interval of time within $[t_{p0}..t_{p1}]$ that precedes/succeeds the subinterval I_i). A symmetrical reasoning may be used for obtaining the intervals I, $\bar{I}$ and I_0 in the case of a Minimum restriction.

The following formal definition, for the narrowing functions associated with the First and Last Maximum and Minimum restrictions, is based on definition 11.2.4-1, but using intervals I, $\bar{I}$ and I_0 as defined above and an auxiliary function $extremeEnclosure_{j,[t_{p0}..t_{p1}]}$. This is similar to function $timeEnclosure_{j,[t_{p0}..t_{p1}]}$ but without the lines 13 to 19 (this is a consequence of working with intervals for representing $TP_j(t_p)$ which give no guarantees regarding its value being a maximum or a minimum value). Moreover, note that the special case for t=a no longer exists since, from the definition of $\bar{I}$, we always have $TP_j(a) \cap \bar{I} = TP_j(a)$.

Definition 11.2.5-1 (First and Last Maximum and Minimum Narrowing Functions).
Consider CSDP $P=(X,D,C)$ as defined in 10.1-1. Let $TR=<TP,TG,TF,TB>$ be the ODE trajectory representing the domain of x_{ODE}, I_i the domain of x_i and $c \in C$ an ODE restriction as defined in 10.1.5-1. If c is a First/Last Maximum restriction, let $k_0=max(leftPoints_{j,[t_{p0}..t_{p1}]}(TR))$, $k_1=max(rightGaps_{j,I_i}(TR))$, $I=[k_1..+\infty]$, $\bar{I}=[-\infty..k_1]$ and $I_0=[k_0..+\infty]$. If c is a First/Last Minimum restriction, let $k_0=min(rightPoints_{j,[t_{p0}..t_{p1}]}(TR))$, $k_1=min(leftGaps_{j,I_i}(TR))$, $I=[-\infty..k_1]$, $\bar{I}=[k_1..+\infty]$ and $I_0=[-\infty.. k_0]$.

 If $c \equiv firstMaximum_{j,[t_{p0}..t_{p1}]}(x_i)$, let $type=first$, $a=left(I_i)$ and $T_{out}=[t_{p0}..a]$.

 If $c \equiv firstMinimum_{j,[t_{p0}..t_{p1}]}(x_i)$, let $type=first$, $a=left(I_i)$ and $T_{out}=[t_{p0}..a]$.

 If $c \equiv lastMaximum_{j,[t_{p0}..t_{p1}]}(x_i)$, let $type=last$, $a=right(I_i)$ and $T_{out}=[a..t_{p1}]$.

 If $c \equiv lastMinimum_{j,[t_{p0}..t_{p1}]}(x_i)$, let $type=last$, $a=right(I_i)$ and $T_{out}=[a..t_{p1}]$.

The ODE restriction c has associated the following pair of narrowing functions:

 (i) $NF_1(<TR,...,I_i,...>) = <TR,..., extremeEnclosure_{j,I_i}(TR,I_0,type),...>$

 (ii) $NF_2(<TR,...,I_i,...>)$
$$\begin{cases} \varnothing & \text{if } \exists_{t_p \neq a \in timePoints_{T_{out}}(TR)} \; TP_j(t_p) \subseteq I \\ < TR',...,I_i,...> & \text{otherwise} \end{cases}$$

where $TR' = TR$ w/ { $\forall_{t_p \neq a \; \in timePoints_{T_{out}}(TR)} \; TP_j(t_p) \leftarrow TP_j(t_p) \cap \bar{I}$,

$\forall_{[t_{pi}..t_{pi+1}] \in timeGaps_{T_{out}}(TR)} \; TG_j([t_{pi}..t_{pi+1}]) \leftarrow TG_j([t_{pi}..t_{pi+1}]) \cap \bar{I}$ } $\square$

11.3 Narrowing Functions for the Uncertainty of the ODE Trajectory

The uncertainty of the ODE trajectory may be decreased by narrowing functions associated with the ODE constraint, based on two basic functions: *pruneGap* and *insertPoint*.

The *pruneGap* function aims at reducing the set of possible ODE solutions that may link two consecutive points along the sequence of time points. To maintain the correctness, this reduction is achieved through a safe interval step method (cf. chapter 9).

The *insertPoint* function introduces a new time point in the sequence of time points. Although not decreasing the ODE trajectory uncertainty on its own, it may lead to important overall reductions when combined with the previous function.

Figure 11.7 shows the pseudocode of function *pruneGap*. It has 4 arguments, the ODE trajectory TR, a vector function $f(S,t)$ defining the ODE system (see definition 9.1), and two F-numbers t_i and t_j (with $t_i \neq t_j$). It returns an ODE trajectory removing from the set of possible ODE solutions of TR, those discarded by the interval step method applied between time points t_i and t_j.

The interval step method is represented in the figure as a procedure *intervalStep*. The first 4 arguments input the vector function $f(S,t)$, the enclosure box $TP(t_i)$ associated in the ODE trajectory for the time t_i, and the specification of the initial and final time points, t_i and t_j respectively. The procedure returns a time point t_k (between t_i and t_j) and safe enclosures S_k and S_{ik} for the possible solution values, at t_k and between t_i and t_k, respectively. It is assumed that the S_{ik} enclosure is the result of some enclosure method (cf. 9.2.2) that firstly tries the validation of the whole time gap between t_i and t_j (with $t_k=t_j$) according to some predefined error tolerances (local and global), and reducing the gap size in case of failure. The S_k enclosure is achieved by the subsequent application of some appropriate interval method (cf. 9.2.3), and the whole process may be repeated until the given tolerance is satisfied.

---

function *pruneGap*(ODE trajectory $TR=<TP,TG,TF,TB>$, vector function $F(S,t)$, F-numbers t_i, t_j)

(1) *intervalStep*($F(S,t),TP(t_i),t_i,t_j,S_{ik},S_k,t_k$);

(2) $T_{ij} \leftarrow [min(t_i,t_j)..max(t_i,t_j)]$; $T_{ik} \leftarrow [min(t_i,t_k)..max(t_i,t_k)]$;

(3) **if** $t_k = t_i$ **then return** TR; **end if;**

(4) **if** $t_k \neq t_i$ **then**

(5)   **if** $t_i < t_j$ **and not** $TF(t_i)=Unbound$ **then** $TF(t_i) \leftarrow TP(t_i)$; **return** TR; **end if;**

(6)   **if** $t_i > t_j$ **and not** $TB(t_i)=Unbound$ **then** $TB(t_i) \leftarrow TP(t_i)$; **return** TR; **end if;**

(7)   **if** *nmbPoints*(TR) = *MaxPoints* **then return** TR; **end if;**

(8) $TR \leftarrow$ *insertPoint*(TR,T_{ij},t_k);

(9) **end if;**

(10) **if** $t_i < t_j$ **then** $TF(t_i) \leftarrow TP(t_i)$; **if** $S_k \subseteq TB(t_k)$ **then** $TB(t_k) \leftarrow S_k$; **end if; end if;**

(11) **if** $t_i > t_j$ **then** $TB(t_i) \leftarrow TP(t_i)$; **if** $S_k \subseteq TF(t_k)$ **then** $TF(t_k) \leftarrow S_k$; **end if; end if;**

(12) **if** $TP(t_k) \cap S_k = \varnothing$ **or** $TG(T_{ik}) \cap S_{ik} = \varnothing$ **then return** $\varnothing$;

(13) $TP(t_k) \leftarrow TP(t_k) \cap S_k$; $TG(T_{ik}) \leftarrow TG(T_{ik}) \cap S_{ik}$;

(14) **return** TR;

end function

Figure 11.7 The definition of the *pruneGap* function.

Firstly, the *intervalStep* procedure is called from t_i to t_j (line 1). If t_i is smaller than t_j, the interval method is applied in the forward time direction, otherwise it is applied in the backward time direction.

Subsequently, the time gaps T_{ij} and T_{ik} are defined (line 2), to represent, respectively, the whole time gap and its subset validated by the interval method.

If the validation procedure fails ($t_k=t_i$), the trajectory is returned unchanged (line 3).

If the validation procedure does not fail but can not validate the whole time gap ($t_k \neq t_j$) then there are 3 possibilities (lines 5-8): either the validation was previously achieved[3] and nothing else is done except updating the respective forward/backward information (lines 5/6); or the maximum number of points (*MaxPoints*) has been reached and the ODE trajectory is returned unchanged (line 7); or a new time point t_k is inserted by the *insertPoint* function (line 8).

Finally, if the procedure did not terminate before, lines 10 through 13 update the ODE trajectory TR according to the enclosures obtained by the interval step method validation from t_i to t_k. Lines 10 and 11 are responsible for updating the forward/backward information (represented by the TF and TB lists, respectively) which keep track of the point enclosures used in the latest application of the interval step method. If the interval method was applied in the forward direction (line 10), point t_i of the TF list is updated with the current enclosure $TP(t_i)$, and point t_k of the TB list is updated if it contains the obtained enclosure S_k[4]. Similarly, line 11 updates the TF and TB lists if the interval method was applied in the backward direction. Line 13 updates the ODE trajectory enclosures at point t_k ($TP(t_k)$) and at the validated time gap ($TG(T_{ik})$) according to the obtained enclosures (after verifying in line 12 that none of these new enclosures becomes empty).

Function *insertPoint* is shown in figure 11.8. It has 3 arguments, the ODE trajectory TR, an F-interval $[t_{pi}..t_{pi+1}]$ and an F-number t_k (with $t_{pi}<t_k<t_{pi+1}$). It returns an ODE trajectory representing the same set of possible ODE solutions as TR, but including a new time point t_k within time gap $[t_{pi}..t_{pi+1}]$, replaced by the new time gaps $[t_{pi}..t_k]$ and $[t_k..t_{pi+1}]$.

---

[3] *Unbound* is used to denote an *n*-ary F-box with all its elements unbounded: $[-\infty..+\infty]$.

[4] This is justified for preventing the subsequent pruning in the opposite direction without an enclosure smaller than S_k.

> **function** *insertPoint*(an ODE trajectory *TR*=<*TP,TG,TF,TB*>, *F*-interval [$t_{pi}..t_{pi+1}$], *F*-number t_k)
> (1) $S_{ij} \leftarrow TG([t_{pi}..t_{pi+1}])$;
> (2) *insert*(*TP*(t_k)=S_{ij}); *insert*(*TF*(t_k)=*Unbound*); *insert*(*TB*(t_k)=*Unbound*);
> (3) *delete*(*TG*([$t_{pi}..t_{pi+1}$])); *insert*(*TG*([$t_{pi}..t_k$])= S_{ij}); *insert*(*TG*([$t_k..t_{pi+1}$])=S_{ij});
> (4) **return** *TR*;
> **end function**

Figure 11.8 The definition of the *insertPoint* function.

The enclosures associated with the new time point and the new time gaps are initialised with the enclosure associated with the previous entire gap. Since no forward or backward interval step method was yet applied starting at the new time point this information is kept unbound.

All narrowing functions described in the next subsections are based on the selection of an appropriate time gap from the ODE trajectory and subsequent application of one or both the above basic functions to such gap.

11.3.1   Propagate Narrowing Function

If an enclosure for the ODE solutions at some time point is reduced by any narrowing function, the reapplication of the interval step method over the adjacent time gaps may further prune these gaps. Moreover, the repeated application of the interval step method triggered by the reduction of the enclosures, propagates this pruning along the ODE trajectory gaps, previously validated with larger starting enclosures.

The propagate narrowing function tries to prune the ODE trajectory through the reapplication of the interval step method over some time gap. This is heuristically chosen to contain the time point with the largest enclosure reduction since the previous application of the interval step method.

Figure 11.9 shows the pseudocode of function *propagateTrajectory*, used for the definition of the propagate narrowing function. It has 2 arguments, the ODE trajectory *TR* and a vector function *F(S,t)* (required by function *pruneGap*). It returns the trajectory *TR* possibly pruned by the interval step method applied from t_i to t_j. These time points are the bounds of a time gap chosen from the existing ODE trajectory by function *choosePropagationGap* (if no time gap is chosen by such function then the ODE trajectory is returned unchanged).

> **function** *propagateTrajectory* (an ODE trajectory *TR*, a vector function *F(S,t)*)
> (1) **if** *choosePropagationGap*(*TR*,t_i,t_j) **then return** *pruneGap*(*TR*,*F(S,t)*,t_i,t_j); **else return** *TR*; **endif**;
> **end function**

Figure 11.9 The definition of the *propagateTrajectory* function.

Function *choosePropagationGap* is shown in figure 11.10. It has an ODE trajectory *TR* as input argument and two *F*-numbers t_i and t_j as output arguments. If it finds some time gap to propagate then it succeeds and outputs t_i and t_j with the time gap bounds, otherwise it fails.

---

function *choosePropagationGap*(an ODE trajectory $TR=\langle TP,TG,TF,TB\rangle$, **out** F-numbers t_i, t_j)

(1) $bestP \leftarrow 1\text{-}MinProp$;

(2) $L_{\text{Gaps}} \leftarrow timeGaps(TR)$;

(3)   **repeat**

(4) $[t_{pi}..t_{pi+1}] \leftarrow L_{\text{Gaps}}.pop\_front()$;

(5)   **if** $TF(t_{pi}) \neq Unbound$ **then**

(6) $w = \prod_{1 \le j \le n} width(TP_j(t_{pi}))/width(TF_j(t_{pi}))$;

(7)   **if** $w < bestP$ **then** $t_i \leftarrow t_{pi}$; $t_j \leftarrow t_{pi+1}$; $bestP \leftarrow w$; **end if;**

(8)   **end if;**

(9)   **if** $TB(t_{pi+1}) \neq Unbound$ **then**

(10) $w = \prod_{1 \le j \le n} width(TP_j(t_{pi+1}))/width(TB_j(t_{pi+1}))$;

(11)   **if** $w < bestP$ **then** $t_i \leftarrow t_{pi+1}$; $t_j \leftarrow t_{pi}$; $bestP \leftarrow w$; **end if;**

(12)   **end if;**

(13) **until** $L_{\text{Gaps}}.size()=0$;

(14) **if** $bestP = 1\text{-}MinProp$ **then return** *false*; **else return** *true*; **end if;**

end function

Figure 11.10 The definition of the *choosePropagationGap* function.

The goal of function *choosePropagationGap* is to chose the best starting point and direction (from t_i to t_j) to reapply the interval step method, comparing the current enclosure at the starting point $TP(t_i)$ and the enclosures $TF(t_i)$ and $TB(t_i)$ at the previous application of the interval step method. The strategy adopted is to choose the starting point and direction with the smallest ratio between the sizes of the current and the previous enclosures.

Firstly (line 1), a threshold for deciding whether an enclosure reduction is propagated is initialised (*MinProp* is some positive value smaller than 1 indicating the minimum enclosure reduction necessary for triggering the propagation). Then, each time gap is analysed within a repeat cycle (lines 3-13) and the function succeeds if the ratio of the best choice is smaller than the defined threshold (line 14).

Each cycle computes for some time gap $[t_{pi}..t_{pi+1}]$ where the interval step method has already been applied (lines 5 and 9 ensure this) the ratios associated with the reapplication of the method starting at t_{pi} in the forward direction (line 6) and starting at t_{pi+1} in the backward direction (line 10). The best choice is updated whenever the associated ratio becomes smaller (lines 7 and 11).

The formal definition of the propagate narrowing function relies on function *propagateTrajectory* to reduce the uncertainty of ODE trajectory. The first argument of such function is the current value of the ODE trajectory TR, and the second argument is an interval vector function $f(S,t)$, which defines the ρ_{ODE} relation of the ODE constraint according to definition 10.1-1.

Definition 11.3.1-1 (Propagate Narrowing Function). Let CSDP $P=(X,D,C)$ be defined as in 10.1-1 and TR be the ODE trajectory representing the domain of x_{ODE}. The ODE constraint $c_{\text{ODE}} \in C$ has associated the following propagate narrowing function:

   (i) $NF_{\text{propagate}}(\langle TR,...\rangle) = \langle propagateTrajectory(TR,f(S,t)),...\rangle$ ❏

11.3.2 Link Narrowing Function

Whereas the propagate narrowing function reapplies the interval step method over some previously validated time gap, the link narrowing function tries to validate (link) some time gap for which the interval step method was never applied in either direction. As a consequence, besides the safe elimination from the ODE trajectory of functions incompatible with the ODE constraint, the time gap may become completely or partially validated (in this case, a new time point is inserted as described earlier).

Figures 11.11 and 11.12 show the pseudocode of function *linkTrajectory* and the auxiliary function *chooseUnlinkedGap* which are quite similar in structure to functions *propagateTrajectory* and *choosePropagationGap*, respectively.

---

function *linkTrajectory*(an ODE trajectory *TR*, a vector function *F(S,t)*)
  (1) **if** *chooseUnlinkedGap*(*TR*,t_i,t_j) **then return** *pruneGap*(*TR*,*F(S,t)*,t_i,t_j); **else return** *TR*; **end if**;
end function

Figure 11.11 The definition of the *linkTrajectory* function.

---

function *chooseUnlinkedGap*(an ODE trajectory *TR*=<*TP,TG,TF,TB*>, **out** *F*-numbers t_i, t_j)
  (1) *bestW1* ← +∞; *bestW2* ← +∞;
  (2) L_{Gaps} ← *timeGaps*(*TR*);
  (3) **repeat**
  (4) $[t_{pi}..t_{pi+1}]$ ←L_{Gaps}.*pop_front*();
  (5)   **if** *TF*(t_{pi})=*Unbound* **and** *TB*(t_{pi+1})=*Unbound* **then**
  (6)   *w1* ← *max*$_{1≤j≤n}$(*width*(*TP_j*(t_{pi}))); *w2* ← *max*$_{1≤j≤n}$(*width*(*TP_j*(t_{pi+1})))
  (7)   **if** *w2*<*w1* **then** *w*←*w2*; *w2* ← *w1*; *w1* ← *w*; *d* ← *backward*; **else** *d* ← *forward*; **end if**;
  (8)   **if** *w1* < *bestW1* **or** (*w1* = *bestW1* **and** *w2* < *bestW2*) **then**
  (9)   *bestW1* ←*w1*; *bestW2* ← *w2*; *direction* ← *d*; *T* ← $[t_{pi}..t_{pi+1}]$;
  (10)   **end if**;
  (11)   **end if**;
  (12) **until** L_{Gaps}.*size*()=0;
  (13) **if** *bestW1* = +∞ **then return** *false*;
  (14) **if** *direction* = *forward* **then** t_i ←*left*(*T*); t_j ← *right*(*T*); **end if**;
  (15) **if** *direction* = *backward* **then** t_i ←*right*(*T*); t_j ← *left*(*T*); **end if**;
  (16) **return** *true*;
end function

Figure 11.12 The definition of the *chooseUnlinkedGap* function.

The main difference is that the time gap chosen by the function *chooseUnlinkedGap* must be some unlinked gap (where no interval step method was ever applied in either direction). The goal of function *chooseUnlinkedGap* is to pick the best time gap (starting point and direction: from t_i to t_j) where to apply, for the first time, the interval step method. The chosen gap is that with the lowest uncertainty on the current enclosures at the starting point *TP*(t_i) and ending point *TP*(t_j) (line 6). The direction is defined accordingly (line 7).

During the repeat cycle (line 3-11) where each non linked (ensured in line 5) time gap is considered, whenever the smallest uncertainty *w1* is smaller than any previously found (or equal but with the other bound uncertainty *w2* smaller) the best choice is updated (lines 8-10).

If at the end of the cycle no time gap was chosen, the function fails (line 13). Otherwise it updates t_i and t_j with the bounds of the chosen gap according to the defined direction (lines 14 and 15) and succeeds (line 16).

The formal definition of the link narrowing function is based on function *linkTrajectory* to reduce the uncertainty of ODE trajectory. As in the case of the propagate narrowing function, the first argument is the current value of the ODE trajectory *TR*, and the second argument is an interval vector function $f(S,t)$, which defines the ρ_{ODE} relation.

Definition 11.3.2-1 (Link Narrowing Function). Let CSDP $P=(X,D,C)$ be defined as in
10.1-1 and TR be the ODE trajectory representing the domain of x_{ODE}. The ODE constraint
$c_{ODE} \in C$ has associated the following link narrowing function:

$\qquad$ (i) $NF_{link}(<TR,...>) = < linkTrajectory(TR,f(S,t)),...>$ $\qquad\qquad$ ❑

11.3.3   Improve Narrowing Functions

When the whole ODE trajectory is completely validated through the application of the
interval step method at every time gap, the precision of the enclosures obtained for each
time point agrees with the error tolerances imposed on the method. However, even with
accurate precision on the time point enclosures, there are no guarantees about the quality of
the enclosures associated with the time gaps that represent the set of ODE solution
functions. The reason is that the representation of the time gap enclosures as intervals
(rectangles in the two dimensional visualization) makes them unsuitable for an accurate
representation of the intermediate function values, in particular if the function is increasing
or decreasing (or both) along the time gap. To minimize this effect it is necessary to
partition the time gap into a set of smaller subintervals and compute the new enclosures
associated with each one (if some of these new enclosures is smaller than the original
enclosure then the uncertainty around the ODE trajectory is reduced).

All the narrowing functions responsible for reducing the domain of a restriction
variable (except the Value narrowing functions) depend on time gap enclosures of the ODE
trajectory (see subsections 11.2.2 through 11.2.5). Therefore, by reducing such time gap
enclosures, the restriction variable domain may eventually be narrowed. This is the goal of
an improve narrowing function, that is, to reduce some time gap enclosure that may later
trigger some other narrowing function associated with an ODE restriction and reduce the
domain of a restriction variable. The reduction of the time gap enclosure is achieved
through the insertion of a new intermediate time point within the gap and the subsequent
application of the interval step method linking this point with its adjacent neighbours.

Within a CSDP there are several improve narrowing functions (one for each ODE
restriction, except Value restrictions) which, according to definition 4.1-1, are associated
with the ODE constraint (in the sense that the values discarded by such functions are those
proved incompatible with the ρ_{ODE} relation through the reapplication of the interval step
method).

In order to decide whether it is worth to improve the ODE trajectory enclosure with
respect to some ODE restriction and to choose the most adequate time gap some heuristic
values are needed. In the following we will use function *heuristicValue* to obtain such
heuristics. This function, which has 3 input arguments, an ODE trajectory TR, an ODE
restriction c and an F-interval T, returns an heuristic value for the domain of the restriction
variable of c that is expected when the segment of TR included in T is enclosed with
maximum precision.

The rational for the implementation of such heuristic function is to consider an
evaluation procedure similar to the procedure used for pruning the domain of the restriction
variable of c from the ODE trajectory TR, which is applied on a changed trajectory TR'.
This trajectory is identical to TR except on the enclosures of the time gaps included in T.
Such enclosures are replaced with smaller enclosures that are expected to be obtained when
T is enclosed with maximum precision (above the lines connecting the lower bounds and
under the lines connecting the upper bounds of consecutive point enclosures). A possible
implementation of the *heuristicValue* function is exemplified in figure 11.13 for the case
where the restriction c is a Maximum restriction (other cases are similar).

---

function *heuristicValue*(an ODE trajectory *TR*, an ODE restriction *c*, an *F*-interval *T*)

(1) **if** $c \equiv Maximum_{j,[t_{p0}..t_{p1}]}(x_i)$ **then**

(2) $a \leftarrow -\infty; b \leftarrow -\infty;$

(3) $L_{\text{Gaps}} \leftarrow timeGaps_{[t_{p0}..t_{p1}]}(TR);$

(4)   **repeat**

(5) $[t_{\text{pi}}..t_{\text{pi+1}}] \leftarrow L_{\text{Gaps}}.pop\_front();$

(6)   **if** $[t_{\text{pi}}..t_{\text{pi+1}}] \subseteq T$ **then** $upper \leftarrow max(right(TR_i(t_{\text{pi}})),right(TR_i(t_{\text{pi+1}})));$

(7)   **else** $upper \leftarrow right(TR_i([t_{\text{pi}}..t_{\text{pi+1}}]));$ **end if;**

(8) $lower \leftarrow max(left(TR_i(t_{\text{pi}})),left(TR_i(t_{\text{pi+1}})));$

(9)   **if** $upper > b$ **then** $b \leftarrow upper;$

(10)   **if** $lower > a$ **then** $a \leftarrow lower;$

(11)   **until** $L_{\text{Gaps}}.size()=0;$

(12) **end if;**

 ⋮   ⋮

(…) **return** $[a..b];$

end function

Figure 11.13 The definition of the *heuristicValue* function.

The only difference with respect to definition 11.2.2-1 is in lines 6-7. Whenever a gap lies within *T* the computed maximum value is given by the maximum of the respective points upper bound instead of the entire gap upper bound.

Figure 11.14 shows the pseudocode of function *improveTrajectory*, used for the definition of the improve narrowing function. It has 4 arguments, the first 2 are the ODE trajectory *TR* and a vector function $F(S,t)$. The third and fourth arguments are an *F*-interval *I* and an ODE restriction *c*, where *I* represents the current domain of the restriction variable of *c*. It returns the ODE trajectory *TR* pruned by the application of the interval step method for linking a new time point that is inserted within some chosen time gap.

---

function *improveTrajectory*(an ODE trajectory *TR*, a vector function $F(S,t)$,
         an *F*-interval *I*, an ODE restriction *c*)

(1) **if** $nmbPoints(TR) = MaxPoints$ **then return** *TR*;

(2) $I_{\text{h}} \leftarrow heuristicValue(TR,c,[t_0..t_1]);$

(3) $\Delta_{\text{left}} \leftarrow left(I_{\text{h}}) - left(I);$

(4) $\Delta_{\text{right}} \leftarrow right(I) - right(I_{\text{h}});$

(5) **if** $\Delta_{\text{left}} \leq \varepsilon$ **and** $\Delta_{\text{right}} \leq \varepsilon$ **then return** *TR*;

(6) **if** $chooseInsertionGap(TR,c,T)$ **then**

(7) $t_i \leftarrow left(T); t_k \leftarrow \lfloor center(T) \rfloor; t_j \leftarrow right(T);$

(8) $TR \leftarrow insertPoint(TR,T,t_k);$

(9) $TR \leftarrow pruneGap(TR,F(S,t),t_i,t_k);$

(10)   **if** $TR \neq \varnothing$ **then** $TR \leftarrow pruneGap(TR,F(S,t),t_j,t_k);$ **end if;**

(11) **end if;**

(12) **return** *TR*;

end function

Figure 11.14 The definition of the *improveTrajectory* function.

Initially, the number of points considered in the ODE trajectory *TR* is checked. If it has already reached its maximum value (*MaxPoints*), *TR* is returned unchanged (line 1).

Lines 2-5 subsequently check whether it is worth to improve the ODE trajectory enclosure with respect to the ODE restriction *c*. The *heuristicValue* function (line 2) predicts the best possible narrowing of the domain I_{h} of the restriction value that would be obtained if the whole ODE trajectory enclosure were known with maximum precision. Then, this interval I_{h} is compared with the current domain *I* of the restriction variable. The improvement procedure is abandoned (line 5) if the maximum predicted gain does not exceed some predefined threshold ε in none of the bounds.

Otherwise, a time gap T is chosen by function *chooseInsertionGap* from the ODE trajectory (line 6), and a new intermediate time point, the mid point of time gap T (line 7), is inserted (line 8), and linked with its left bound (line 9). If the trajectory does not become empty, the time point is also linked to its right bound (line 10).

Function *chooseInsertionGap* is illustrated in figure 11.15. Besides the ODE trajectory TR, it has an ODE restriction c as input argument and outputs an F-interval T. If successful, T is updated with the time gap from the ODE trajectory which, according to the *heuristicValue* function, would narrow the most the domain of the restriction variable of c. It fails if any expected improvement on the ODE trajectory uncertainty is not able to sufficiently narrow the restriction variable domain.

---

function *chooseInsertionGap*(an ODE trajectory TR, an ODE restriction c, **out** an F-interval T)

(1) $bestW \leftarrow 1; T \leftarrow \varnothing;$

(2) $I \leftarrow heuristicValue(TR,c,\varnothing);$

(3) $L_{\text{Gaps}} \leftarrow timeGaps(TR);$

(4) **repeat**

(5) $\quad G \leftarrow L_{\text{Gaps}}.pop\_front();$

(6) $\quad$ **if not** $isCanonical(G)$ **then**

(7) $\quad\quad I_{\text{G}} \leftarrow heuristicValue(TR,c,G);$

(8) $\quad\quad w \leftarrow width(I_{\text{G}})/width(I);$

(9) $\quad\quad$ **if** $w < bestW$ **then** $T \leftarrow G; bestW \leftarrow w;$ **end if;**

(10) $\quad$ **end if;**

(11) **until** $L_{\text{Gaps}}.size()=0;$

(12) **if** $T = \varnothing$ **then return** *false*; **else return** *true*; **end if;**

end function

Figure 11.15 The definition of the *choosePropagationGap* function.

The value I for the restriction variable domain, that may be obtained from the ODE trajectory without any further information, is computed by the *heuristicValue* function with the empty set as third argument (line 2). The heuristic value I_{G} for the restriction variable domain, obtained from the ODE trajectory when considering the best predicted enclosure for some time gap G, is computed by the *heuristicValue* function with the interval G as third argument (line 7). Since the heuristic value I_{G} must be included within the current value I, a measure for the best predicted narrowing associated with G is given by the ratio between the widths of I_{G} and I (line 8). Analysing all the non canonical time gaps (line 6) within a repeat cycle (lines 4-11), the gap that minimises such ratio is chosen to update the output variable T.

The formal definition of an improve narrowing function regarding an ODE restriction c is based on function *improveTrajectory* where the first argument is the current value of the ODE trajectory TR, the second argument is the interval vector function $f(S,t)$, the third argument is the current domain I_i of the constraint variable and the fourth argument is the ODE restriction c_r.

Definition 11.3.3-1 (Improve Narrowing Functions). Let CSDP $P=(X,D,C)$ be defined as in 10.1-1, TR be the ODE trajectory representing the domain of x_{ODE} and I_i be the domain of x_i. For each non Value ODE restriction $c_r=(\langle x_{\text{ODE}},x_i\rangle,\rho_r)\in C$ the ODE constraint $c_{\text{ODE}}\in C$ has associated the following improve narrowing function:

$\quad$ (i) $NF_{\text{improve}(c_r)}(\langle TR,\ldots,I_i,\ldots\rangle) = \langle improveTrajectory(TR,F(S,t),I_i,c_r),\ldots,I_i,\ldots\rangle$ $\quad\square$

11.4 The Constraint Propagation Algorithm for CSDPs

The constraint propagation algorithm for pruning the domains of the CSDP variables is derived from the generic propagation algorithm for pruning the domains of the variables of a CCSP (see function *prune*, illustrated in figure 4.1).

Since there are no guarantees of monotonicity for the narrowing functions associated with the CSDP constraints, the order of their application may be crucial, not only for the efficiency of the propagation but also for the pruning achieved.

The strategy followed by the algorithm is to propagate as soon as possible any information related with the restriction variables and delay as much as possible the application of the narrowing functions for reducing the ODE trajectory uncertainty. The reason is that whereas the former are easy to deal with and may provide fast domain pruning, the latter may be computationally more expensive as they require the application of the interval step method.

Among the narrowing functions for the ODE trajectory uncertainty, the selection criterion favours the propagate narrowing function for spreading as soon as possible any domain reduction achieved by any other narrowing function. Moreover, since it does not make sense to try to improve an ODE trajectory that is not completely validated, the link narrowing function is always preferred to any of the improve narrowing functions.

Figure 11.16 shows the constraint propagation algorithm for CSDPs represented by function *pruneCSDP*. The first argument Q is a set of narrowing functions composed of all the narrowing functions associated with the constraints of the CSDP (see subsections 11.2 and 11.3). The second argument A is an element of the domains lattice representing the original variable domains (before applying the propagation algorithm). The result is a smaller (or equal) element of the domains lattice.

---

function *pruneCSDP*(a set Q of narrowing functions, an element A of the domains lattice)

(1) $Q_2 \leftarrow \{NF \in Q: NF$ is a propagate, link or improve narrowing function$\}$; $Q_1 \leftarrow Q \setminus Q_2$;

(2) $S_1 \leftarrow \varnothing$; $S_2 \leftarrow \varnothing$;

(3)   **while** $Q_1 \cup Q_2 \neq \varnothing$ **do**

(4)   **if** $Q_1 \neq \varnothing$ **then** choose $NF \in Q_1$;

(5)   **else if** $NF_{\text{propagate}} \in Q_2$ **then** $NF \leftarrow NF_{\text{propagate}}$;

(6)   **else if** $NF_{\text{link}} \in Q_2$ **then** $NF \leftarrow NF_{\text{link}}$;

(7)   **else** choose $NF \in Q_2$;

(8)   **end if**;

(9) $A' \leftarrow NF(A)$;

(10)   **if** $A' = \varnothing$ **then return** $\varnothing$;

(11) $P_1 \leftarrow \{ NF' \in S_1: \exists_{x \in \text{Relevant}_{NF'}} A[x] \neq A'[x] \}$;

(12) $P_2 \leftarrow \{ NF' \in S_2: \exists_{x \in \text{Relevant}_{NF'}} A[x] \neq A'[x] \}$;

(13) $Q_1 \leftarrow Q_1 \cup P_1$; $S_1 \leftarrow S_1 \setminus P_1$;

(14) $Q_2 \leftarrow Q_2 \cup P_2$; $S_2 \leftarrow S_2 \setminus P_2$;

(15)   **if** $NF \in Q_1$ **then** $Q_1 \leftarrow Q_1 \setminus \{NF\}$; $S_1 \leftarrow S_1 \cup \{NF\}$;

(16)   **else if** $A' = A$ **then** $Q_2 \leftarrow Q_2 \setminus \{NF\}$; $S_2 \leftarrow S_2 \cup \{NF\}$;

(17)   **end if**;

(18) $A \leftarrow A'$;

(19) **end while**

(20) **return** A ;

end function

Figure 11.16 The constraint propagation algorithm for CSDPs.

The structure of the algorithm is identical to the constraint propagation algorithm described in chapter 4. The main difference is that the set Q of narrowing functions is subdivided into two sets, Q_2 containing all the narrowing functions associated with the ODE constraint and Q_1 containing the remainder (line 1). Similarly, two different sets, S_1 associated with Q_1 and S_2 with Q_2, are maintained to keep track of the narrowing functions for which A is a fixed-point (lines 2,11-16). The selection criterion (line 4-8) only chooses a narrowing function from Q_2 if Q_1 is empty. In this case, it tries first the propagate narrowing function (line 5), then the link narrowing function (line 6), and only as a last option chooses it some improve narrowing function (line 7). After applying some narrowing function NF from Q_1 it is assumed that the obtained element A' is a fixed-point of NF since all the narrowing functions of Q_1 are necessarily idempotent (line 15). Such assumption is not made for the narrowing functions of Q_2 and so, it must be verified if the new element A' is equal to the previous element A (line 16).

The algorithm is correct and terminates. The correctness of the algorithm derives from the correctness of each narrowing function as proved in the case of the constraint propagation algorithm for CCSPs. The termination of the algorithm is only guaranteed by the imposition of a maximum number of points (*MaxPoints*) to consider for the ODE trajectory and can be proved by contradiction. Suppose that the algorithm does not terminate. Then, there are always some narrowing functions for which the current element A is not a fixed-point. If no new point is introduced in the ODE trajectory TR (only the link and improve narrowing functions may insert new points in TR) then the number of applications of the narrowing functions must be finite. This is justified because, in this case, either the current element A is a fixed-point of a narrowing function (and it is not applied) or its application will narrow some F-interval domain in the representation of A. Without considering new points, such representation is always the same finite set of F-intervals, whose lattice is finite and, as in the CCSP case, the process of obtaining a smaller element will necessary stop. So, the termination of the algorithm is only problematic if new points are inserted in the ODE trajectory TR. However the maximum number of points is limited to *MaxPoints* since, when this number is reached, both the link and the improve narrowing functions do not insert new points. Consequently, if the algorithm did not terminate before, then after reaching the maximum number of points no new point can be further introduced and, as proved above, the number of additional applications of the narrowing functions must be finite and the algorithm terminates.

The goal of a solving procedure for a CSDP is to prune as much as possible the initial domains of the restriction variables $I_1,\ldots,I_m$. This goal is achieved through function *pruneCSDP* with the ODE trajectory TR previously initialised by function *initializeTrajectory* (as described in section 11.2). This is illustrated in figure 11.17 in the pseudocode of function *solveCSDP*. From the initial domains of the restriction variables $I_1,\ldots,I_m$ it either returns the empty set, or narrows them further through the constraint propagation algorithm for CSDPs.

---

function *solveCSDP*($<I_1,\ldots,I_m>$)

  (1) $TR \leftarrow$ *initializeTrajectory*();

  (2) $A \leftarrow$ *pruneCSDP*($Q,< TR,I_1,\ldots,I_m>$);

  (3)   **if** $A = \varnothing$ **then return** $\varnothing$;

  (4)   **if** $A = < TR',I_1',\ldots,I_m'>$ **then return** $<I_1',\ldots,I_m'>$;

end function

Figure 11.17 The solving function associated with an CSDP.

11.5 Summary

In this chapter a procedure was described for solving a CSDP based on constraint propagation over a set of narrowing functions associated with the CSDP. An ODE trajectory was defined and represents an enclosure of the ODE solution set. All the narrowing functions, either for enforcing the ODE restrictions or for reducing the uncertainty of the ODE trajectory, were fully characterised. In the next chapter the extended interval constraints framework is used for solving several problems in different biomedical domains.

Chapter 12

Biomedical Decision Support with ODEs

Biomedical models provide a representation of the functioning of living organisms, making it possible to reason about them and eventually to take decisions about their state or adequate actions regarding some intended goals. Parametric differential equations are general and expressive mathematical means to model systems dynamics, and are suitable to express the deep modelling of many biophysical systems. Notwithstanding its expressive power, reasoning with such models may be quite difficult, given their complexity.

Analytical solutions are available only for the simplest models. Alternative numerical simulations require precise numerical values for the parameters involved, which are usually impossible to gather given the uncertainty on available data. This may be an important drawback since, given the usual non-linearity of the models, small differences on the input parameters may cause important differences on the output produced.

To overcome this limitation, Monte Carlo methods rely on a large number of simulations, that may be used to estimate the likelihood of the different options under study. However, they cannot provide safe conclusions regarding these options, given the various sources of errors that they suffer from, both input precision errors and round-of errors accumulated in the simulations.

In contrast with such methods, constraint reasoning assumes the uncertainty of numerical variables within given bounds and propagates such knowledge through a network of constraints on these variables, in order to decrease the underlying uncertainty. To be effective it must rely on advanced safe methods so that uncertainty is sufficiently bound as to be possible to make safe decisions.

The extended CCSP framework offers an alternative approach for modelling system dynamics with uncertain data as a set of constraints and provides reliable reasoning methods for supporting safe decisions. In this chapter, the expressive power of the extended CCSP framework is illustrated for decision support in several examples from biomedicine: diagnosis of diabetes (section 12.1), tuning of drug design (section 12.2) and epidemic studies (section 12.3).

12.1   A Differential Model for Diagnosing Diabetes

Diabetes mellitus prevents the body from metabolising glucose due to an insufficient supply of insulin. A glucose tolerance test (GTT) is frequently used for diagnosing diabetes. The patient ingests a large dose of glucose after an overnight fast and in the subsequent hours, several blood tests are made. From the evolution of the glucose concentration a diagnosis is made by the physicians.

Ackerman and al [1] proposed a well-known model for the blood glucose regulatory system during a GTT, with the following parametric differential equations:

$$\frac{dg(t)}{dt} = -p_1 g(t) - p_2 h(t) \qquad\qquad \frac{dh(t)}{dt} = -p_3 h(t) + p_4 g(t)$$

where g is the deviation of the glucose blood concentration from its fasting level;

h is the deviation of the insulin blood concentration from its fasting level;

p_1, p_2, p_3 and p_4 are positive, patient dependent, parameters.

Let $t=0$ be the instant immediately after the absorption of a large dose of glucose, g_0, when the deviation of insulin from the fasting level is still negligible. According to the model, the evolution of glucose and insulin blood concentrations is described by the trajectory of the above ODE system, with initial values $g(0)=g_0$ and $h(0)=0$, and depends on the parameter values p_1 to p_4.

Figure 12.1 shows the evolution of the glucose concentration for two patients with a glucose fasting level concentration of 110 mg glucose/100 ml blood. Immediately after the ingestion of an initial dose of glucose, the glucose concentration increases to 190 (i.e. $g_0 = 190\text{-}110 = 80$). The different trajectories are due to different parameters.

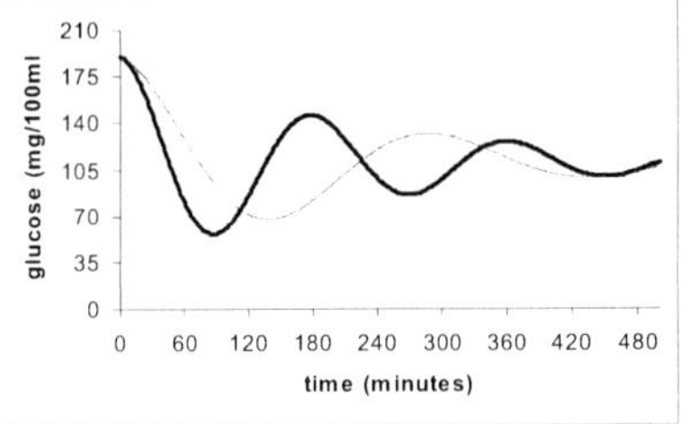

Figure 12.1 Evolution of the blood glucose concentration.

The general behaviour of the glucose trajectory (and insulin trajectory as well) oscillates around, and eventually converges to, the fasting concentration level. The natural period T of such trajectory is given (in minutes) by: $T = \dfrac{2\pi}{\sqrt{p_1 p_3 + p_2 p_4}}$

A criterion used for diagnosing diabetes is based on the natural period T, which is increased in diabetic patients. It is generally accepted that a value for T higher than 4 hours is an indicator of diabetes, otherwise normalcy is concluded.

We next show how the extended CCSP framework can be used to support the diagnosis of diabetes, possibly interrupting the sequence of blood tests if a safe decision can be made.

12.1.1   *Representing the Model and its Constraints with an Extended CCSP*

The decision problem regarding the diagnosis of diabetes may be modelled by an extended CCSP with a CSDP constraint and a numerical constraint.

The CSDP constraint relates the evolution of the glucose and insulin concentrations with the trajectory values obtained through the blood tests. It is associated with the following ODE system S based on the original system of differential equations but with the parameters included as new components with null derivatives:

$$S \equiv \begin{cases} s_1'(t) = & -s_3(t)s_1(t) - s_4(t)s_2(t) \\ s_2'(t) = & -s_5(t)s_2(t) + s_6(t)s_1(t) \\ s_3'(t) = & s_4'(t) = s_5'(t) = s_6'(t) = 0 \end{cases} \quad \text{where } s_1 \equiv g,\ s_2 \equiv h,\ s_3 \equiv p_1,\ s_4 \equiv p_2,\ s_5 \equiv p_3,\ s_6 \equiv p_4$$

If n blood tests were made at times $t_1,\ldots,t_n$, the constraint is defined by CSDP P_{Sn}, which includes the ODE constraint enforcing the trajectories to satisfy ODE system S between $t=0.0$ and $t=t_n$, together with Value restrictions representing each known trajectory component value. Variables g_0, h_0, p_1, p_2, p_3 and p_4, are the initial values and variables $g_{t1},\ldots,g_{tn}$, the glucose values at times $t_1,\ldots,t_n$.

CSDP $P_{Sn}=(X_n,D_n,C_n)$ where:

$$X_n = <x_{ODE},\quad g_0,\quad h_0,\quad p_1,\quad p_2,\quad p_3,\quad p_4,\quad g_{t1},\dots,\quad g_{tn}>$$

$$D_n = <D_{ODE}, Dg_0, Dh_0, Dp_1, Dp_2, Dp_3, Dp_4, Dg_{t1}, \dots, Dg_{tn}>$$

$$C_n = \{ \quad ODE_{S,[0.0\,..\,tn]}(x_{ODE}),\quad Value_{1,0.0}(g_0),\quad Value_{2,0.0}(h_0),$$
$$Value_{3,0.0}(p_1),\quad Value_{4,0.0}(p_2),\quad Value_{5,0.0}(p_3),\quad Value_{6,0.0}(p_4),$$
$$Value_{1,t1}(g_{t1}),\dots,\quad Value_{1,tn}(g_{tn}) \}$$

The numerical constraint is a simple equation relating the natural period with the ODE parameters according to its defining expression. With n blood tests performed, the resulting extended CCSP is:

CSDP $P_n=(X,D,C)$ where:

$$X = <\quad g_0,\quad h_0,\quad p_1,\quad p_2,\quad p_3,\quad p_4,\quad g_{t1},\dots,\quad g_{tn},\quad T>$$

$$D = < Dg_0, Dh_0, Dp_1, Dp_2, Dp_3, Dp_4, Dg_{t1}, \dots, Dg_{tn}, DT>$$

$$C = \{ P_{Sn}(g_0, h_0, p_1, p_2, p_3, p_4, g_{t1},\dots, g_{tn}),\ T = 2\pi/sqrt(p_1p_3+p_2p_4) \}$$

Here and in the remainder of this chapter we will use the following denotation: if a CSDP constraint is defined by the CSDP $P=(<x_{ODE},x_1,\dots,x_n><D_{ODE},Dx_1,\dots,Dx_n>,C)$ then it will be referred in an extended CCSP as $P\ (<x'_1,\dots,x'_n>)$ where $x'_1,\dots,x'_n$ are the CCSP variables that are shared by the CSDP.

12.1.2 *Using the Extended CCSP for Diagnosing Diabetes*

By solving the extended CCSP P_n with the initial variable domains set up to the available information, the natural period T will be safely bounded, and a guaranteed diagnosis can be made if T is clearly above or below the threshold of 240 minutes.

In the following we assume that the acceptable bounds for the parameter values are 50% above/below the typical normal values (p_1=0.0044, p_2=0.04, p_3=0.0045, p_4=0.03) and study two different patients, A and B, whose observed values agree with Figure 12.1.

The first blood test on patient A, performed 1 hour after the glucose ingestion, indicates a glucose deviation from the fasting level concentration of –29.8 (an error of ±0.05 is always considered with regard to the precision of the measuring process).

The extended CCSP P_1 (with a single blood test) is solved by enforcing Global Hull-consistency on the following initial variable domains:

Dp_1=[0.0022..0.0066], Dp_2=[0.0200..0.0600], Dp_3=[0.0022..0.0068], Dp_4=[0.0150..0.0450], Dg_0=[80.0], Dh_0=[0.0], Dg_{60}=[–29.85..–29.75], DT=[–∞..+∞]

Table 12.1 shows results for T obtained after 10, 30 and 60 minutes of CPU execution time[1] (with 10^{-6} precision).

Table 12.1 Narrowing results obtained for patient A from the information of the first blood test.

	10 minutes	30 minutes	60 minutes
T	[140.5..233.3]	[149.6..213.9]	[154.9..206.0]

After 10 minutes of CPU time (in fact after 7 minutes), the natural period is proved to be smaller than 240 minutes and a normal diagnosis can be guaranteed with no need of further examinations. When the next blood test were due, 60 minutes later, T was proved to be under 206, much less than the threshold for diagnosing diabetes.

---

[1] The tests of this chapter were all performed on a Pentium III with 128MBytes RAM running at 500MHz. The CPU execution times were divided by a factor of 3 to provide more realistic real time results that can be easily obtained by up-to-date computer configurations.

In patient B, the observed glucose deviation at the same first blood examination is 17.9. The initial domains for the variables of P_1 are thus the same of the previous case, except for the observed glucose value Dg_{60}=[17.85.. 17.95].

Enforcing Global Hull-consistency on P_1 with such information alone, no safe diagnosis can be attained before the next blood test (1 hour later). After 60 minutes of CPU time, T was proved to be within [236.4..327.9] and both diagnoses, normal or diabetic, are still possible, though diabetes is quite likely. Further information is required, and a second test is performed, indicating a glucose concentration of -38.9. The extended CCSP P_2 (two blood tests) is solved with the initial domains:

$$Dg_{60}=[17.85..17.95], \qquad Dg_{120}=[-38.95..-38.85], \qquad DT=[236.4..327.9]$$

In less than 20 minutes, T was proved to be above 240. One hour later, when the next examination would be due, T is clearly above such threshold ($T \in [245.0..323.8]$), and the patient is safely diagnosed as diabetic, requiring no further blood tests.

Note the importance of using a strong consistency requirement such as Global Hull-consistency. Table 12.2 presents the narrowing results, for patient A and B, obtained by enforcing Global Hull-consistency, together with those obtained by other strong consistency requirements such as 3B- and 4B-consistency (all with 10^{-6} precision). Each row shows the narrowing of T domain achieved by considering the number of blood tests specified in the first column, when the next blood examination is due.

Table 12.2 Narrowing of T domain achieved by enforcing 3B-, 4B- and Global Hull-consistency.

tests	3B	4B	Global Hull
1	[126.4..257.7]	[144.8..222.0]	[154.9..206.0]
2	[126.4..257.5]	—	—
3	[126.4..257.5]	—	—

Patient A

tests	3B	4B	Global Hull
1	[191.2..344.9]	[211.8..340.7]	[236.4..327.9]
2	[192.6..344.9]	[228.9..340.7]	[245.0..323.8]
3	[192.6..344.9]	[232.1..339.9]	—

Patient B

Clearly, if 3B-consistency is enforced, the results obtained are insufficient to make any safe decision. In fact, for both patients, even after the 3rd blood examination, no safe diagnosis can be made since, in either cases, the interval obtained for the T domain contains the diagnosis threshold of 240 minutes.

Enforcing 4B-consistency (stronger than 3B-consistency) patient A can be safely diagnosed with the information of the first examination alone, requiring no further blood tests. For this patient case, and comparing with the Global Hull-consistency results, the reduction of the T domain is about 3 times slower (only after 23 minutes can it be proved that the value of T is under 240) and 50% wider, but a safe diagnosis still can be made. However, in the case of patient B, the pruning achieved (with T domain about 40% wider than when enforcing Global Hull-consistency) is again insufficient to make a safe diagnosis in time for avoiding further blood examinations.

12.2   A Differential Model for Drug Design

Pharmacokinetics studies the time course of drug concentrations in the body, how they move around it and how quickly this movement occurs. Oral drug administration is a widespread method for the delivery of therapeutic drugs to the blood stream. This section is based on the following two-compartment model of the oral ingestion/gastro-intestinal absorption process (see [130] and [138] for details):

$$\frac{dx(t)}{dt} = -p_1 x(t) + D(t) \qquad\qquad \frac{dy(t)}{dt} = p_1 x(t) - p_2 y(t)$$

where x is the concentration of the drug in the gastro-intestinal tract;

y is the concentration of the drug in the blood stream;

D is the drug intake regimen; p_1 and p_2 are positive parameters.

The model considers two compartments, the gastro-intestinal tract and the blood stream. The drug enters the gastro-intestinal tract according to a drug intake regimen, described as a function of time $D(t)$. It is then absorbed into the blood stream at a rate, p_1, proportional to its gastro-intestinal concentration and independently from its blood concentration. The drug is removed from the blood through a metabolic process at a rate, p_2, proportional to its concentration there.

The drug intake regimen $D(t)$ depends on several factors related with the production of the drug by the pharmaceutical company. We assume that the drug is taken on a periodic basis (every six hours), providing a unit dosage that is uniformly dissolved into the gastro-intestinal tract during the first half hour. Consequently, for each period of six hours the intake regimen is defined as:

$$D(t) = \begin{cases} 2 & \textit{if } 0.0 \leq t \leq 0.5 \\ 0 & \textit{if } 0.5 < t \leq 6.0 \end{cases}$$

The effect of the intake regimen on the concentrations of the drug in the blood stream during the administration period is determined by the absorption and metabolic parameters, p_1 and p_2. Maintaining the above intake regimen, the solution of the ODE system asymptotically converges to a six hours periodic trajectory called the limit cycle, shown in figure 12.2 for specific values of the ODE parameters.

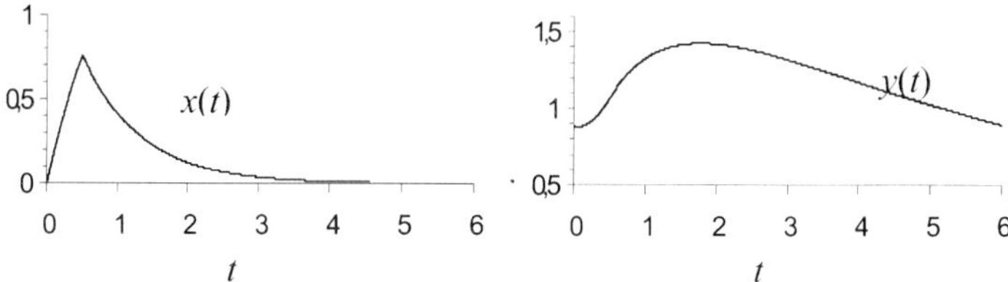

Figure 12.2 The periodic limit cycle with $p_1=1.2$ and $p_2=\ln(2)/5$.

In designing a drug, it is necessary to adjust the ODE parameters to guarantee that the drug concentrations are effective, but causing no significant side effects. In general, it is sufficient to guarantee some constraints on the concentrations over a limit cycle.

One constraint is to keep the drug concentration at the blood within predefined bounds, namely to prevent its maximum value (the Peak Concentration) to exceed a threshold associated with a side effect. Other constraint imposes bounds on the area under the curve of the drug blood concentration (known as AUC) guaranteeing that the accumulated dosage is high enough to be effective. Finally, bounding the total time that such concentration remains above or under some threshold is an additional requirement for controlling drug concentration during the limit cycle. Figure 12.3 shows maximum, minimum, area (≥ 1.0) and time (≥ 1.1) values for the limit cycle of figure 12.2.

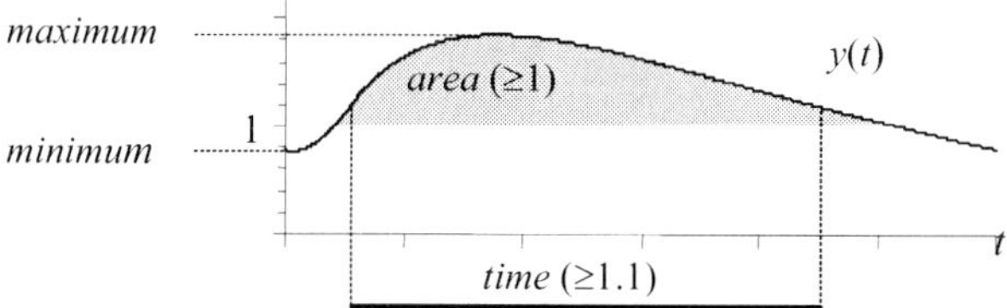

Figure 12.3 Maximum, minimum, area and time values at the limit cycle ($p_1=1.2$ and $p_2=\ln(2)/5$).

We show below how the extended CCSP framework can be used for supporting the drug design process. We will focus on the absorption parameter, p_1, which may be adjusted by appropriate time release mechanisms (the metabolic parameter p_2, tends to be characteristic of the drug itself and cannot be easily modified). The tuning of p_1 should satisfy the following requirements on the drug concentration in the blood during the limit cycle, namely:

 (i) Its "instantaneous" value bounded between 0.8 and 1.5;
 (ii) Its area under the curve (and above 1.0) bounded between 1.2 and 1.3;
 (iii) Its value above 1.1 should not last more than 4 hours.

12.2.1 *Representing the Model and its Constraints with an Extended CCSP*

The expressive power of the extended CCSP framework allows its use for representing the limit cycle and the different requirements illustrated in figure 12.3. Due to the intake regimen definition $D(t)$, the ODE system has a discontinuity at time $t=0.5$, and is represented by two CSDP constraints, P_{S1} and P_{S2}, in sequence.

The first, P_{S1}, ranges from the beginning of the limit cycle ($t=0.0$) to time $t=0.5$, and a second P_{S2}, is associated to the remaining trajectory of the limit cycle (until $t=6.0$). S1 and S2 are the corresponding ODE systems, where p_1 and p_2 are included as new components with null derivatives and the intake regimen $D(t)$ is a constant:

$$S1 \equiv \begin{cases} s'_1(t) = & -s_3(t)s_1(t) + 2 \\ s'_2(t) = & s_3(t)s_1(t) - s_4(t)s_2(t) \\ s'_3(t) = & s'_4(t) = 0 \end{cases} \qquad S2 \equiv \begin{cases} s'_1(t) = & -s_3(t)s_1(t) \\ s'_2(t) = & s_3(t)s_1(t) - s_4(t)s_2(t) \\ s'_3(t) = & s'_4(t) = 0 \end{cases}$$

The CSDP constraints for P_{S1} are defined as shown below (P_{S2} is similar). Besides the ODE constraint, Value, Maximum, Minimum, Area and Time restrictions associate variables with different trajectory properties relevant in this problem. Variables x_{init}, y_{init}, p_1 and p_2 are the initial trajectory values, and x_{fin} and y_{fin} are the final trajectory values of the 1st and 2nd components. Variables y_{max} and y_{min} are the maximum and minimum trajectory values of the 2nd component (drug concentration in the blood stream) for this period. Variables y_a and y_t denote the area above 1.0 and the time above 1.1 of the 2nd component in this same period.

 CSDP $P_{S1} = (X_1, D_1, C_1)$ where:

$X_1 = <\; x_{ODE}, \quad x_{init}, \quad y_{init}, \quad p_1, \quad p_2, \quad x_{fin}, \quad y_{fin}, \quad y_{max}, \quad y_{min}, \quad y_a, \quad y_t\;>$

$D_1 = <D_{ODE}, Dx_{init}, Dy_{init}, Dp_1, Dp_2, Dx_{fin}, Dy_{fin}, Dy_{max}, Dy_{min}, Dy_a, Dy_t>$

$C_1 = \{ODE_{S1,\,[0.0\,..\,0.5]}(x_{ODE}),$

  $Value_{1,\,0.0}(x_{init}),\; Value_{2,\,0.0}(y_{init}),\; Value_{3,\,0.0}(p_1),\; Value_{4,\,0.0}(p_2),$

  $Value_{1,\,0.5}(x_{fin}),\; Value_{2,\,0.5}(y_{fin}),$

  $Maximum_{2,\,[0.0\,..\,0.5]}(y_{max}),\; Minimum_{2,\,[0.0\,..\,0.5]}(y_{min}),$

  $Area_{2,\,[0.0\,..\,0.5],\,\geq 1.0}(y_a),\; Time_{2,\,[0.0\,..\,0.5],\,\geq 1.1}(y_t)\}$

The extended CCSP P connects the two ODE segments in sequence by assigning the same variables x_{05} and y_{05} to both the final values of P_{S1} and the initial values of P_{S2} (parameters p_1 and p_2 are shared by both constraints). Moreover, the 6 hours period is guaranteed by the assignment of the same variables x_0 and y_0 to both the initial values of P_{S1} and the final values of P_{S2}. Besides considering all the restriction variables $(y_{max},\ldots,y_t)$ of each ODE segment, new variables for the whole trajectory y_{area} and y_{time} sum the values in each segment.

CCSP $P=(X,D,C)$ where:

$$X = <\ x_0,\ y_0,\ p_1,\ p_2,\ x_{05},\ y_{05},\ y_{max1},\ y_{max2},\ y_{min1},\ y_{min2},\ y_{a1},\ y_{a2},\ y_{area},\ y_{t1},\ y_{t2},\ y_{time}>$$

$$D=\ <Dx_0,Dy_0,Dp_1,Dp_2,Dx_{05},Dy_{05},Dy_{max1},Dy_{max2},Dy_{min1},Dy_{min2},Dy_{a1},Dy_{a2},Dy_{area},Dy_{t1},Dy_{t2},Dy_{time}>$$

$$C = \{\,P_{S1}(x_0,\ y_0,\ p_1,\ p_2, x_{05}, y_{05},\ y_{max1},\ y_{min1}, y_{a1}, y_{t1}),$$
$$P_{S2}(x_{05}, y_{05}, p_1, p_2,\ x_0,\ y_0,\ y_{max2},\ y_{min2}, y_{a2}, y_{t2}),$$
$$y_{area} = y_{a1} + y_{a2},\qquad y_{time} = y_{t1} + y_{t2}\,\}$$

12.2.2 Using the Extended CCSP for Parameter Tuning

The tuning of drug design may be supported by solving P with the appropriate set of initial domains for its variables. We will assume p_2 to be fixed to a five-hour half live ($Dp_2=[ln(2)/5]$) and p_1 to be adjustable up to about ten-minutes half live ($Dp_1=[0..4]$). The initial value x_0, always very small, is safely bounded in interval $Dx_0=[0.0..0.5]$.

The assumptions about the parameter ranges together with the bounds imposed by the above requirements justify the following initial domains for the variables of P (all the remaining variable domains are unbounded):

$$Dx_0=\ [0.0 .. 0.5],\qquad Dy_0 =\ [0.8 .. 1.5],\qquad Dp_1 =\ [0.0 .. 4.0],\qquad Dp_2 =\ [ln(2)/5],$$
$$Dy_{max1}=[0.8 .. 1.5],\qquad Dy_{max2}=[0.8 .. 1.5],\qquad Dy_{min1}=[0.8 .. 1.5],\qquad Dy_{min2}=[0.8..1.5],$$
$$Dy_{area}= [1.2 .. 1.3],\qquad Dy_{time}= [0.0 .. 4.0]$$

Solving the extended CCSP P (enforcing Global Hull-consistency), with a precision of 10^{-3}, narrows the original p_1 interval to $[1.191..1.543]$ in less than 3 minutes. Hence, for p_1 outside this interval the set of requirements cannot be satisfied.

This may help to adjust p_1 but offers no guarantees on specific choices within the obtained interval. For instance, the two extreme canonical solutions for p_1, $[1.191.. 1.192]$ and $[1.542..1.543]$, contain no real solution, since when solving the problem with a higher precision (10^{-6} - which took about 18 minutes of CPU time), the domain of p_1 is narrowed to $[1.209233..1.474630]$ that does not include the above canonical solutions (obtained with the lower 10^{-3} precision).

Nevertheless, using CCSP P with different initial domains, may produce guaranteed results for particular choices of the p_1 parameter values. For example, for $p_1 \in [1.3..1.4]$ (an acceptable uncertainty in the manufacturing process), and the following initial domains (the remaining are unbounded):

$$Dx_0=[0.0..0.5],\qquad Dy_0=[0.8..1.5],\qquad Dp_1=[1.3..1.4],\qquad Dp_2=[ln(2)/5]$$

Global Hull-consistency on P (with 10^{-3} precision) narrows the following, initially unbounded, domains to:

$$y_{min1} \in [0.881..0.891],\qquad y_{max1} \in [1.090..1.102],\qquad y_{area} \in [1.282..1.300],$$
$$y_{min2} \in [0.884..0.894],\qquad y_{max2} \in [1.447..1.462],\qquad y_{time} \in [3.908..3.967].$$

Notwithstanding the uncertainty, these results do prove that with p_1 within $[1.3..1.4]$, all limit cycle requirements are safely guaranteed (the obtained bounds are well within the requirements). Moreover, they offer some insight on the requirements showing, for instance, the area requirement to be the most critical constraint.

The above bounds were obtained in about 13 minutes. However, faster results may be obtained if the goal is simply to check whether the requirements are met. Since Global Hull-consistency is enforced by an any time algorithm, its execution may be interrupted as soon as the requirements are satisfied (10 minutes in this case).

A better approach in this case would be to prove that the CCSP P with the initial domains $Dx_0=[0.0..0.5]$, $Dy_0=[0.8..1.5]$, $Dp_1=[1.3..1.4]$ and $Dp_2=[ln(2)/5]$ together with each of the following domains cannot contain any solution (again, the remaining domains are kept unbound):

$$Dy_{\max 1}=[1.5..+\infty], \quad Dy_{\max 2}=[1.5..+\infty], \quad Dy_{\min 1}=[-\infty..0.8], \quad Dy_{\min 2}=[-\infty..0.8],$$
$$Dy_{\text{area}}=[1.3..+\infty], \quad Dy_{\text{area}}=[-\infty..1.2], \quad Dy_{\text{time}}=[4.0..+\infty].$$

By independently proving that no solutions exist for the above 7 problems, which cover all non satisfying possibilities, it is proved that all the requirements are necessarily satisfied. This was achieved in less than 5 minutes.

Again the requirement of a strong consistency is important for obtaining good pruning results. A possible alternative would be to enforce 3B-consistency. In this case, the initial domain of parameter p_1 would be narrowed from [0..4] to [1.158..1.577] which is 20% wider than the obtained with Global Hull-consistency, but could be obtained faster (in about 1 minute and half). Despite providing the same p_1 domain reduction, 4B-consistency is not a good alternative for Global Hull-consistency since it is 4 times slower (about 14 minutes of CPU execution time).

12.3   The SIR Model of Epidemics

The time development of epidemics is the subject of many mathematical models that have been proved useful for the understanding and control of infectious diseases. The SIR model [102] is a well known model of epidemics which divides a population into three classes of individuals and is based of the following parametric ODE system:

$$\frac{dS(t)}{dt} = -rS(t)I(t) \qquad\qquad \frac{dI(t)}{dt} = rS(t)I(t) - aI(t) \qquad\qquad \frac{dR(t)}{dt} = aI(t)$$

where S are the susceptibles - individuals who can catch the disease;

I are the infectives - individuals who have the disease and can transmit it;

R are the removed - individuals who had the disease and are immune or died;

r and a are positive parameters.

The model assumes that the total population N is constant ($N=S(t)+I(t)+R(t)$) and the incubation period is negligible. Parameter r accounts for the efficiency of the disease transmission (proportional to the frequency of contacts between susceptibles and infectives). Parameter a measures the recovery (removing) rate from the infection.

Important questions in epidemic situations are: whether the infection will spread or not; what will be the maximum number of infectives; when will it start to decline; when will it ends; and how many people will catch the disease.

Figure 12.4 shows the number of susceptibles, infectives and removed as a function of time, as predicted by the SIR model with $S(0)=762$, $I(0)=1$, $R(0)=0$, $r=0.00218$ and $a=0.44$. In this case, the infection will spread up to a maximum number of infected of about 294 individuals ($i_{\max}$), starting to decline after 6.5 days ($t_{\max}$), ending after 22.2 days (t_{end}) and affecting a total of 744 individuals (r_{end}).

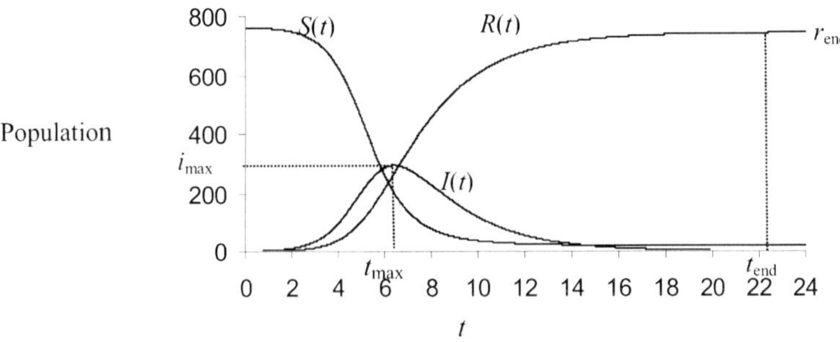

Figure 12.4 SIR model predictions with $S(0)=762$, $I(0)=1$, $R(0)=0$, $r=0.00218$ and $a=0.44036$.

Frequently, there is information available about the spread of a disease on a particular population. This is usually gathered as series of time-infectives (t_i,I_i) or time-removed (t_i,R_i) data points together with the values (t_0,S_0), (t_0,I_0) or (t_0,R_0) that initiated the epidemics on the population. An important problem is to predict the behaviour of a similar disease (with similar parameter values) when occurring in a different environment, namely with a different population size or a different number of initial infectives.

The following study is based on data reported in the British Medical Journal (4th March 1978) from an influenza epidemic that occurred in an English boarding school (taken from [102]): a single boy (from a total population of 763) initiated the epidemics and the evolution of the number of infectives, available daily, from day 3 to day 14, is shown in table 12.3.

Table 12.3 Infectives reported during an epidemics in an English boarding school.

t	0	3	4	5	6	7	8	9	10	11	12	13	14
I_t	1	22	78	222	300	256	233	189	128	72	28	11	6

The goal of our study is to predict what would happen if a similar disease occurs in a different place, say a small town with a population of about 10000 individuals. Moreover, if there is a vaccine to that disease, what would be the vaccination rate necessary to guarantee that the maximum number of infectives never exceeds some predefined threshold, for example, half of the total population.

12.3.1 Using the Extended CCSP for Predicting the Epidemic Behaviour

The first step for solving the above problem is to characterize an epidemic disease which is similar to the one reported in the boarding school. The classical approach would be to perform a numerical best fit approximation to compute the parameter values r' and a' that minimize the residual:

$$\sum_{j=1}^{m}\left(I(t_j)-It_j\right)^2$$

where $I_{t_1},...,I_{t_m}$ are the infectives observed at times $t_1,...,t_m$, and $I(t_1),...,I(t_m)$ their respective values predicted by the SIR model with $r=r'$ and $a=a'$. In [102] this method is used to compute $r=0.00218$ and $a=0.44036$ with a residual of about 4221 (figure 12.4 shows the best fit solution).

However, generating a single value for each parameter does not capture the essence of the problem which is not to determine the most similar disease but rather to reason with a set of similar enough diseases. Moreover such approach does not provide any sensitive analysis about the quality of the data fitting, namely on the effects of small changes on the parameter values.

An alternative, possible in a constraints framework, is to relax the imposition of the "best" fit and merely impose a "good" fit. This can be achieved either by considering acceptable errors ε_j for each observed data and computing ranges for the parameters such that the distance between the model predictions and the observed data does not exceed these errors or by imposing some upper bound on the residual value (or any other measure of the unfitness of the model).

Either the first approach, known as the data driven inverse problem, or the second approach, denoted here as the maximum residual problem, cannot be solved by classical constraint approaches since the epidemic model has no analytical solution form.

However, both problems can be represented as extended CCSPs, P_1 and P_2, respectively, which include a CSDP constraint P_S, representing the evolution of the susceptibles and infectives during the reported period of time (the first 14 days). The associated ODE system S is composed by the first two components of the SIR model together with two extra components with null derivatives for representing the parameters[2]:

$$S \equiv \begin{cases} s_1'(t) = & -0.01 s_3(t) s_1(t) s_2(t) \\ s_2'(t) = & 0.01 s_3(t) s_1(t) s_2(t) - s_4(t) s_2(t) \\ s_3'(t) = & s_4'(t) = 0 \end{cases}$$

CSDP P_S contains several Value restrictions for associating variables with: the initial values of the susceptible (s_0) and infective (i_0); the parameter values (r and a); and the values of the infective at times $3,\dots,14$ ($i_3,\dots,i_{14}$).

CSDP $P_S = (X,D,C)$ where:

$$X = <\ x_{ODE},\quad s_0,\quad i_0,\quad r,\quad a,\quad i_3,\ \dots,\quad i_{14}\ >$$
$$D = <\ D_{ODE},\ Ds_0,\ Di_0,\ Dr,\ Da,\ Di_3,\ \dots,\ Di_{14}\ >$$
$$C = \{\ ODE_{S,\ [0.0\ ..\ 14.0]}(x_{ODE}),$$
$$Value_{1,\ 0.0}(s_0),\ Value_{2,\ 0.0}(i_0),\ Value_{3,\ 0.0}(r),\ Value_{4,\ 0.0}(a),$$
$$Value_{2,\ 3.0}(i_3),\ \dots,\ Value_{2,\ 14.0}(i_{14})\ \}$$

The extended CCSP P_1, which represents the data driven inverse problem, contains a single constraint defined as CSDP P_S. The extended CCSP P_2, which represents the maximum residual problem, besides CSDP constraint P_S, contains also a numerical constraint defining the residual (R) from the variables $i_3,\dots,i_{14}$ and the observed values (represented as constants $k_3,\dots,k_{14}$).

CCSP $P_1=(X_1,D_1,C_1)$ where:

$$X_1 =<\ s_0,\quad i_0,\quad r,\quad a,\quad i_3,\dots,\quad i_{14}>$$
$$D_1 =<\ Ds_0,\ Di_0,\ Dr,\ Da,\ Di_3,\dots,\ Di_{14}>$$
$$C_1 = \{\ P_S(s_0,\ i_0,\ r,\ a,\ i_3,\dots,\ i_{14})\ \}$$

CCSP $P_2=(X_2,D_2,C_2)$ where:

$$X_2 =<\ s_0,\quad i_0,\quad r,\quad a,\quad i_3,\dots,\quad i_{14},\quad R >$$
$$D_2 =<\ Ds_0,\ Di_0,\ Dr,\ Da,\ Di_3,\dots,\ Di_{14},\ DR >$$
$$C_2 = \{\ P_S(s_0,\ i_0,\ r,\ a,\ i_3,\dots,\ i_{14}),\ R = \Sigma(i_j - k_j)^2\ \}$$

Assuming very wide initial parameter ranges ($Dr=Da=[0..1]$), the "good" fit requirement can now be enforced by solving either P_1 or P_2 with appropriate initial domains for the remaining variables (the values of the susceptible and infective are initialized accordingly to the report, $Ds_0=[762]$ and $Di_0=[1]$). In the case of P_1, each Di_j should be initialized with the interval $[\lfloor k_j - \varepsilon_j \rfloor .. \lceil k_j + \varepsilon_j \rceil]$ (for example with $\varepsilon_j=30$). In the case of P_2, all Di_j are kept unbounded, but the residual initial domain DR must be upper bounded (for example with $DR=[0..4800]$).

Enforcing Global Hull-consistency (with precision 10^{-6}) on P_1 with $\varepsilon_j=30$, the parameter ranges are narrowed from $[0..1]$ to $r \in [0.214..0.222]$ and $a \in [0.425..0.466]$ in about 50 minutes. Identical narrowing would be obtained by enforcing Global Hull-consistency (with precision 10^{-6}) on P_2 with $DR=[0..4800]$: $r \in [0.213..0.224]$ and $a \in [0.423..0.468]$.

As in the case of the preceding problems, the enforcing of less strong consistency requirements such as 3B- or 4B-consistency is not a good alternative. For example, enforcing 3B-consistency on P_1 the r domain is only reduced to $[0.143..0.495]$ (43 times wider than with Global Hull-consistency) and the domain of a cannot be pruned at all. Enforcing 4B-consistency on P_1 the domains reduction is similar to the obtained with Global Hull-consistency, but the execution time is much slower (the execution was interrupted after 5 hours of CPU time).

---

[2] In the equations r is multiplied by 0.01 re-scaling it to the interval $[0..1]$ (its best fit value 0.00218 is re-scaled to 0.218).

Once obtained the parameter ranges that may be considered acceptable to characterize epidemic diseases similar to the one observed, the next step is to use them for making predictions in the new context of a population of 10000 individuals.

In this case a single CSDP constraint $P_{S'}$ represents the first two components of the model together with ODE restrictions associating variables with the predicted values (besides the Value restrictions to associate variables with the parameter values r and a and the initial values s_0 and i_0). A Maximum restriction represents the infectives maximum value i_{max} and a First Maximum restriction represents the time of such maximum t_{max}. A Last Value restriction represents the duration t_{end} of the epidemics as the last time that the number of infectives exceeds 1. Finally a Value restriction represents the number of people s_{25} that are still susceptible at a time (25) safely after the end of the epidemics.

CSDP $P_{S'} = (X,D,C)$ where:

$$X = < x_{ODE}, \quad s_0, \quad i_0, \quad r, \quad a, \quad i_{max}, \quad t_{max}, \quad t_{end}, \quad s_{25} >$$
$$D = <D_{ODE}, Ds_0, Di_0, Dr, Da, Di_{max}, Dt_{max}, Dt_{end}, Ds_{25} >$$
$$C = \{ ODE_{S,[0.0\,..\,25.0]}(x_{ODE}),$$
$$Value_{1,0.0}(s_0), Value_{2,0.0}(i_0), Value_{3,0.0}(r), Value_{4,0.0}(a),$$
$$Maximum_{2,[0.0..25.0]}(i_{max}), firstMaximum_{2,[0.0..25.0]}(t_{max}),$$
$$lastValue_{2,[0.0..25.0],\geq 1.0}(t_{end}), Value_{1,25.0}(s_{25}) \}$$

Solving such problem with the parameters ranging within the previously obtained intervals (for example, $Dr=[0.213..0.224]$ and $Da=[0.423..0.468]$), the initial value domains $Ds_0=[9999]$ and $Di_0=[1]$, and all the other variable domains unbounded, the results obtained for these domains indicated that:

$i_{max} \in [8939..9064]$ clearly suggesting the spread of a severe epidemics;

$t_{max} \in [0.584..0.666]$ and $t_{end} \in [20.099..22.405]$ predicting that the maximum will occur during the first 14 to 16 hours, starting then to decline and ending before the 10th hour of day 22;

$s_{25} \in [0..0.001]$ showing that everyone will eventually catch the disease.

If the administration of a vaccine is considered at a rate λ proportional to the number of susceptibles then, the differential model must be modified into:

$$\frac{dS(t)}{dt} = -rS(t)I(t) - \lambda S(t) \qquad \frac{dI(t)}{dt} = rS(t)I(t) - aI(t) \qquad \frac{dR(t)}{dt} = aI(t) + \lambda S(t)$$

The requirement that the maximum number of infectives cannot exceed half of the population is represented by adding the numerical constraint $i_{max} \leq 5000$. Solving this new CCSP with the λ initial domain $[0..1.5]$, its lower bound is raised up to 0.985 indicating that at least such vaccination rate is necessary to satisfy the requirement.

12.4   Summary

In this chapter the potentiality of the extended interval constraints framework was tested for solving decision problems based on differential models. The diagnosis of diabetes, the tuning of drug design and an epidemic study were effectively supported. Such examples illustrated the expressive power of CSDP restrictions strengthening the importance of our contribution for the integration of biophysical models in decision support. In the next chapter the contributions of this work are analysed and directions for future work are suggested.

Chapter 13

Conclusions and Future Work

In this chapter we overview each individual contribution of this work, discuss its usefulness and applicability, and identify related directions for further research. In the last section we analyse the global value of our contribution to the more ambitious goal of integrating biophysical models within decision support.

13.1   Interval Constraints for Differential Equations

The interval constraints framework was extended to handle Constraint Satisfaction Differential Problems (CSDPs), which, for the first time, provide the full integration of ordinary differential equations in constraint reasoning. Many existing ordinary differential models may be represented as CSDP constraints and combined with other constraints in a constraint model. Moreover, its expressive power, enhanced with non-conventional restrictions, may encourage the development of new differential models that cannot be handled by traditional techniques.

Previous versions of our approach were already proposed in [36, 37, 38], but they were specially developed for handling initial value problems, and lacked the formalism and adequate methodology for the integration of differential models into constraint reasoning. Only recently such an integration could be achieved and evaluated by both the mathematical community [41] and the constraint community [42][1], with positive results.

Despite its encouraging results the approach can still be further improved and several research directions are worth exploring.

Firstly, new kinds of ODE restrictions could be considered. The type of restrictions considered in the CSDP formalism were derived from our experience in the biomedical context, namely from the representation needs evidenced by the underlying differential models. However, the expressive power can be improved by allowing new types of restrictions more adequate for different kinds of differential models. For instance, a new restriction could be considered for representing the period of an ODE.

Secondly, different representations of the ODE system could be supported. An assumption of our approach is the continuity of the right hand side of the differential equations. When there are discontinuities, a sequence of CSDP constraints is considered, new variables are added for representing the restrictions at each continuous segment, and new constraints are included to combine them and obtain its global value (see subsection 12.2.1). However, maximum, minimum, first and last restrictions cannot be easily combined in numerical constraints without comparison operators. An useful direction for research would be to extend the CSDP framework to allow the definition by segments of the ODE system.

Thirdly, alternative solving procedures could be implemented. The procedure developed for solving a CSDP relies on an Interval Taylor Series (ITS) method which was originally conceived for solving initial value problems (IVPs), obtaining reliable enclosures

---

[1] "This paper is an interesting contribution to the handling of differential equations in constraint programming that goes beyond simple initial values problems" (a citation from a reviewer of CP'2003).

along the ODE trajectory. One limitation of our approach is a direct consequence of such a method, which requires that at least one time point should be completely confined (even if within wide bounds). Research could be accomplished for allowing other complementary solving methods. For instance, methods used for solving boundary problems could be studied. Moreover, from the different reliable approaches for solving IVPs, the ITS direct method (see subsection 9.2.3) was selected due to its simplicity of implementation rather than its efficiency. The application of more competitive approaches, such as the Hermite-Obreschkoff interval method [103] or the Hermite interpolation constraint method [81], should be considered, or alternatively, new Runge-Kutta interval methods could be developed.

13.2   Global Hull-consistency

The interval constraints framework was extended with the new consistency criterion of Global Hull-consistency, for which several enforcing algorithms were developed. Among such algorithms, the Tree Structured Algorithm (TSA) presented the best performances offering the advantages of keeping a tree-based compact description of the feasible space and providing any-time results. Constraint problems, for which weaker consistency criteria provide insufficient pruning of variable domains, may be solved, with reasonable computing costs, by enforcing Global Hull-consistency. Such improvement on domains pruning is especially important in problems that include differential equations (see chapter 12). In general, due to its complexity, the applicability of the Global Hull-consistency criterion is not suitable for problems with a large number of variables.

The criterion of Global Hull-consistency was firstly introduced in [39], where only preliminary results on a simple example were presented. The complete description of the various Global Hull-consistency enforcing algorithms (presented in chapter 6) and their experimental results (as discussed in chapter 8) was published in [40].

Several research directions can be envisaged for improving the constraint pruning techniques in continuous domains and, in particular, for ameliorating our Global Hull-consistency approach.

Firstly, the underlying algorithm for enforcing Local-consistency can be further improved. The local consistency criterion currently required is Box-consistency, which is enforced by the constraint Newton method (see subsection 4.2.2). The method associates a narrowing function to each variable of each constraint for reducing its domain bounds accordingly to the constraint. Since only a single smaller box is required, the narrowing algorithm is exclusively concerned with the outer limits of the variable domains. However, during the narrowing process unfeasible regions within each variable domain may be detected. Such knowledge could be incorporated in the tree-based description of the feasible space. Other related research directions would be the development of alternative or complementary narrowing functions. For instance, any of the complementary approaches described in 4.2.3 could be considered for improving the domains pruning achieved through constraint propagation.

Secondly, the branching strategy of the Global Hull-consistency enforcing algorithms may be refined. Currently, whenever a box is subdivided, two sub-boxes are considered, sharing all variable domains of the original box, except the one with largest width, which is split by its mid point. Better strategies could be implemented taking into account other domains to split, other split points, or even more subdivisions. Such decision should be equated in the context of both the overall enforcing algorithm requiring the subdivision and the underlying constraint propagation algorithm that will eventually prune each sub-box.

Thirdly, the advantages of having a tree-based compact description of the feasible space could be further explored. The tree structure representation of the search space is exclusively used by the *TSA* algorithm for supporting the search without losing any previously obtained pruning information. An interesting research direction would be to develop a visual interface tool for user interaction, based on the current state of the domains tree. Such a tool could provide the user with a general perspective on the feasible space, allowing for the interactive specification of sub-regions of interest which would be further pruned by reinforcing Global Hull-consistency. A direct consequence of the generalisation of this idea is the definition of even stronger consistency criteria based on the recursive enforcement of Global Hull-consistency. Another appealing research area related with the domains tree and with both of the above extensions is the parallel exploration of the different branches of the tree.

13.3   Local Search for Interval Constraint Reasoning

A local search procedure was developed for integration with constraint reasoning in continuous domains. The local search is confined to specific boxes of the search space, relying on the generic branch and bound strategy of the constraint reasoning algorithm to overcome local minimum. It can be used for accelerating the finding of canonical solutions in CCSPs. In particular, the integration of local search with the enforcement of Global Hull-consistency may be advantageous for reducing both the overall execution time and the memory storage required.

The local search procedure was introduced in association with the Global Hull-consistency criterion and described as an optional tool for enhancing its enforcing algorithms [39, 40]. Not enough testing was yet performed for identifying the kind of constraint problems where the application of local search is advantageous. Future work should be done to evaluate the impact of local search in continuous constraint programming and, in particular, for enforcing Global Hull-consistency.

The integration of local search in continuous constraint programming is a widely open research area. In particular, our local search approach may be improved in several ways.

Firstly, the line search procedure may be modified. In the proposed approach, new points are obtained through minimisation along the Newton vector, as long as they are kept inside the search box. Since local minima are overcome by a generic branch and bound strategy, a natural variation could be to skip the minimisation procedure and to take complete Newton steps bounded within the search box. Another alternative that is worth further research could be to allow the search to evolve outside the original search box.

Secondly, different kinds of local search procedures could be explored. Any of the alternatives described in section 7.2 could be considered. For instance, the alternatives based on strategies for solving constrained optimisation problems, such as Penalty methods and Lagrange-Multiplier methods, seem to be the best candidates.

Thirdly, the integration of local search with the CSDP constraints can be improved. In the Jacobian matrix, necessary for computing the Newton vector, the elements associated with variables of the solving base of a CSDP constraint are computed approximately from the derivative definition (see definition 10.2-7). Such a method implies an extra ODE evaluation on a nearby point, and is subject to errors depending on the distance of such a point from the current one. A better alternative could be to use the available information about the ODE system to obtain the partial derivatives with respect to those variables.

13.4   Prototype Implementation: Applications to Biophysical Modelling

All the proposed extensions to the interval constraints framework were implemented in a prototype application. The usefulness of the application for supporting decisions based on differential models was tested in diverse biophysical domains. It proved to be a valuable supporting tool for the diagnosis of diabetes, the tuning of drug design, and the study of epidemics. In general, its ability to handle parametric differential models makes the approach potentially applicable on a wide range of problems, namely, those requiring the integration of a dynamics model into decision support.

The practical application of our approach combines constraint reasoning with mathematical modelling and biomedical knowledge. Consequently, an evaluation of the approach should verify its correctness on these complementary perspectives. The results on biomedical decision support with ODEs (chapter 12) were published in a constraint programming conference [42], a mathematical conference [41], and a biomedical conference [43].

With respect to the implementation of our prototype application many improvements are conceivable. The application was developed for testing the feasibility of our approach and does not pretend to be competitive with other well established constraint programming systems. Consequently, an important practical issue is the full revision of the code for efficiency proposes.

Another important topic for further research is the tuning of the prototype parameters. Several decisions with respect to the functioning of the underlying algorithms were left as parameters of the prototype application and may be adjusted for each specific problem. Examples of such parameters are the order of the Interval Taylor Series method, the error tolerances accepted, the maximum number of time points considered, etc. A thorough study should be made to try to understand which parameter values are more suitable for each particular problem.

Finally, the application of our approach for modelling other domains is an open area with a variety of research possibilities. In particular, we plan to apply it to predator-prey models, neurophysiology models, reaction kinetic models, satellite localisation models, and aerodynamic models.

13.5   Conclusions

The integration of deep biophysical models into decision support is a challenging goal, difficult to accomplish, but fundamental for the development of model-based reasoning in biomedical domains. Our work is a contribution towards that. Several differential models can now be used for decision support through constraint reasoning. The constraint reasoning techniques were extended for a better adequacy to such a demanding context of decision support based on complex non linear models.

However, there are still important research areas that may provide valuable contributions for adapting constraint reasoning to such context.

The expressive power on the constraints framework can be further extended. Given the hybrid nature of many biophysical models, combining real variables with integer and boolean variables, integer constraints and conditional constraints should also be supported. Another possibility is to provide a new domain type for representing functions as primary objects, allowing to reason about their properties. This is an idea borrowed from the CSDP framework, where there is a variable for representing a function and restrictions on its properties. Constraints on the maximum or minimum values of a real function over some interval are not specific of differential equations, and should also be handled in the case where the function is defined analytically.

Additionally, the representation of other kinds of differential equations could be supported by the CSDP framework, broadening the spectrum of its potential applications. Ordinary differential equations are the present scope of the CSDP framework. However, many differential models (e.g. econometric models, flow models) cannot be represented by

ODEs. An important area for future research is the extension of the framework for handling other kinds of differential equations such as Partial Differential Equations (PDEs) or Delay Differential Equations (DDEs).

Finally, constraint reasoning could be extended with probabilistic reasoning. In decision problems there are often several possible options, all of them consistent with the constraint set. In practice, whenever a decision is required, the choice is made based on the probability/likelihood of the possible options. Uncertainty may be naturally represented in the constraints framework as intervals of possible values. However, there is often also information about the distribution of the different value possibilities, some values being more likely than others. Defining a distribution function for each variable range and knowing how each individual contribution would be combined (for example, assuming its independence), a global probability/likelihood value could be computed. Such knowledge could be subsequently included in the constraint model and used for supporting probabilistic reasoning based on the ranking of each canonical solution.

We believe that the relative unpopularity of constraint reasoning in continuous domains (at least compared with finite domains) is essentially due to not providing what is needed in practice. Problems in continuous domains are very demanding and solutions are required to be obtained efficiently. Users are willing to sacrifice safety for speed. They are used to the traditional mathematical tools capable of "quickly" providing approximations that are usually good enough for their needs. To change this situation constraint programming must impose itself as a convincing better alternative. It must be able to provide modelling and reasoning capabilities that go beyond what is traditionally offered by the competing alternatives, namely supporting safe decisions with acceptable computational resources. We hope that this work has been a valuable contribution in such a direction.

References

[1] E. Ackerman, L. Gatewood, J. Rosevar, and G. Molnar. Blood Glucose Regulation and Diabetes. In: Concepts and Models of Biomathematics, Chapter 4:131-156, F. Heinmets (Ed.), Marcel Dekker, 1969.

[2] G. Alefeld. Intervallrechnung uber den Komplexen Zahlen und einige Anwendungen. Ph.D. Thesis, University of Karlsrube, 1968.

[3] G. Alefeld, and J. Herzberger. Introduction to Interval Computations. Academic Press, New York, USA, 1983.

[4] K.R. Apt. The essence of constraint propagation. Theoretical Computer Science, 221(1-2):179-210, 1999.

[5] U.M. Ascher, and L.R. Petzold. Computer Methods for Ordinary Differential Equations and Differential-Algebraic Equations. Society for Industrial and Applied Mathematics, Philadelphia, USA, 1998.

[6] C. Bendsten, and O. Stauning. FADBAD, a Flexible C++ Package for Automatic Differentiation Using the Forward and Backward Methods, Technical Report 1996-x5-94, Department of Mathematical Modelling, Technical University of Denmark, Lyngby, Denmark, 1996.

[7] C. Bendsten, and O. Stauning. TADIFF, a Flexible C++ Package for Automatic Differentiation Using Taylor Series, Technical Report 1997-x5-94, Department of Mathematical Modelling, Technical University of Denmark, Lyngby, Denmark, 1997.

[8] F. Benhamou. Interval Constraint Logic Programming. In: Constraint Programming: Basics and Trends, LNCS 910, 1-21, A. Podelski (Ed.), Springer-Verlag, 1995.

[9] F. Benhamou. Heterogeneous Constraint Solving. In: Proceedings of 5th International Conference on Algebraic and Logic programming (ALP'96), LNCS 1139, 62-76, M. Hanus, and M. Rodriguez-Artalejo (Eds.), Springer-Verlag, Aachen, Germany, 1996.

[10] F. Benhamou, F. Goualard, and L. Granvilliers. An Extension of the WAM for Cooperative Interval Solvers. Technical Report, Department of Computer Science, University of Nantes, France, 1998.

[11] F. Benhamou, F. Goualard, L. Granvilliers, and J.-F. Puget. Revising Hull and Box Consistency. In: Proceedings of International Conference on Logic Programming (ICLP'99), 230-244, D. De Schreye (Ed.), MIT Press, Las Cruces, New Mexico, USA, 1999.

[12] F. Benhamou, and L. Granvilliers. Combining Local Consistency, Symbolic Rewriting and Interval Methods. In: Proceedings of 3rd International Conference on Artificial Intelligence and Symbolic Mathematical Computation (AISMC-3), LNCS 1138, 144-159, J. Calmet, J.A. Campbell, and J. Pfalzgraf (Eds.), Springer-Verlag, Steyr, Austria, 1996.

[13] F. Benhamou, and L. Granvilliers. Automatic Generation of Numerical Redundancies for Non-Linear Constraint Solving. Reliable Computing, 3(3):335-344, 1997.

[14] F. Benhamou, D. McAllester, and P. Van Hentenryck.. CLP(Intervals) revisited. In: Proceedings of International Logic Programming Symposium (ILPS'94), 124-138, M. Bruynooghe (Ed.), MIT Press, Ithaca, New York, USA, 1994.

[15] F. Benhamou, and W.J. Older. Applying Interval Arithmetic to Real, Integer and Boolean Constraints. Journal of Logic Programming, 32(1):1-24, 1997.

[16] D.P. Bertsekas. Constrained Optimisation and Lagrange Multiplier Methods, Academic Press, New York, USA, 1982.

[17] D.P. Bertsekas. Nonlinear Programming. (2nd Edition) Athena Scientific, Belmont, 1999.

[18] M. Berz. COSY INFINITY version 8 reference manual. Technical Report MSUCL-1088, National Superconducting Cyclotron Lab., Michigan State University, East Lansing, Michigan, USA, 1997.

[19] M. Berz, and K. Makino. Verified Integration of ODEs and Flows Using Differential Algebraic Methods on High-Order Taylor Models. Reliable Computing, 4:361-369, 1998.

[20] L. Bordeaux, E. Monfroy, and F. Benhamou. Improved bounds on the complexity of kB-consistency. In: Proceedings of 17th International Joint Conference on Artificial Inteligence (IJCAI'2001), Volume 1, 640-650, B. Nebel (Ed.), M. Kaufmann Publishers Inc., Seattle, Washington, USA, 2001.

[21] R.W. Brankin, I. Gladwell, and L.F. Shampine. RKSUITE: a Suite of Runge-Kutta Codes for the Initial Value Problem of ODEs. Softreport 92-S1, Department of Mathematics, Southern Methodist University, Dallas, Texas, USA, 1992.

[22] R.P. Brent. Algorithms for Minimization without Derivatives. Prentice-Hall, Englewood Cliffs, New Jersey, USA, 1973.

[23] P.N. Brown, G.D. Byrne, and A.C. Hindmarsh. VODE: a Variable-coefficient ODE Solver. SIAM Journal on Scientific Computing, 10(5):1038-1051, 1989.

[24] C.G. Broyden. A Class of Methods for Solving Nonlinear Simultaneous Equations. Mathematics of Computation. 19: 577-593, 1965.

[25] R. Bulirsch, and J. Stoer. Numerical treatment of ordinary differential equations by extrapolation methods. Numerische Mathematik, 8(1): 1-13, 1966.

[26] O. Caprani, B.Godthaab, and K. Madsen. Use of a Real-Valued Local Minimum in Parallel Interval Global Optimization. Interval Computations, 2:71-82, 1993.

[27] J.G. Cleary. Logical Arithmetic. Future Computing Systems, 2(2):125-149, 1987.

[28] H. Collavizza, F. Delobel, and M. Rueher. A Note on Partial Consistencies over Continuous Domains. In: Proceedings of 4th International Conference on Principles and Practice of Constraint Programming (CP'98), LNCS 1520, 147-161, M.J. Maher, and J.-F. Puget (Eds.), Springer Verlag, Pisa, Italy, 1998.

[29] H. Collavizza, F. Delobel, and M. Rueher. Comparing Partial Consistencies. *Reliable Computing*, 5:1-16, 1999.

[30] [CDR99b]   H. Collavizza, F. Delobel, and M. Rueher. Extending Consistent Domains of Numeric CSP. In: Proceedings of 16th International Joint Conference on Artificial Inteligence (IJCAI'99), 406-413, T. Dean (Ed.), Morgan Kaufmann, Stockholm, Sweden, 1999.

[31] G.F. Corliss, and R. Rihm. Validating an A Priori Enclosure Using High-Order Taylor Series. In: Scientific Computing, Computer Arithmetic and Validated Numerics: Proceedings of the International Symposium on Scientific Computing, Computer Arithmetic and Validated Numerics (SCAN'95), 228-238, G. Alefeld, A. Frommer, and B. Lang (Eds.), Akademie Verlag, Wuppertal, Germany, 1996.

[32] J. Cruz. Um Modelo Causal-Funcional para o Diagnóstico de Doenças Neuromusculares, Masters Thesis in Informatics Engineering, Science and Technology Faculty of the New University of Lisbon, Portugal, 1995

[33] J. Cruz, P. Barahona, and F. Benhamou. Integrating Deep Biomedical Models into Medical Decision Support Systems: an Interval Constraint Approach. In: Proceedings of the 7th Joint European Conference on Artificial Intelligence in Medicine and Medical Decision Making (AIMDM'99), LNAI 1620, 185-194, W. Horn, Y. Shahar, G. Lindberg, S. Andreassen, J. Wyatt (Eds.), Springer, Aalborg, Denmark, 1999.

[34] J. Cruz, P. Barahona, A. Figueiredo, M. Veloso, and M. Carvalho, DARE: A Knowledge-Based System for the Diagnosis of Neuromuscular Disorders, Applications of Artificial Intelligence, Advanced Manufacturing Forum, 1:29-40, J. Mamede, and C. Pinto-Ferreira (Eds.), N. Scitec Publications, 1996.

[35] J. Cruz, and P. Barahona. A Causal-Functional Model Applied to EMG Diagnosis. In: Proceedings of the 6th Conference on Artificial Intelligence in Medicine in Europe (AIME'97). LNCS 1211, 249-260, E.T. Keravnou, C. Garbay, R.H. Baud, and J.C. Wyatt (Eds.), Springer, Grenoble, France, 1997.

[36] [CB99a]J. Cruz, and P. Barahona. An Interval Constraint Approach to Handle Parametric Ordinary Differential Equations for Decision Support. In: Proceedings of the 5th International Conference on Principles and Practice of Constraint Programming (CP'99), LNCS 1713, 478-479, J. Jaffar (Ed.), Springer, Alexandria, Virginia, USA, 1999.

[37] [CB99b]J. Cruz, and P. Barahona. An Interval Constraint Approach to Handle Parametric Ordinary Differential Equations for Decision Support. In: Proceedings of the 2nd International Workshop on Extraction of Knowledge from Data Bases (EKDB'99), associated with 9th Portuguese Conference on Artificial Intelligence (EPIA'99), 93-108, F. Moura Pires, G. Guimarães and A. Jorge (Eds.), Springer, Évora, Portugal, 1999.

[38] J. Cruz, and P. Barahona. Handling Differential Equations with Constraints for Decision Support. In: Proceedings of the 3rd International Workshop on Frontiers of Combining Systems (FroCoS'2000), LNAI 1794, 105-120, H. Kirtchner, and C. Ringeissen (Eds.), Springer, Nancy, France, 2000.

[39] J. Cruz, and P. Barahona. A Global Hull Consistency with Local Search for Continuous Constraint Solving. In: Proceedings of the 10th Portuguese Conference on Artificial Intelligence (EPIA'2001). LNCS 2258, 349-362, P. Brazdil, and A. Jorge (Eds.), Springer, Porto, Portugal, 2001.

[40] J. Cruz, and P. Barahona. Maintaining Global-Hull Consistency with Local Search for Continuous CSPs. In: Proceedings of the 1st International Workshop on Global Constrained Optimisation and Constraint Satisfaction (Cocos'02), Springer, Valbonne, Sophia-Antipolis, France, 2002.

[41] [CB03a]J. Cruz, and P. Barahona. Constraint Reasoning with Differential Equations. In: Proceedings of the International Conference on Applied Numerical Analysis & Computational Mathematics (NaCoM-2003), NaCoM-2003 Extended Abstracts, 38-41, G. Psihoyios (Ed.), Wiley, Cambridge, UK, 2003.

[42] [CB03b]J. Cruz, and P. Barahona. Constraint Satisfaction Differential Problems. In: Proceedings of the 9th International Conference on Principles and Practice of Constraint Programming (CP'03), LNCS, Springer, Cork, Ireland, 2003 (accepted for publication).

[43] [CB03c]J. Cruz, and P. Barahona. Constraint Reasoning in Deep Biomedical Models. In: Proceedings of the 9th Conference on Artificial Intelligence in Medicine (AIME'03), LNAI, Springer, Cyprus, 2003 (accepted for publication).

[44] E. Davis. Constraint Propagation with Interval Label. Artificial Intelligence, 32:281-331, 1987.

[45] J.W. Demmel. Applied Numerical Linear Algebra. SIAM Publications, Philadelphia, 1997.

[46] J.E. Dennis, and R.B. Schnabel. Numerical Methods for Unconstrained Optimization and Nonlinear Least Squares. Prentice-Hall, Englewood Cliffs, New Jersey, USA, 1983.

[47] Y. Deville, M. Janssen, and P. Van Hentenryck. Consistency Techniques for Ordinary Differential Equations. In: Proceedings of the 4th International Conference on Principles and Practice of Constraint Programming (CP'98), LNCS 1520, 162-176, M. Maher, and J.-F. Puget (Eds.), Springer, Pisa, Italy,1998.

[48] P. Eijgenraam. The Solution of Initial Value Problems Using Interval Arithmetic. Mathematical Centre Tracts N°144. Stichting Mathematisch Centrum, Amsterdam, The Netherlands, 1981.

[49] A. V. Fiacco, and G. P. McCormick. Nonlinear Programming: Sequential Unconstrained Minimization Techniques. SIAM Publications, Philadelphia, USA, 1990.

[50] R. Fletcher. A New Approach to Variable Metric Algorithms. Computer Journal, 13:317-322, 1970.

[51] R. Fletcher. Practical Methods of Optimization, 2nd Edition, Wiley, Chichester & New York, 1987.

[52] R. Fletcher, and M. Powell. A Rapidly Convergent Descent Method for Minimization. Computer Journal, 6:163-168, 1963.

[53] R. Fletcher, and C. Reeves. Function Minimization by Conjugate Gradients. Computer Journal, 7:149-154, 1964.

[54] G.E. Forsythe, M.A. Malcolm, and C.B. Moler. Computer Methods for Mathematical Computations. Prentice-Hall, Englewood Cliffs, New Jersey, USA, 1977.

[55] E.C. Freuder. Synthesizing Constraint Expressions. In: Communications of the ACM (CACM), 21(11):958-966, 1978.

[56] C.W. Gear. Numerical Initial Value Problems in Ordinary Differential Equations, Prentice-Hall, Englewood Cliffs, New Jersey, USA, 1971.

[57] P.E. Gill, and W. Murray. Numerical Methods for Constrained Optimization, Academic Press, 1975.

[58] P.E. Gill, W. Murray, and M. Wright. Practical Optimization, Academic Press, New York, USA, 1981.

[59] G.H. Golub, and C. Reinsch. Singular Values Decomposition and Least Squares Solution. Contribution I/10 in [WR71], 1971.

[60] G.H. Golub, and C.F. Van Loan. Matrix Computations, 3th Edition, Johns Hopkins University Press, Baltimore and London, 1996.

[61] F. Goualard. Langages et environnements en programmation par contraintes d'intervalles. Ph.D. Thesis, Institut de Recherche en Informatique de Nantes, Université Nantes, Nantes, France, 2000.

[62] M. Gu, J.W. Demmel, and I. Dhillon. Efficient Computation of the Singular Value Decomposition with Applications to Least Squares Problems. Technical Report LBL-96201, Lawrence Berkeley National Laboratory, 1994.

[63] E. Hairer, S.P. Nørsett, and G. Wanner. Solving Ordinary Differential Equations I: Nonstiff Problems, 2nd Edition, Springer-Verlag, Berlin, Germany, 1991.

[64] E. Hansen. Topics in Interval Analysis, Oxford University Press, London, UK, 1969.

[65] E. Hansen. A Globally Convergent Interval Method for Computing and Bounding Real Roots. BIT 18:415-424, 1978.

[66] E. Hansen. Global Optimization Using Interval Analysis. 2nd Edition, Marcel Dekker, New York, USA, 1992.

[67] E. Hansen, and S. Sengupta. Bounding Solutions of Systems of Equations Using Interval Analysis. BIT 21:203-211, 1981.

[68] R.J. Hanson. Interval Arithmetic as a Closed Arithmetic System on a Computer. Technical Report 197, Jet Propulsion Laboratory, 1968.

[69] D. Sam-Haroud, and B.V. Faltings. Consistency Techniques for Continuous Constraints. Constraints, 1(1-2):85-118, 1996.

[70] P. Hartman. Ordinary Differential Equations. John Wiley and Sons, New York, USA, 1964.

[71] M. Hartmann. Runge-Kutta Methods for the Validated Solution of ODEs. In: Proceedings of the 4th International Congress on Industrial and Applied Mathematics (ICIAM'99), 202, Edinburgh, Scotland, UK, 1999.

[72] P. Henrici. Discrete Variable Methods in Ordinary Differential Equations. John Wiley & Sons, New York, USA, 1962.

[73] T.J. Hickey. CLP(F) and Constrained ODEs. In: Proceedings of the Workshop on Constraint Languages and their use in Problem Modelling, ECRC Technical Report ECRC-94-38, 69-79, Jourdan, Lim, and Yap (Eds.), 1994.

[74] T.J. Hickey, M.H. van Emden, and H. Wu. A Unified Framework for Interval Constraints and Interval Arithmetic. In: Proceedings of the 4th International Conference on Principles and Practice of Constraint Programming (CP'98), LNCS 1520, 250-264, M. Maher, and J.-F. Puget (Eds.), Springer, Pisa, Italy, 1998.

[75] T.J. Hickey, Q. Ju, and M.H. van Emden. Interval Arithmetic: from Principles to Implementation. Technical Report CS-99-202, Brandeis University, USA, 1999.

[76] J. Hoefkens, M. Berz, and K. Makino. Verified High-Order Integration of DAEs and Higher-Order ODEs. In: Scientific Computing, Validated Numerics and Interval Methods, 281-292, W. Kraemer, and J.W.V. Gudenberg (Eds.), Kluwer Academic Publishers, Dordrecht, The Netherlands, 2001.

[77] E. Hyvönen. Constraint Reasoning based on Interval Arithmetic: the Tolerance Propagation Approach. Artificial Intelligence, 58(1-3):71-112, 1992.

[78] K. Ichida, and Y. Fugii. An Interval Arithmetic Method for Global Optimization. Computing, 23(1):85-97, 1979.

[79] M. Janssen, Y. Deville, and P. Van Hentenryck. Multistep Filter Operators for Ordinary Differential Equations. Proceedings of the 5th International Conference on Principles and Practice of Constraint Programming (CP'99), LNCS 1713, 246-260, J. Jaffar (Ed.), Springer, Alexandria, Virginia, USA, 1999.

[80] M. Janssen, P. Van Hentenryck, and Y. Deville. A Constraint Satisfaction Approach to Parametric Differential Equations. In: Proceedings of the 17th International Joint Conference on Artificial Intelligence (IJCAI'2001), Volume 1, 297-302, B. Nebel (Ed.), M. Kaufmann Publishers Inc., Seattle, Washington, USA, 2001.

[81] M. Janssen, P. Van Hentenryck, and Y. Deville. Optimal Pruning in Parametric Differential Equations. In: Proceedings of 7th International Conference on Principles and Practice of Constraint Programming (CP'01), LNCS 2239, 539-562, T. Walsh (Ed.), Springer-Verlag, Paphos, Cyprus, 2001.

[82] C. Jansson. A Global Optimization Method Using Interval Arithmetic. In: Proceeding of the 3rd IMACS-GAMM Symposium on Computer Arithmetic and Scientific Computing (SCAN'92), 259-268, L. Atanassova (Ed.), 1992.

[83] P. Jeavons. Constructing Constraints. In: Proceedings of 4th International Conference on Principles and Practice of Constraint Programming (CP'98), LNCS 1520, 2-16, M.J. Maher, and J.-F. Puget (Eds.), Springer-Verlag, Pisa, Italy, 1998

[84] W.M. Kahan. A More Complete Interval Arithmetic. Technical Report, University of Toronto, Canada, 1968.

[85] R. Klatte, U. Kulisch, M. Neaga, D. Ratz, and C. Ullrich. Pascal-XSC: Language Reference with Examples. Springer-Verlag, Berlin, Germany, 1992.

[86] R. Krawczyk. Newton-Algorithmen zur Bestimmung von Nullstellen mit Fehlershranken, Computing, 4:187-201, 1969.

[87] L. Krippahl, and P. Barahona. Applying Constraint Propagation to Protein Structure Determination. In: Proceedings of the 5th International Conference on Principles and Practice of Constraint Programming (CP'99), LNCS 1713, 289-302, J. Jaffar (Ed.), Springer, Alexandria, Virginia, USA, 1999.

[88] L. Krippahl, and P. Barahona. PSICO: Solving Protein Structures with Constraint Programming and Optimisation, Constraints, 7(3-4):317-331, 2002.

[89] F. Krückeberg. Ordinary Differential Equations. In: Topics in Interval Analysis, 91-97, E. Hansen (Ed.), Clarendon Press, Oxford, UK, 1969.

[90] J.D. Lambert. Numerical Methods for Ordinary Differential Systems. Wiley, London, UK, 1991.

[91] O. Lhomme. Consistency Techniques for Numeric CSPs. In: Proceedings of the 13th International Joint Conference on Artificial Intelligence (IJCAI'1993), 232-238, R. Bajcsy (Ed.), Morgan Kaufmann, Chambéry, France, 1993.

[92] O. Lhomme, A. Gotlieb, M. Rueher, and P. Taillibert. Boosting the Interval Narrowing Algorithm. In: Proceedings of the Joint International Conference and Symposium on Logic Programming (JICSLP'96), 378-392, M.J. Maher (Ed.), MIT Press, Bonn, Germany, 1996.

[93] O. Lhomme, A. Gotlieb, and M. Rueher. Dynamic Optimization of Interval Narrowing Algorithms. Journal of Logic Programming, 37(1-3):165-183, 1998.

[94] R.J. Lohner. Enclosing the solutions of ordinary initial and boundary value problems. In: Computer Arithmetic: Scientific Computation and Programming Languages. 255-286, E.W. Kaucher, U.W. Kulisch, and C. Ullrich, (Eds.), Wiley-Teubner Series in Computer Science, Stuttgart, Germany, 1987.

[95] R.J. Lohner. Einschließung der Lösung Gewöhnlicher Anfangs- und Randwertaufgaben und Anwendungen, Ph.D. Thesis, Universität Karlsruhe, Germany, 1988.

[96] R.J. Lohner. Step Size and Order Control in the Verified Solution of IVP with ODEs. In: Proceedings of International Conference on Scientific Computation and Differential Equations (SciCADE'95), Stanford, California, USA, 1995.

[97] A.K. Mackworth. Consistency in Network of Relations. Artificial Intelligence, 8(1):99-118, 1977.

[98] U. Montanari. Networks of Constraints: Fundamental Properties and Applications to Picture Processing. Information Science, 7(2):95-132, 1974.

[99] R.E. Moore. Interval Analysis. Prentice-Hall, Englewood Cliffs, New Jersey, USA, 1966.

[100] R.E. Moore. Methods and Applications of Interval Analysis. SIAM, Studies in Applied Mathematics 2, Philadelphia, USA, 1979.

[101] J.J. Moré, and S.J. Wright. Optimization Software Guide. SIAM, Frontiers in Applied Mathematics 14, Philadelphia, USA, 1993.

[102] J.D. Murray. Mathematical Biology, 2nd Edition, Springer, 1991.

[103] N.S. Nedialkov. Computing Rigorous Bounds on the Solution of an Initial Value Problem for an Ordinary Differential Equation. Ph.D. Thesis, Department of Computer Science, University of Toronto, Canada, 1999.

[104] N.S. Nedialkov, and K.R. Jackson. An Interval Hermite-Obreschkoff method for Computing Rigorous Bounds on the Solution of an Initial Value Problem for an Ordinary Differential Equation. Reliable Computing 5(3), 289-310, 1999.

[105] N.S. Nedialkov, and K.R. Jackson. ODE Software that Computes Guaranteed Bounds on the Solution. In: Advances in Software Tools for Scientific Computing, 197-224, H.P. Langtangen, A.M. Bruaset and E. Quak (Eds.), Springer-Verlag, 2000.

[106] N.S. Nedialkov, and K.R. Jackson. A New Perspective on the Wrapping Effect in Interval Methods for Initial Value Problems for Ordinary Differential Equations. In: Perspectives on Enclosure Methods, 219-264, A. Facius, U. Kulisch, and R. Lohner (Eds.), Springer-Verlag, Vienna, Austria, 2001.

[107] N.S. Nedialkov, and K.R. Jackson. The Design and Implementation of an Object-Oriented Validated ODE Solver. Submitted to Advances of Computational Mathematics, 2002.

[108] N.S. Nedialkov, K.R. Jackson, and G.F. Corliss. Validated Solutions of Initial Value Problems for Ordinary Differential Equations. Applied Mathematics and Computation 105(1):21-68, 1999.

[109] N.S. Nedialkov, K.R. Jackson, and J.D. Pryce. An Effective High-Order Interval Method for Validating Existence and Uniqueness of the Solution of an IVP for an ODE. Reliable Computing 7(6):1-17, 2001.

[110] M. Novoa. Theory of Preconditioners for the Interval Gauss-Siedel Method and Existence/Uniqueness Theory with Interval Newton Methods. Department of Mathematics, University of Southwestern Louisiana, USA, 1993.

[111] W. Older. Interval Arithmetic Specification. Technical Report, BNR Computing Research Laboratory, 1989.

[112] W. Older. Application of Relational Interval Arithmetic to Ordinary Differential Equations. In: Proceedings of the Workshop on Constraint Languages and their use in Problem Modelling, International Logic Programming Symposium, M. Bruynooghe (Ed.), Ithaca, New York, USA, 1994.

[113] W. Older, and F. Benhamou. Programming in CLP(BNR). In: Proceedings of the International Conference on Principles and Practice of Constraint Programming (PPCP'93), 228-238, Newport, Rode Island, USA, 1993.

[114] W. Older, and A. Vellino. Constraint Arithmetic on Real Intervals. In: Constraint Logic Programming: Selected Research, 175-196, A. Colmerauer, and F. Benhamou (Eds.), MIT Press, London, UK, 1993.

[115] J. Ortega, and W. Rheinboldt. Iterative Solution of Nonlinear Equations in Several Variables. Academic Press, New York, USA, 1970.

[116] K. Petras. Validating Runge-Kutta Methods for ODEs with Analytic Right-Hand Side. Oral Communication at the International Symposium on Scientific Computing, Computer Arithmetic and Validated Numerics (SCAN'00), Karlsruhe, Germany, 2000.

[117] E. Polak, and G. Ribiere. Note sur la Convergence de Methods de Directions Conjuges. Revue Francaise Informat, Recherche Operationelle, 16:35-43, 1969.

[118] W.H. Press, S.A. Teukolsky, W.T. Vetterling, and B.P. Flannery. Numerical Recipes in C: The Art of Scientific Computing, 2nd Edition, Cambridge University Press, 1992.

[119] J-F. Puget, and P. Van Hentenryck. A Constraint Satisfaction Approach to a Circuit Design Problem. Journal of Global Optimization, 13:75-93, MIT Press, 1997.

[120] H. Ratschek, and J. Rokne. Computer Methods for the Range of Functions. Ellis Horwood Limited, Chichester, UK, 1984.

[121] H. Ratschek, and J. Rokne. New Computer Methods for Global Optimization. Wiley, New York, USA, 1988.

[122] D. Ratz. On Extended Interval Arithmetic and Inclusion Monotonicity. Institut fur Angewandte Mathematik, University of Karlsrube, Germany, 1996.

[123] R. Rihm. Interval Methods for Initial Value Problems in ODEs. In: Topics in Validated Computations: Proceedings of the IMACS-GAMM International Workshop on Validated Computations, 173-207, University of Oldenburg, J. Herzberger, (Ed.), Elsevier Studies in Computational Mathematics, Elsevier, Amsterdam, New York, USA, 1994.

[124] R. Rihm. Implicit Methods for Enclosing Solutions of ODEs. Journal of Universal Computer Science, 4(2): 202-209, 1998.

[125] A. Semenov, T. Kashevarova, A. Leshchenko, and D. Petunin. Combining Various Techniques with the Algorithm of Subdefinite Calculations. In: Proceedings of the 3rd International Conference on the Practical Application of Constraint Technology (PACT'97), 287-306, London, UK, 1997.

[126] L.F. Shampine, and M.K. Gordon. Computer Solution of Ordinary Differential Equations: the Initial Value Problem. W.H. Freeman and Company, San Francisco, USA, 1975.

[127] L.F. Shampine. Numerical Solution of Ordinary Differential Equations. Chapman and Hall, New York, USA, 1994.

[128] G. Sidebottom, and W.S. Havens. Hierarchical Arc Consistency for Disjoint Real Intervals in Constraint Logic Programming. Computational Intelligence, 8(4):601-623, 1992.

[129] S. Skelboe. Computation of Rational interval Functions. BIT, 14:87-95, 1974.

[130] E. Spitznagel. Two-Compartment Pharmacokinetic Models. Consortium for Ordinary Differential Equations Experiments Newsletter (C-ODE-E). Harvey Mudd College, Claremont, California, USA, (1992).

[131] O. Stauning. Enclosing Solutions of Ordinary Differential Equations.Technical Report IMM-REP-1996-18, Department of Mathematical Modelling, Technical University of Denmark, Lyngby, Denmark, 1996.

[132] O. Stauning. Automatic Validation of Numerical Solutions. Ph.D. Thesis, Technical University of Denmark, Lyngby, Denmark, 1997.

[133] J. Stoer, and R. Bulirsch. Introduction to Numerical Analysis. 2nd Edition, Springer, New York, USA, 1992.

[134] P. Van Hentenryck, D. McAllester, and D. Kapur. Solving Polynomial Systems Using a Branch and Prune Approach. SIAM Journal of Numerical Analysis, 34(2): 797-827, 1997.

[135] P. Van Hentenryck, L. Michel, and Y. Deville. Numerica: A Modeling Language for Global Optimization. MIT Press, 1997.

[136] D.L. Waltz. Generating Semantic Descriptions from Drawings of Scenes with Shadows. Techical Report AI-TR-271, MIT, Cambridge, USA, 1972.

[137] J.H. Wilkinson, and C. Reinsch. Linear Algebra. Handbook for Automatic Computation II. Springer, Berlin, Germany, 1971.

[138] E.K. Yeargers, R.W. Shonkwiler, and J.V. Herod. An Introduction to the Mathematics of Biology: with Computer Algebra Models. Birkhäuser, Boston, USA, 1996.

Appendix A

Interval Analysis Theorems

The demonstrations of the Interval Arithmetic theorems are based on several assumptions about the basic interval arithmetic operators and their approximate evaluation. These assumptions are in accordance with the original interval arithmetic proposal [99] where division by an interval containing zero was not considered. However to integrate the interval arithmetic theorems within the broader context of extended interval arithmetic it would be necessary to extend the definitions of an interval arithmetic operator and of the interval expression evaluation to handle multiple intervals[1].

Assumption A-1 If Φ is an m-ary basic interval arithmetic operator then, for the real intervals $I_1,\ldots,I_m$, $\Phi(I_1,\ldots,I_m)$ is the smallest real interval enclosing the set S obtained by applying it to m-tuples of real numbers, one from each of the m intervals:

$$S=\{\Phi(r_1,\ldots,r_m) \mid r_1 \in I_1 \wedge \ldots \wedge r_m \in I_m\} \subseteq \Phi(I_1,\ldots,I_m) \quad \wedge \quad \forall_{I \supseteq S}\ \Phi(I_1,\ldots,I_m) \subseteq I$$

In particular, if Φ is said to be able to compute the exact ranges within its interval arguments then:

$$S=\{\Phi(r_1,\ldots,r_m) \mid r_1 \in I_1 \wedge \ldots \wedge r_m \in I_m\} = \Phi(I_1,\ldots,I_m) \qquad \Box$$

Assumption A-2 The basic interval arithmetic operators are all inclusion monotonic. If Φ is an m-ary basic interval arithmetic operator then, for any real intervals $I_1,\ldots,I_m$ and $I_1',\ldots,I_m'$, the following property holds:

$$\forall_{1 \leq i \leq m}\ I_i' \subseteq I_i \Rightarrow \Phi(I_1',\ldots,I_m') \subseteq \Phi(I_1,\ldots,I_m) \qquad \Box$$

Assumption A-3 If Φ is an m-ary basic interval arithmetic operator then, for the real intervals $I_1,\ldots,I_m$, its approximate evaluation $\Phi_{apx}(I_1,\ldots,I_m)$ with the outward evaluation rules is defined as:

$$\Phi_{apx}(I_1,\ldots,I_m)=I_{apx}(\Phi(I_1,\ldots,I_m))$$

In particular, when the interval arithmetic evaluation is said to be performed with infinite precision:

$$\Phi_{apx}(I_1,\ldots,I_m)=\Phi(I_1,\ldots,I_m) \qquad \Box$$

The following lemmas will be used in the demonstrations of the Interval Arithmetic theorems.

Lemma A-1 Let I_1 and I_2 be two real intervals. If $I_1 \cup I_2$ is a real interval then:

$$I_{apx}(I_1) \cup I_{apx}(I_2) = I_{apx}(I_1 \cup I_2) \qquad \Box$$

Proof:

Let $I_1=\langle a..b\rangle$ and $I_2=\langle c..d\rangle$ with $a \leq d$ (this is always possible by choosing appropriately I_1 and I_2). If $I_1 \cup I_2$ is a real interval then there cannot exist any real value r such that $b < r < c$ (otherwise there would be a gap within $I_1 \cup I_2$). Moreover, the smallest of the left bounds

---

[1] As discussed in subsection 3.1.1 multiple intervals may be originated from a division by an interval containing zero.

and the largest of the right bounds of both intervals I_1 and I_2 must bound the real interval $I_1 \cup I_2$. Consequently its *RF*-interval approximation is (see definition 2.2.3-1):

$$I_{apx}(I_1 \cup I_2) = [\lfloor \min(a,c) \rfloor .. \lceil \max(b,d) \rceil] \tag{1}$$

On the other hand, the *RF*-interval approximations of each interval I_1 and I_2 are:

$$I_{apx}(I_1) = [\lfloor a \rfloor .. \lceil b \rceil]$$
$$I_{apx}(I_2) = [\lfloor c \rfloor .. \lceil d \rceil]$$

Their union is necessarily a real interval because if it cannot exist any real value r such that $b<r<c$ then it also cannot exist any real value r such that $b \leq \lceil b \rceil < r < \lfloor c \rfloor \leq c$.

And so their union is:

$$I_{apx}(I_1) \cup I_{apx}(I_2) = [\lfloor a \rfloor .. \lceil b \rceil] \cup [\lfloor c \rfloor .. \lceil d \rceil] = [\min(\lfloor a \rfloor, \lfloor c \rfloor) .. \max(\lceil b \rceil, \lceil d \rceil)] \tag{2}$$

If $a \leq c$ then $\lfloor a \rfloor \leq \lfloor c \rfloor$ and so $\min(\lfloor a \rfloor, \lfloor c \rfloor) = \lfloor a \rfloor = \lfloor \min(a,c) \rfloor$ else $\min(\lfloor a \rfloor, \lfloor c \rfloor) = \lfloor c \rfloor = \lfloor \min(a,c) \rfloor$. So, in both cases:

$$\min(\lfloor a \rfloor, \lfloor c \rfloor) = \lfloor \min(a,c) \rfloor \tag{3}$$

If $b \leq d$ then $\lceil b \rceil \leq \lceil d \rceil$ and so $\max(\lceil b \rceil, \lceil d \rceil) = \lceil d \rceil = \lceil \max(b,d) \rceil$ else $\max(\lceil b \rceil, \lceil d \rceil) = \lceil b \rceil = \lceil \max(b,d) \rceil$ So, in both cases:

$$\max(\lceil b \rceil, \lceil d \rceil) = \lceil \max(b,d) \rceil \tag{4}$$

From (3) and (4) it follows that the right sides of equations (1) and (2) are equal thus:

$$I_{apx}(I_1) \cup I_{apx}(I_2) = I_{apx}(I_1 \cup I_2) \qquad\qquad \blacksquare$$

Lemma A-2 Let F_E be an interval expression. Let B, B_1 and B_2 be *n*-ary *R*-boxes. If $B = B_1 \cup B_2$ then $F_E(B_1) \cup F_E(B_2)$ is a real interval. ❏

Proof:

Definition 3.2-1 gives an inductive definition for the set of all possible interval expressions: the two base clauses (i and ii) specify that interval constants (I) and interval variables (X_i) are the basic elements of the defined set; the inductive clause (iii) specifies how to generate additional elements from existing elements $E_1,\ldots,E_m$ and an *m*-ary basic interval arithmetic operator Φ.

We will present an inductive proof that for every interval expression F_E belonging to this set, the property that $F_E(B_1) \cup F_E(B_2)$ is a real interval must hold (where B, B_1 and B_2 be *n*-ary *R*-boxes and $B = B_1 \cup B_2$). The proof consists on a basis step which shows that the property holds for the basic elements ($F_E(B_1) \cup F_E(B_2)$ is a real interval with $F_E \equiv I$ and with $F_E \equiv X_i$) and an inductive step which shows that if the property holds for some elements $E_1,\ldots,E_m$ (inductive hypothesis: $E_i(B_1) \cup E_i(B_2)$ is a real interval with $1 \leq i \leq m$) then it holds for any elements generated from them by the inductive clause ($F_E(B_1) \cup F_E(B_2)$ is a real interval with $F_E \equiv \Phi(E_1,\ldots,E_m)$).

Basis Step

① Proof that $F_E(B_1) \cup F_E(B_2)$ is a real interval with $F_E \equiv I$ (where I is a real interval):
Accordingly to definition 3.2-4, in the case of $F_E \equiv I$:

$$F_E(B_1) = F_E(B_2) = I_{apx}(I)$$

Consequently:

$$F_E(B_1) \cup F_E(B_2) = I_{apx}(I)$$

which is a real interval (see definition 2.2.3-1 of *RF*-interval approximation) ❶

② Proof that $F_E(B_1) \cup F_E(B_2)$ is a real interval with $F_E \equiv X_i$ (where X_i is an interval valued variable):
Accordingly to definition 3.2-4, in the case of $F_E \equiv X_i$:

$$F_E(B_1) = I_{apx}(B_1[X_i])$$

$F_E(B_2)=I_{apx}(B_2[X_i])$

Consequently:

$F_E(B_1)\cup F_E(B_2) = I_{apx}(B_1[X_i]) \cup I_{apx}(B_2[X_i])$ (1)

If $B=B_1\cup B_2$ and B, B_1 and B_2 are n-ary R-boxes then:

$\forall_{1\leq i\leq m} B[X_i]=B_1[X_i]\cup B_2[X_i]$

So, for any $1\leq i\leq m$, $B_1[X_i]\cup B_2[X_i]$ is a real interval since $B[X_i]$ is a real interval (see definition 2.2.4-1 of an R-box). Because $B_1[X_i]$ and $B_2[X_i]$ and both real intervals and their union is a real interval then, by Lemma A1 it follows:

$I_{apx}(B_1[X_i]) \cup I_{apx}(B_2[X_i]) = I_{apx}(B_1[X_i]\cup B_2[X_i])$ (2)

From (1) and (2) it follows that:

$F_E(B_1)\cup F_E(B_2) = I_{apx}(B_1[X_i]\cup B_2[X_i])$

which is a real interval.   ❷

Inductive Step

③   Proof that $F_E(B_1)\cup F_E(B_2)$ is a real interval with $F_E\equiv\Phi(E_1,\ldots,E_m)$ and $E_i(B_1)\cup E_i(B_2)$ real intervals ($1\leq i\leq m$):

Accordingly to definition 3.2-4 in the case of $F_E\equiv\Phi(E_1,\ldots,E_m)$ and Assumption A3:

$F_E(B_1)= \Phi_{apx}(E_1(B_1),\ldots, E_m(B_1)) = I_{apx}(\Phi(E_1(B_1),\ldots, E_m(B_1)))$

$F_E(B_2)= \Phi_{apx}(E_1(B_2),\ldots, E_m(B_2)) = I_{apx}(\Phi(E_1(B_2),\ldots, E_m(B_2)))$

Consequently:

$F_E(B_1) \cup F_E(B_2) = I_{apx}(\Phi(E_1(B_1),\ldots, E_m(B_1))) \cup I_{apx}(\Phi(E_1(B_2),\ldots, E_m(B_2)))$ (3)

But by Assumption A1:

$S_1=\{\Phi(r_1,\ldots,r_m) \mid r_1\in E_1(B_1) \wedge\ldots\wedge r_m\in E_m(B_1)\} \subseteq \Phi(E_1(B_1),\ldots, E_m(B_1))$ (4)

$S_2=\{\Phi(r_1,\ldots,r_m) \mid r_1\in E_1(B_2) \wedge\ldots\wedge r_m\in E_m(B_2)\} \subseteq \Phi(E_1(B_2),\ldots, E_m(B_2))$ (5)

The union of these two sets gives:

$S_1\cup S_2 = \{\Phi(r_1,\ldots,r_m)|r_1\in E_1(B_1)\cup E_1(B_2)\wedge\ldots\wedge r_m\in E_m(B_1)\cup E_m(B_2)\}\subseteq \Phi(E_1(B_1),\ldots, E_m(B_1))$

$\cup\Phi(E_1(B_2),\ldots, E_m(B_2))$

From the inductive hypothesis we know that each $E_i(B_1)\cup E_i(B_2)$ is a real interval denoted as IR_i:

$S_1\cup S_2 = \{\Phi(r_1,\ldots,r_m) \mid r_1\in IR_1\wedge\ldots\wedge r_m\in IR_m\} \subseteq \Phi(E_1(B_1),\ldots,E_m(B_1))\cup\Phi(E_1(B_2),\ldots, E_m(B_2))$

And again by Assumption A1, $\Phi(IR_1,\ldots,IR_m)$ must be the smallest real interval enclosing $S_1\cup S_2$:

$S_1\cup S_2\subseteq \Phi(IR_1,\ldots,IR_m) \subseteq \Phi(E_1(B_1),\ldots, E_m(B_1)) \cup \Phi(E_1(B_2),\ldots, E_m(B_2))$ (6)

From (4), (5) and (6) we may conclude that $\Phi(E_1(B_1),\ldots,E_m(B_1))\cup\Phi(E_1(B_2),\ldots,E_m(B_2))$ is necessarily a real interval. This fact will be proved by contradiction where it is assumed that $\Phi(E_1(B_1),\ldots,E_m(B_1))\cup\Phi(E_1(B_2),\ldots,E_m(B_2))$ is not a real interval and a contradiction is proved:

Suppose $\Phi(E_1(B_1),\ldots,E_m(B_1))\cup\Phi(E_1(B_2),\ldots,E_m(B_2))$ is not a real interval then $\Phi(IR_1,\ldots,IR_m)$ which is an interval must be a subset of $\Phi(E_1(B_1),\ldots,E_m(B_1))$ or a subset of $\Phi(E_1(B_2),\ldots, E_m(B_2))$. The reason for this is that there is no way of some elements of the interval $\Phi(IR_1,\ldots,IR_m)$ be within $\Phi(E_1(B_1),\ldots,E_m(B_1))$ and other elements be within $\Phi(E_1(B_2),\ldots, E_m(B_2))$ without these two intervals be connected (and their union be a real interval).

If $\Phi(IR_1,\ldots,IR_m) \subseteq \Phi(E_1(B_1),\ldots, E_m(B_1))$ then from (6):

$S_1\cup S_2\subseteq\Phi(E_1(B_1),\ldots, E_m(B_1))$

But in (4) we said that $\Phi(E_1(B_1),\ldots, E_m(B_1))$ is the smallest real interval enclosing S_1, so $S_1\cup S_2=S_1$, which means that:

$S_2\subseteq S_1$

And from (4) and (5) because they are the smallest real intervals enclosing these sets:

$\Phi(E_1(B_2),\dots, E_m(B_2)) \subseteq \Phi(E_1(B_1),\dots, E_m(B_1))$
which is contradictory with the assumption that $\Phi(E_1(B_1),\dots,E_m(B_1)) \cup \Phi(E_1(B_2),\dots, E_m(B_2))$ is not a real interval.

If $\Phi(IR_1,\dots,IR_m) \subseteq \Phi(E_1(B_2),\dots, E_m(B_2))$ a similar reasoning may reach to the analogous conclusion that $\Phi(E_1(B_1),\dots,E_m(B_1)) \subseteq \Phi(E_1(B_2),\dots,E_m(B_2))$ which is contradictory with the assumption that $\Phi(E_1(B_1),\dots,E_m(B_1)) \cup \Phi(E_1(B_2),\dots, E_m(B_2))$ is not a real interval.

Consequently $\Phi(E_1(B_1),\dots,E_m(B_1)) \cup \Phi(E_1(B_2),\dots,E_m(B_2))$ is necessarily a real interval. But if it is a real interval and both, $\Phi(E_1(B_1),\dots,E_m(B_1))$ and $\Phi(E_1(B_2),\dots,E_m(B_2))$, are real intervals, then from Lemma A1:

$I_{apx}(\Phi(E_1(B_1),\dots,E_m(B_1))) \cup I_{apx}(\Phi(E_1(B_2),\dots,E_m(B_2)))=I_{apx}(\Phi(E_1(B_1),\dots,E_m(B_1)) \cup \Phi(E_1(B_2),\dots,E_m(B_2)))$

And from (3) it follows that:

$F_E(B_1) \cup F_E(B_2) = I_{apx}(\Phi(E_1(B_1),\dots,E_m(B_1)) \cup \Phi(E_1(B_2),\dots,E_m(B_2)))$
which is a real interval.
    ❸
    ■

Lemma A-3 The *RF*-interval approximation of a real interval is inclusion monotonic, that is, for any real intervals I and I' the following property holds:

$I' \subseteq I \Rightarrow I_{apx}(I') \subseteq I_{apx}(I)$ ❑

Proof:

Let $I=<a..b>$ and $I'=<a'..b'>$ be two real intervals. If $I' \subseteq I$ then $a \leq a'$ and $b' \leq b$.

From definition 2.2.3-1 of *RF*-interval approximation we know that:

$I_{apx}(I')=I_{apx}(<a'..b'>)=[\lfloor a' \rfloor .. \lceil b' \rceil]$ (1)

$I_{apx}(I)=I_{apx}(<a..b>)=[\lfloor a \rfloor .. \lceil b \rceil]$ (2)

But $\lfloor a \rfloor$ is the largest *F*-number not greater than a and so, because $a \leq a'$, it must also be not greater than a'. Consequently:

$\lfloor a \rfloor \leq \lfloor a' \rfloor$ (3)

Similarly, $\lceil b \rceil$ is the smallest *F*-number not lower than b and so, because $b' \leq b$, it must also be not lower than b'. Consequently:

$\lceil b' \rceil \leq \lceil b \rceil$ (4)

From (1) and (2) together with (3) and (4) it follows:

$I_{apx}(I') \subseteq I_{apx}(I)$ ■

Lemma A-4 If Φ is an *m*-ary basic interval arithmetic operator then its approximate evaluation is inclusion monotonic, that is, for any real intervals $I_1,\dots,I_m$ and $I_1',\dots,I_m'$, the following property holds:

$\forall_{1 \leq i \leq m} I_i' \subseteq I_i \Rightarrow \Phi_{apx}(I_1',\dots,I_m') \subseteq \Phi_{apx}(I_1,\dots,I_m)$ ❑

Proof:

If $\forall_{1 \leq i \leq m} I_i' \subseteq I_i$, then from Assumption A3 we know that:

$\Phi_{apx}(I_1',\dots,I_m')=I_{apx}(\Phi(I_1',\dots,I_m'))$ (1)

$\Phi_{apx}(I_1,\dots,I_m)=I_{apx}(\Phi(I_1,\dots,I_m))$ (2)

But due to the inclusion monotonicity property of the basic interval arithmetic operator Φ (Assumption A3):

$\Phi(I_1',\dots,I_m') \subseteq \Phi(I_1,\dots,I_m)$

From the inclusion monotonicity property of the *RF*-interval approximation (Lemma A3):

$$I_{apx}(\Phi(I_1',\ldots,I_m')) \subseteq I_{apx}(\Phi(I_1,\ldots,I_m)) \qquad (3)$$

From (1) and (2) together with (3) it follows:

$$\Phi_{apx}(I_1',\ldots,I_m') \subseteq \Phi_{apx}(I_1,\ldots,I_m) \qquad \blacksquare$$

Lemma A-5 If F_E be an interval expression representing the n-ary interval function F then its interval arithmetic evaluation is inclusion monotonic, that is, for any two n-ary R-boxes B and B', the following property holds:

$$B' \subseteq B \Rightarrow F_E(B') \subseteq F_E(B) \qquad \square$$

Proof:

Definition 3.2-1 gives an inductive definition for the set of all possible interval expressions: the two base clauses (i and ii) specify that interval constants (I) and interval variables (X_i) are the basic elements of the defined set; the inductive clause (iii) specifies how to generate additional elements from existing elements $E_1,\ldots,E_m$ and an m-ary basic interval arithmetic operator Φ.

We will present an inductive proof that for every interval expression F_E belonging to this set, the inclusion monotonicity property must hold. The proof consists on a basis step which shows that the property holds for the basic elements ($F_E{\equiv}I$ and $F_E{\equiv}X_i$) and an inductive step which shows that if the property holds for some elements $E_1,\ldots,E_m$ (inductive hypothesis: $\forall_{1\leq i\leq m} E_i$ is inclusion monotonic) then it holds for any elements generated from them by the inductive clause ($F_E{\equiv}\Phi(E_1,\ldots,E_m)$).

Basis Step

① Proof that $F_E{\equiv}I$ is inclusion monotonic (where I is a real interval):

Accordingly to definition 3.2-4, in the case of $F_E{\equiv}I$, for any two R-boxes B and B':

$$F_E(B) = F_E(B') = I_{apx}(I)$$

Consequently:

$$B' \subseteq B \Rightarrow F_E(B') \subseteq F_E(B) \qquad ❶$$

② Proof that $F_E{\equiv}X_i$ is inclusion monotonic (where X_i is an interval valued variable):

Accordingly to definition 3.2-4, in the case of $F_E{\equiv}X_i$, for any two R-boxes B and B':

$$F_E(B')=I_{apx}(B'[X_i]) \qquad (1)$$
$$F_E(B)=I_{apx}(B[X_i]) \qquad (2)$$

But if $B' \subseteq B$ then for every $1\leq i\leq m$ it follows:

$$B'[X_i] \subseteq B[X_i]$$

And from the inclusion monotonicity property of the RF-interval approximation (Lemma A3):

$$I_{apx}(B'[X_i]) \subseteq I_{apx}(B[X_i]) \qquad (3)$$

From (1) and (2) together with (3) it follows:

$$B' \subseteq B \Rightarrow F_E(B') \subseteq F_E(B) \qquad ❷$$

Inductive Step

③ Proof that $F_E{\equiv}\Phi(E_1,\ldots,E_m)$ is inclusion monotonic (where $\forall_{1\leq i\leq m} E_i$ is inclusion monotonic):

Accordingly to definition 3.2-4, in the case of $F_E{\equiv}\Phi(E_1,\ldots,E_m)$, for any two R-boxes B and B':

$$F_E(B')= \Phi_{apx}(E_1(B'),\ldots, E_m(B')) \qquad (4)$$
$$F_E(B)= \Phi_{apx}(E_1(B),\ldots, E_m(B)) \qquad (5)$$

But if $B' \subseteq B$ and for every $1\leq i\leq m$ E_i is inclusion monotonic it follows:

$\forall_{1\le i\le m}\ E_i(B')\subseteq E_i(B)$

And from the inclusion monotonicity property of approximate evaluation of the basic interval arithmetic operator (Lemma A4):

$$\Phi_{apx}(E_1(B'),\ldots,E_m(B')) \subseteq \Phi_{apx}(E_1(B),\ldots,E_m(B)) \qquad (6)$$

From (4) and (5) together with (6) it follows:

$B'\subseteq B \Rightarrow F_E(B') \subseteq F_E(B)$ ❸

∎

Lemma A-6 If F is an n-ary interval function represented by interval expression F_E then F is inclusion monotonic, that is, for any two n-ary R-boxes B and B', the following property holds:

$$B'\subseteq B \Rightarrow F(B') \subseteq F(B) \qquad \square$$

Proof:

Let B' and B be the two n-ary R-boxes $<I_1',\ldots,I_n'>$ and $<I_1,\ldots,I_n>$ respectively (with $B'\subseteq B$). Accordingly to definition 3.2-3, $F(B')$ and $F(B)$ are the smallest real intervals containing respectively the ranges $f^*(<I_1',\ldots,I_n',I_{n+1},\ldots,I_{n+k}>)$ and $f^*(<I_1,\ldots,I_n,I_{n+1},\ldots,I_{n+k}>)$ (where $I_{n+1},\ldots,I_{n+k}$ are all the interval constants appearing in F_E) and f is expressed as $f_E \equiv \Phi(e_1,\ldots,e_m)$ (obtained by replacing in F_E each interval variable X_i by the real variable x_i, each interval constant I_{n+j} by the real variable x_{n+j} and each interval operator by the corresponding real operator).

In this case the ranges $f^*(<I_1',\ldots,I_n',I_{n+1},\ldots,I_{n+k}>)$ and $f^*(<I_1,\ldots,I_n,I_{n+1},\ldots,I_{n+k}>)$ are:

$$f^*(<I_1',\ldots,I_n',I_{n+1},\ldots,I_{n+k}>)= \{f(d) \mid d\in<I_1',\ldots,I_n',I_{n+1},\ldots,I_{n+k}>\} \subseteq F(B') \qquad (4)$$

$$f^*(<I_1,\ldots,I_n,I_{n+1},\ldots,I_{n+k}>)= \{f(d) \mid d\in<I_1,\ldots,I_n,I_{n+1},\ldots,I_{n+k}>\} \subseteq F(B) \qquad (5)$$

But if $B'\subseteq B$ then for every $1\le i\le m$ it follows:

$I_i' \subseteq I_i$

And so:

$$\forall_{1\le i\le m}\ d\in<I_1',\ldots,I_n',I_{n+1},\ldots,I_{n+k}> \Rightarrow d\in<I_1,\ldots,I_n,I_{n+1},\ldots,I_{n+k}>$$

Consequently:

$$\{f(d) \mid d\in<I_1',\ldots,I_n',I_{n+1},\ldots,I_{n+k}>\}\subseteq\{f(d) \mid d\in<I_1,\ldots,I_n,I_{n+1},\ldots,I_{n+k}>\} \qquad (6)$$

Because $F(B')$ is the smallest real interval enclosing the range $\{f(d) \mid d\in<I_1',\ldots,I_n',I_{n+1},\ldots,I_{n+k}>\}$ (from (4)) and $F(B)$ is also a real interval enclosing it (from (5) and (6)), it follows:

$F(B') \subseteq F(B)$ ∎

Theorem 3.2-1 (Soundness of the Interval Expression Evaluation). Let F_E be an interval expression representing the n-ary interval function F. Let B be an n-ary R-box. The interval arithmetic evaluation of F_E with respect to B is sound:

$$F(B) \subseteq F_E(B) \qquad \square$$

Proof:

Definition 3.2-1 gives an inductive definition for the set of all possible interval expressions: the two base clauses (i and ii) specify that interval constants (I) and interval variables (X_i) are the basic elements of the defined set; the inductive clause (iii) specifies how to generate additional elements from existing elements $E_1,\ldots,E_m$ and an m-ary basic interval arithmetic operator Φ.

We will present an inductive proof that for every interval expression F_E belonging to this set, the property $F(B) \subseteq F_E(B)$ must hold (where F is the n-ary interval function represented by F_E and B is an n-ary R-box). The proof consists on a basis step which shows that the property holds for the basic elements ($F(B){\subseteq}F_E(B)$ with $F_E{\equiv}I$ and with $F_E{\equiv}X_i$) and an inductive step which shows that if the property holds for some elements $E_1,\ldots,E_m$ (inductive hypothesis: $\forall_{1\leq i\leq m} F_i(B){\subseteq}E_i(B)$ where E_i represents F_i) then it holds for any elements generated from them by the inductive clause ($F(B){\subseteq}F_E(B)$ with $F_E{\equiv}\Phi(E_1,\ldots,E_m)$).

Basis Step

① Proof that $F(B){\subseteq}F_E(B)$ with $F_E{\equiv}I$ (where I is a real interval):

Accordingly to definition 3.2-3 (extended for allowing the representation of interval constants – see footnote), in the case of $F_E{\equiv}I$, $F(B)$ is the smallest real interval containing the range $f^*(I)$ with f expressed as $f_E{\equiv}x_1$. By the definition of the range of a real function f over I:

$$f^*(I){=}\{f(r_1) \mid r_1{\in}I\}$$

and in this particular case:

$$f^*(I){=}\{r_1 \mid r_1{\in}I\}{=}I$$

Since I is a real interval, it is the smallest real interval containing the range $f^*(I)$ and so:

$$F(B){=}I \qquad\qquad (1)$$

Accordingly to definition 3.2-4, in the case of $F_E{\equiv}I$:

$$F_E(B){=}I_{apx}(I) \qquad\qquad (2)$$

Because $IR{\subseteq}I_{apx}(IR)$ for any real interval IR (see definition 2.2.3-1), from (1) and (2) it follows:

$$F(B){\subseteq}F_E(B) \qquad\qquad\qquad ❶$$

② Proof that $F(B){\subseteq}F_E(B)$ with $F_E{\equiv}X_i$ (where X_i is an interval valued variable):

Accordingly to definition 3.2-3, in the case of $F_E{\equiv}X_i$, $F(B)$ is the smallest real interval containing the range $f^*(I)$ with $I{=}B[X_i]$ and f expressed as $f_E{\equiv}x_i$. So, in this particular case:

$$f^*(I){=}\{r_i \mid r_i{\in}I\}{=} I {=}B[X_i]$$

Since $B[X_i]$ is a real interval, it is the smallest real interval containing the range $f^*(I)$ and so:

$$F(B){=}B[X_i] \qquad\qquad (3)$$

Accordingly to definition 3.2-4, in the case of $F_E{\equiv}X_i$:

$$F_E(B){=}I_{apx}(B[X_i]) \qquad\qquad (4)$$

From (3) and (4) it follows:

$$F(B){\subseteq}F_E(B) \qquad\qquad\qquad ❷$$

Inductive Step

③ Proof that $F(B){\subseteq}F_E(B)$ with $F_E{\equiv}\Phi(E_1,\ldots,E_m)$ and $\forall_{1\leq i\leq m} F_i(B){\subseteq}E_i(B)$ (where E_i represents F_i):

Let B be the n-ary R-box $<I_1,\ldots,I_n>$. Accordingly to definition 3.2-3, if $F_E{\equiv}\Phi(E_1,\ldots,E_m)$ then $F(B)$ is the smallest real interval containing the range $f^*(B')$ with $B'{=}<I_1,\ldots,I_n,I_{n+1},\ldots,I_{n+k}>$ (where $I_{n+1},\ldots,I_{n+k}$ are all the interval constants appearing in F_E) and f is expressed as $f_E{\equiv}\Phi(e_1,\ldots,e_m)$ (obtained by replacing in F_E each interval variable X_i by the real variable x_i, each interval constant I_{n+j} by the real variable x_{n+j} and each interval operator by the corresponding real operator).

Let f_i be the real function represented by e_i, and s_i the variables appearing in e_i (with $1\leq i\leq m$). In this case the range $f^*(B')$ is:

$$f^*(B'){=} \{ f(d) \mid d{\in}B'\} = \{\Phi(f_1(d[s_1]),\ldots,f_m(d[s_m])) \mid d{\in}B'\}$$

Moreover, the range of each f_i over $B'[s_i]$ (with $1\leq i\leq m$) is:

$f_i^*(B'[s_i])=\{ f_i(d_i) \mid d_i \in B'[s_i]\}$

and accordingly to definition 3.2-3, $F_i(B)$ is the smallest real interval containing it:

$f_i^*(B'[s_i]) \subseteq F_i(B)$

Consequently:

$f^*(B') = \{\Phi(f_1(d[s_1]),\ldots,f_m(d[s_m]))|d \in B'\} \subseteq \{\Phi(r_1,\ldots,r_m) \mid r_1 \in F_1(B) \wedge \ldots \wedge r_m \in F_m(B)\}$

But by Assumption A1:

$f^*(B') \subseteq \{\Phi(r_1,\ldots,r_m) \mid r_1 \in F_1(B) \wedge \ldots \wedge r_m \in F_m(B)\} \subseteq \Phi(F_1(B),\ldots, F_m(B))$

Because $F(B)$ is the smallest real interval enclosing the range $f^*(B')$ and $\Phi(F_1(B),\ldots, F_m(B))$ is also a real interval enclosing it:

$F(B) \subseteq \Phi(F_1(B),\ldots, F_m(B))$

Assuming the inductive hypothesis, $F_i(B) \subseteq E_i(B)$ (with $1 \le i \le m$), and from the inclusion monotonicity of the basic interval operators (Assumption A2), it follows that:

$F(B) \subseteq \Phi(F_1(B),\ldots, F_m(B)) \subseteq \Phi(E_1(B),\ldots, E_m(B))$ $\qquad$ (5)

On the other hand, in the case of $F_E \equiv \Phi(E_1,\ldots,E_m)$ (see definition 3.2-4 and Assumption A3):

$F_E(B) = \Phi_{apx}(E_1(B),\ldots, E_m(B)) = I_{apx}(\Phi(E_1(B),\ldots, E_m(B)))$ $\qquad$ (6)

From (5) and (6) it follows:

$F(B) \subseteq F_E(B)$ $\qquad\qquad$ ❸
$\qquad\qquad\qquad\qquad\qquad\qquad$ ∎

Theorem 3.2.1-1 (Soundness of the Evaluation of an Interval Extension). Let F be an interval extension of an n-ary real function f. Let F_E be an interval expression representing F. Let B be an n-ary R-box. Then, both $F(B)$ and $F_E(B)$, enclose the range of f over B:

$f^*(B) \subseteq F(B) \subseteq F_E(B)$ $\qquad\qquad$ ❏

Proof:

By the definition of the range $f^*(B)$ of a real function f over the n-ary R-box B:

$f^*(B) = \{ f(<r_1,\ldots,r_n>) \mid <r_1,\ldots,r_n> \in B\}$ $\qquad$ (1)

By definition 3.2.1-1, if F is an interval extension of f within B then:

$\forall_{<r_1,\ldots,r_n> \in B}\ f(<r_1,\ldots,r_n>) \in F(<[r_1..r_1],\ldots,[r_n..r_n]>)$

If F_E is an interval expression representing F then F must be inclusion monotonic (Lemma A6), so:

$<[r_1..r_1],\ldots,[r_n..r_n]> \subseteq B \Rightarrow F(<[r_1..r_1],\ldots,[r_n..r_n]>) \subseteq F(B)$

thus:

$\forall_{<r_1,\ldots,r_n> \in B}\ f(<r_1,\ldots,r_n>) \subseteq F(B)$ $\qquad$ (2)

From (1) and (2) it follows:

$f^*(B) = \{ f(<r_1,\ldots,r_n>) \mid <r_1,\ldots,r_n> \in B\} \subseteq F(B)$

and from theorem 3.2-1, $F(B) \subseteq F_E(B)$, consequently:

$f^*(B) \subseteq F(B) \subseteq F_E(B)$ $\qquad\qquad\qquad\qquad$ ∎

Theorem 3.2.1-2 (Natural Interval Extension). Let f_E be a real expression representing the real function f. Let F_n be the natural interval expression of f_E. The interval function F represented by F_n is the smallest interval enclosing for the range of f and the interval arithmetic evaluation of F_n is an interval extension of f denominated Natural interval extension with respect to f_E. $\qquad$ ❏

Proof:

The intended meaning of an interval expression F_E, as defined in 3.2-3, is to represent an interval function F which applying to an R-box B obtains the smallest real interval containing the range, within this box, of an associated real function. In the particular case of the Natural interval extension F_n with respect to f_E, the associated real function is the function represented by the real expression f_E. Thus by definition 3.2-3, F obtains the smallest interval enclosing for the range of f. Moreover, from theorem 3.2-1, $F(B) \subseteq F_n(B)$, and so interval arithmetic evaluation of F_n is also an interval extension of f.

∎

Theorem 3.2.1-3 (Intersection of Interval Extensions). Let F_1 and F_2 be two n-ary interval functions and B an n-ary R-box. Let F be an n-ary interval function defined by: $F(B)=F_1(B) \cap F_2(B)$. If F_1 and F_2 are interval extensions of the real function f, then F is also an interval extension of f. ❏

Proof:

By definition 3.2.1-1, if F_1 and F_2 are interval extensions of f within B then:

$$\forall_{<r_1,\ldots,r_n> \in B}\ f(<r_1,\ldots,r_n>) \in F_1(<[r_1..r_1],\ldots,[r_n..r_n]>)$$

$$\forall_{<r_1,\ldots,r_n> \in B}\ f(<r_1,\ldots,r_n>) \in F_2(<[r_1..r_1],\ldots,[r_n..r_n]>)$$

Consequently:

$$\forall_{<r_1,\ldots,r_n> \in B}[f(<r_1,\ldots,r_n>) \in F_1(<[r_1..r_1],\ldots,[r_n..r_n]>) \wedge f(<r_1,\ldots,r_n>) \in F_2(<[r_1..r_1],\ldots,[r_n..r_n]>)]$$

which is equivalent to:

$$\forall_{<r_1,\ldots,r_n> \in B}\ [f(<r_1,\ldots,r_n>) \in F_1(<[r_1..r_1],\ldots,[r_n..r_n]>) \cap F_2(<[r_1..r_1],\ldots,[r_n..r_n]>)]$$

and if F is the interval function defined by $F(B)=F_1(B) \cap F_2(B)$ then:

$$\forall_{<r_1,\ldots,r_n> \in B}\ f(<r_1,\ldots,r_n>) \in F(<[r_1..r_1],\ldots,[r_n..r_n]>)$$

which proves (definition 3.2.1-1) that F is also an interval extension of f. ∎

Theorem 3.2.1-4 (Decomposed Evaluation of an Interval Extension). Let F be an interval extension of the n-ary real function f. Let F_E be an interval expression representing F. Let B, B_1 and B_2 be n-ary R-boxes. If $B=B_1 \cup B_2$ then:
$$F(B) \subseteq F_E(B_1) \cup F_E(B_2) \subseteq F_E(B)$$
❏

Proof:

The proof of the above theorem is divided in two sub-proofs. In the first it is demonstrated that $F_E(B_1) \cup F_E(B_2) \subseteq F_E(B)$ whereas in the second sub-proof $F(B) \subseteq F_E(B_1) \cup F_E(B_2)$ is asserted.

① Proof that $F_E(B_1) \cup F_E(B_2) \subseteq F_E(B)$ with $B=B_1 \cup B_2$ and F_E an interval expression:

If $B=B_1 \cup B_2$ then $B_1 \subseteq B$ and $B_2 \subseteq B$, and from the inclusion monotonicity property of the evaluation of an interval expression F_E (Lemma A5):

$$F_E(B_1) \subseteq F_E(B)$$
$$F_E(B_2) \subseteq F_E(B)$$

and consequently (carrying out the union of the two left sides and the two right sides):

$$F_E(B_1) \cup F_E(B_2) \subseteq F_E(B) \cup F_E(B) = F_E(B)$$
❶

② Proof that $F(B) \subseteq F_E(B_1) \cup F_E(B_2)$ with $B=B_1 \cup B_2$ and F_E an interval expression representing F which is an interval extension of the n-ary real function f:

Let B be the n-ary R-box $<I_1,\ldots,I_n>$. Accordingly to definition 3.2-3, if F_E is an interval expression representing F then $F(B)$ is the smallest real interval containing the range $g^*(B')$

of a real function (not necessarily f) with $B'=<I_1,\ldots,I_n,I_{n+1},\ldots,I_{n+k}>$ (where $I_{n+1},\ldots,I_{n+k}$ are all the interval constants appearing in F_E)

If $B=B_1\cup B_2$ (with $B_1=<I1_1,\ldots,I1_n>$ and $B_2=<I2_1,\ldots,I2_n>$) then the range $g^*(B')$ is:

$$g^*(B') = \{g(<r_1,\ldots,r_n,r_{n+1},\ldots,r_{n+k}>) \mid <r_1,\ldots,r_n>\in B \wedge <r_{n+1},\ldots,r_{n+k}>\in<I_{n+1},\ldots,I_{n+k}>\}$$

$$g^*(B') = \{g(<r_1,\ldots,r_n,r_{n+1},\ldots,r_{n+k}>) \mid <r_1,\ldots,r_n>\in B_1 \wedge <r_{n+1},\ldots,r_{n+k}>\in<I_{n+1},\ldots,I_{n+k}>\}$$

$$\cup\{g(<r_1,\ldots,r_n,r_{n+1},\ldots,r_{n+k}>)\mid <r_1,\ldots,r_n>\in B_2 \wedge<r_{n+1},\ldots,r_{n+k}>\in<I_{n+1},\ldots,I_{n+k}>\}$$

$$= g^*(B_1') \cup g^*(B_2') \quad (B_1'=<I1_1,\ldots,I1_n,I_{n+1},\ldots,I_{n+k}> \text{ and } B_2'=<I2_1,\ldots,I2_n,I_{n+1},\ldots,I_{n+k}>)$$

But again, accordingly to definition 3.2-3, $F(B_1)$ and $F(B_2)$ are the smallest real intervals enclosing $g^*(B_1')$ and $g^*(B_2')$ respectively, thus:

$$g^*(B') = g^*(B_1') \cup g^*(B_2') \subseteq F(B_1) \cup F(B_2)$$

By theorem 3.2.1-1 $F(B_1) \subseteq F_E(B_1)$ and $F(B_2) \subseteq F_E(B_2)$, so:

$$g^*(B') \subseteq F(B_1) \cup F(B_2) \subseteq F_E(B_1) \cup F_E(B_2)$$

From Lemma A2 we know that if B, B_1 and B_2 are n-ary R-boxes with $B=B_1\cup B_2$ then $F_E(B_1)\cup F_E(B_2)$ is a real interval. Because $F(B)$ is the smallest real interval enclosing the range $g^*(B')$ and $F_E(B_1)\cup F_E(B_2)$ is also a real interval enclosing it:

$$F(B) \subseteq F_E(B_1)\cup F_E(B_2)$$

❷

■

Theorem 3.2.1-5 (No Overestimation Without Multiple Variable Occurrences). Let F_E be an interval expression representing the n-ary interval function F. Let B be an n-ary R-box. If F_E is an interval expression in which each variable occurs only once, then:

$F(B) = F_E(B)$ (with exact interval operators and infinite precision arithmetic evaluation)❑

Proof:

Definition 3.2-1 gives an inductive definition for the set of all possible interval expressions: the two base clauses (i and ii) specify that interval constants (I) and interval variables (X_i) are the basic elements of the defined set; the inductive clause (iii) specifies how to generate additional elements from existing elements $E_1,\ldots,E_m$ and an m-ary basic interval arithmetic operator Φ.

We will present an inductive proof that for every interval expression F_E belonging to this set, if F_E is an interval expression in which each variable occurs only once then $F(B)=F_E(B)$ (where F is the n-ary interval function represented by F_E, B is an n-ary R-box). During this proof it is assumed that each basic interval operator appearing in F_E is able to compute the exact ranges within its interval arguments and all the interval arithmetic evaluations are performed with infinite precision. The proof consists on a basis step which shows that the property holds for the basic elements ($F(B)=F_E(B)$ with $F_E\equiv I$ and with $F_E\equiv X_i$) and an inductive step which shows that if the property holds for some elements $E_1,\ldots,E_m$ (inductive hypothesis: $\forall_{1\le i\le m} F_i(B)=E_i(B)$ where E_i represents F_i) then it holds for any elements generated from them by the inductive clause ($F(B)=F_E(B)$ with $F_E\equiv\Phi(E_1,\ldots,E_m)$).

Basis Step

① Proof that $F(B)=F_E(B)$ with $F_E\equiv I$ (where I is a real interval):

Accordingly to definition 3.2-3 (extended for allowing the representation of interval constants – see footnote), in the case of $F_E\equiv I$, $F(B)$ is the smallest real interval containing the range $f^*(I)$ with f expressed as $f_E\equiv x_1$. By the definition of the range of a real function f over I:

$$f^*(I)=\{f(r_1) \mid r_1\in I\}$$

and in this particular case:

$$f^*(I)=\{r_1 \mid r_1 \in I\}=I$$

Since I is a real interval, it is the smallest real interval containing the range $f^*(I)$ and so:

$\quad F(B)=I$ $\qquad\qquad\qquad\qquad$ (1)

Accordingly to definition 3.2-4, in the case of $F_E\equiv I$:

$\quad F_E(B)=I_{apx}(I)$

Because we are assuming infinite precision arithmetic $I_{apx}(I)=I$ and so:

$\quad F_E(B)=I_{apx}(I)=I$ $\qquad\qquad$ (2)

From (1) and (2) it follows:

$\quad F(B)=F_E(B)$ $\qquad\qquad\qquad\qquad$ ❶

② Proof that $F(B)=F_E(B)$ with $F_E\equiv X_i$ (where X_i is an interval valued variable):

Accordingly to definition 3.2-3, in the case of $F_E\equiv X_i$, $F(B)$ is the smallest real interval containing the range $f^*(I)$ with $I=B[X_i]$ and f expressed as $f_E\equiv x_i$. So, in this particular case:

$$f^*(I)=\{r_i \mid r_i \in I\}= I =B[X_i]$$

Since $B[X_i]$ is a real interval, it is the smallest real interval containing the range $f^*(I)$ and so:

$\quad F(B)=B[X_i]$ $\qquad\qquad\qquad\qquad$ (3)

Accordingly to definition 3.2-4, in the case of $F_E\equiv X_i$:

$\quad F_E(B)=I_{apx}(B[X_i])$

Because we are assuming infinite precision arithmetic $I_{apx}(B[X_i])=B[X_i]$ and so:

$\quad F_E(B)=I_{apx}(B[X_i])=B[X_i]$ $\qquad$ (4)

From (3) and (4) it follows:

$\quad F(B)=F_E(B)$ $\qquad\qquad\qquad\qquad$ ❷

Inductive Step

③ Proof that $F(B)=F_E(B)$ with $F_E\equiv\Phi(E_1,\ldots,E_m)$ and $\forall_{1\leq i\leq m}\ F_i(B)=E_i(B)$ (where E_i represents F_i):

Let B be the n-ary R-box $<I_1,\ldots,I_n>$. Accordingly to definition 3.2-3, if $F_E\equiv\Phi(E_1,\ldots,E_m)$ then $F(B)$ is the smallest real interval containing the range $f^*(B')$ with $B'=<I_1,\ldots,I_n,I_{n+1},\ldots,I_{n+k}>$ (where $I_{n+1},\ldots,I_{n+k}$ are all the interval constants appearing in F_E) and f is expressed as $f_E\equiv\Phi(e_1,\ldots,e_m)$ (obtained by replacing in F_E each interval variable X_i by the real variable x_i, each interval constant I_{n+j} by the real variable x_{n+j} and each interval operator by the corresponding real operator).

Let f_i be the real function represented by e_i, and s_i the variables appearing in e_i (with $1\leq i\leq m$):

$$f^*(B')= \{ f(d) \mid d \in B'\} = \{\Phi(f_1(d[s_1]),\ldots,f_m(d[s_m])) \mid d \in B'\}$$

Because there are no multiple occurrences of the X_i variables, each s_i must contain a different subset (with no common variables) of the all set of variables. Consequently, if d_i denotes a tuple of real values associated with the subset s_i, the range $f^*(B')$ may be rewritten as:

$$f^*(B')=\{\Phi(f_1(d[s_1]),\ldots,f_m(d[s_m]))|d\in B'\}=\{\Phi(f_1(d_1),\ldots,f_m(d_m))|d_1\in B'[s_1]\wedge\ldots\wedge d_m\in B'[s_m]\}$$

The range of each f_i over $B'[s_i]$ (with $1\leq i\leq m$) is:

$$f_i^*(B'[s_i])=\{ f_i(d_i) \mid d_i\in B'[s_i]\}$$

Note that in the basis step it was proved that the range associated with each basic element ($F_E\equiv I$ and with $F_E\equiv X_i$) is a real interval. In the following we will prove that if the ranges f_i^* associated with each $E_1,\ldots,E_m$ are all real intervals then the range f^* associated with $F_E\equiv\Phi(E_1,\ldots,E_m)$ must also be a real interval. This will prove by induction that the range associated with any interval expression (in the conditions of the theorem) is a real interval.

If the range $f_i^*(B'[s_i])$ is a real interval then, accordingly to definition 3.2-3, $F_i(B)$ must be exactly this range since it is the smallest real interval containing it:

$f_i^*(B'[s_i])=F_i(B)$

Consequently:

$f^*(B')=\{\Phi(f_1(d_1),\ldots,f_m(d_m))|d_1\in B'[s_1]\wedge\ldots\wedge d_m\in B'[s_m]\}=\{\Phi(r_1,\ldots,r_m)|r_1\in F_1(B)\wedge\ldots\wedge r_m\in F_m(B)\}$

But by Assumption A1, in the particular case that Φ is an m-ary basic interval operator which is able to compute the exact ranges within its interval arguments:

$f^*(B') = \{\Phi(r_1,\ldots,r_m) \mid r_1\in F_1(B) \wedge\ldots\wedge r_m\in F_m(B)\} = \Phi(F_1(B),\ldots, F_m(B))$

$f^*(B')$ must be a real interval because it is defined as the result of a basic interval operation and the result of a basic interval operation is a real interval (Assumption A1). This completes the proof that the range associated with any interval expression (in the conditions of the theorem) is a real interval.

Because $F(B)$ is the smallest real interval enclosing $f^*(B')$ and $\Phi(F_1(B),\ldots, F_m(B))$ is a real interval:

$F(B) = \Phi(F_1(B),\ldots, F_m(B))$

Assuming the inductive hypothesis, $F_i(B)=E_i(B)$ (with $1\leq i\leq m$) it follows that:

$F(B) = \Phi(F_1(B),\ldots, F_m(B)) = \Phi(E_1(B),\ldots, E_m(B))$ (5)

On the other hand, and accordingly to definition 3.2-4, in the case of $F_E\equiv\Phi(E_1,\ldots,E_m)$:

$F_E(B)= \Phi_{apx}(E_1(B),\ldots, E_m(B))$

Because we are assuming infinite precision arithmetic, with Assumption A3:

$F_E(B) = \Phi_{apx}(E_1(B),\ldots, E_m(B)) = \Phi(E_1(B),\ldots, E_m(B))$ (6)

From (5) and (6) it follows:

$F(B)=F_E(B)$ ❸

■

Appendix B

Constraint Propagation Theorems

The demonstrations of the first three Constraint Propagation theorems (4.1-1, 4.1-2 and 4.1-3) rely on one of the following assumptions about the restrictions on the representation of the domains of the variables of a CCSP (see subsection 2.2.5).

Assumption B-1 Let $P=(X,D,C)$ be a CCSP where X is the n-ary tuple of variables $<x_1,\ldots,x_n>$ and D is the Cartesian Product of their respective original domains $D_1 \times D_2 \times \ldots \times D_n$. The domains of the variables of the CCSP must be represented by unions of n-ary F-boxes. The domain of any narrowing function associated with any constraint of C is the set of all the elements within $I_{apx}(D_1) \times I_{apx}(D_2) \times \ldots \times I_{apx}(D_n)$ that are representable by the union of n-ary F-boxes. ❏

Assumption B-2 Let $P=(X,D,C)$ be a CCSP where X is the n-ary tuple of variables $<x_1,\ldots,x_n>$ and D is the Cartesian Product of their respective original domains $D_1 \times D_2 \times \ldots \times D_n$. The domains of the variables of the CCSP must be represented by single n-ary F-boxes. The domain of any narrowing function associated with any constraint of C is the set of all the elements within $I_{apx}(D_1) \times I_{apx}(D_2) \times \ldots \times I_{apx}(D_n)$ that are representable by a single n-ary F-box. ❏

The following assumption is used in the demonstration of theorem 4.2.1-1.

Assumption B-3 Let $c=(s,\rho)$ be an n-ary primitive constraint of a CCSP $P=(X,D,C)$ with $s=<x_1,\ldots,x_n>$. Let Φ be an m-ary basic operator and e_i (with $0 \le i \le m$) a variable from s or the real constant k_{e_i}. Let v_{e_i} be either r_j if e_i is the variable x_j or k_{e_i} if e_i is the constant k_{e_i}. Let $\diamond \in \{\le,=,\ge\}$ and K be a set of real numbers defined as:

$$K = \begin{cases} [0..+\infty] & \text{if } \diamond \equiv \le \\ [0] & \text{if } \diamond \equiv = \\ [-\infty..0] & \text{if } \diamond \equiv \ge \end{cases}$$

If c is expressed in the form $e_1 \diamond e_0$ then it represents the relation (with $n \le 2$):
$$\rho = \{<r_1,\ldots,r_n> \mid r_1 \in D[x_1] \wedge \ldots \wedge r_n \in D[x_n] \wedge v_{e_1} = v_{e_0} + k \wedge k \in K\}$$

If c is expressed in the form $\Phi(e_1,\ldots,e_m) \diamond e_0$ then it represents the relation (w/ $n \le m+1$):
$$\rho = \{<r_1,\ldots,r_n> \mid r_1 \in D[x_1] \wedge \ldots \wedge r_n \in D[x_n] \wedge \Phi(v_{e_1},\ldots, v_{e_m}) = v_{e_0} + k \wedge k \in K\}$$

and is assumed that for any $1 \le i \le m$ there is a basic Φ_{e_i} operator such that:
$$\rho = \{<r_1,\ldots,r_n> \mid r_1 \in D[x_1] \wedge \ldots \wedge r_n \in D[x_n] \wedge \Phi_{e_i}(v_{e_0}+k,v_{e_1},\ldots,v_{e_{i-1}},v_{e_{i+1}},\ldots,v_{e_m}) = v_{e_i} \wedge k \in K\} ❏$$

The following lemmas will be used in the demonstrations of the Constraint Propagation theorems.

Lemma B-1 If the union of two fixed points, A_1 and A_2, of a monotonic narrowing function NF is an element of its domain then it is also a fixed point:
$$\forall_{A_1,A_2 \in \text{Domain}_{NF}} NF(A_1)=A_1 \wedge NF(A_2)=A_2 \wedge A_1 \cup A_2 \in \text{Domain}_{NF} \Rightarrow NF(A_1 \cup A_2)=A_1 \cup A_2 ❏$$

Proof:

If *NF* is a monotonic narrowing function, because $A_1 \subseteq A_1 \cup A_2$ and both A_1 and $A_1 \cup A_2$ are elements of its domain then, from definition 4.1-2 (property P3) it follows:

$NF(A_1) \subseteq NF(A_1 \cup A_2)$

Since A_1 is a fixed point of the narrowing function, $NF(A_1)=A_1$ and, rewriting the left side:

$A_1 \subseteq NF(A_1 \cup A_2)$ (1)

If *NF* is a monotonic narrowing function, because $A_2 \subseteq A_1 \cup A_2$ and both A_2 and $A_1 \cup A_2$ are elements of its domain then, from definition 4.1-2 (property P3) it follows:

$NF(A_2) \subseteq NF(A_1 \cup A_2)$

Since A_2 is a fixed point of the narrowing function $NF(A_2)=A_2$, rewriting the left side:

$A_2 \subseteq NF(A_1 \cup A_2)$ (2)

From (1) and (2), carrying out the union of the two left sides and the two right sides, it follows:

$A_1 \cup A_2 \subseteq NF(A_1 \cup A_2) \cup NF(A_1 \cup A_2) = NF(A_1 \cup A_2)$ (3)

On the other hand, due to the contractance property of any narrowing function (definition 4.1-1, property P1):

$NF(A_1 \cup A_2) \subseteq A_1 \cup A_2$ (4)

Consequently, from (3) and (4), $A_1 \cup A_2$ must be a fixed point of *NF*:

$NF(A_1 \cup A_2) = A_1 \cup A_2$ ∎

Lemma B-2 If the elements of the domain of a monotonic narrowing function *NF* are those representable by unions of *n*-ary *F*-boxes then the union of any two fixed points, A_1 and A_2, is also a fixed point:

$$\forall_{A_1,A_2 \in \text{Domain}_{NF}} NF(A_1)=A_1 \wedge NF(A_2)=A_2 \Rightarrow NF(A_1 \cup A_2)= A_1 \cup A_2 \quad \square$$

Proof:

If A_1 and A_2, are representable by unions of *n*-ary *F*-boxes then they can be rewritten as:

$A_1 = B_1' \cup \ldots \cup B_k'$

$A_2 = B_1'' \cup \ldots \cup B_m''$

Where each B_i' (with $1 \leq i \leq k$) and each B_j'' (with $1 \leq i \leq k$) are *n*-ary *F*-boxes.

Consequently, the union of these elements is:

$A_1 \cup A_2 = B_1' \cup \ldots \cup B_k' \cup B_1'' \cup \ldots \cup B_m''$

Which is also representable by unions of *n*-ary *F*-boxes and so, is an element of the domain of the narrowing function:

$A_1 \cup A_2 \in \text{Domain}_{NF}$

Therefore, from Lemma B1, it follows that $A_1 \cup A_2$ is also a fixed point of the narrowing function:

$NF(A_1 \cup A_2) = A_1 \cup A_2$ ∎

Lemma B-3 If the elements of the domain of a monotonic narrowing function *NF* are those representable by unions of *n*-ary *F*-boxes then the union of all its fixed points within an element *A* of its domain is its greatest fixed point:

$\cup \text{Fixed-Points}_{NF}(A) \in \text{Fixed-Points}_{NF}(A)$

$\forall_{A_i \in \text{Fixed-Points}_{NF}(A)} A_i \subseteq \cup \text{Fixed-Points}_{NF}(A)$ $\square$

Proof:

Consider the set of all fixed points of *NF* within an element *A* of its domain:

Fixed-Points$_{NF}(A)$ = { $A_i \in$ Domain$_{NF}$ | $A_i \subseteq A \wedge NF(A_i) = A_i$ }

This set must be a finite set because the number of elements A_i of the domain of *NF* which are subsets of *A* must be finite (see subsection 2.2.5). Thus if the number of its elements is *m* the set may be represented as:

Fixed-Points$_{NF}(A)$ = { $A_i \in$ Domain$_{NF}$ | $A_i \subseteq A \wedge NF(A_i) = A_i$ } = { $A_1,..., A_m$ }

Consider the union of all the elements of the above set:

$\cup$Fixed-Points$_{NF}(A) = A_1 \cup ... \cup A_m$

Let *A'* be the element of the domain of *NF* resulting from the union of A_{m-1} and A_m:

$A' = A_{m-1} \cup A_m$

As seen in Lemma B2 the union of these two elements is also representable by unions of *n*-ary *F*-boxes and so *A'* must be an element of the domain of *NF*. *A'* must be within *A* because both A_{m-1} and A_m are within *A* and, since A_{m-1} and A_m are fixed points of *NF*, from Lemma B2, the element *A'* is also a fixed point of *NF*. Consequently it must be a member of the set of all fixed points of *NF* within *A*:

$A' \in \{ A_1,..., A_m \}$

So if we add the *F*-box *A'* to the union of all fixed points of *NF*, it will have no effect:

$\cup$Fixed-Points$_{NF}(A) = A_1 \cup ... \cup A_m = A_1 \cup ... \cup A_m \cup A'$

But if $A' = A_{m-1} \cup A_m$ then the elements A_{m-1} and A_m may be removed from the union of all fixed points of *NF* without changing its result:

$\cup$Fixed-Points$_{NF}(A) = A_1 \cup ... \cup A_m \cup A' = A_1 \cup ... \cup A_{m-2} \cup A'$

This way, the union of all fixed points of *NF*, which was represented by a union of *m* fixed points of *NF*, is now equivalently represented by a union of *m*-1 fixed points of *NF*.

Repeating the above procedure *m*-1 times, the union of all fixed points of *NF* will be represented by a single element of the domain of *NF*. This element must be a fixed point of *NF* (all the elements in the union are fixed points of *NF*) and must be its greatest fixed point since it includes all the other fixed points (it results from the union of all of them). ■

Lemma B-4 If the elements of the domain of a monotonic narrowing function *NF* are those representable by a single *n*-ary *F*-box then the smallest *F*-box *B* enclosing any two fixed points, B_1 and B_2, is also a fixed point:

$$\forall B_1,B_2 \in \text{Domain}_{NF}\ NF(B_1)=B_1 \wedge NF(B_2)=B_2 \Rightarrow NF(B)=B \qquad \square$$

Proof:

Let the fixed points B_1 and B_2, be the *n*-ary *F*-boxes $<I_1',...,I_n'>$ and $<I_1'',...,I_n''>$ respectively. If *B* is the smallest *F*-box enclosing B_1 and B_2 then:

$B=<I_1' \uplus I_1'',...,I_n' \uplus I_n''>$

If *NF* is a monotonic narrowing function then, because $B_1 \subseteq B$ and $B_2 \subseteq B$, it follows:

$NF(B_1) \subseteq NF(B)$

$NF(B_2) \subseteq NF(B)$

Because B_1 and B_2 are fixed points of *NF*, $NF(B_1)=B_1$ and $NF(B_2)=B_2$, hence:

$B_1 \subseteq NF(B)$ (1)

$B_2 \subseteq NF(B)$ (2)

We will prove by contradiction that the above *F*-box *B* must be a fixed point of *NF*. We will assume that *B* is not a fixed point of *NF* and prove a contradiction.

Assuming that *B* is not a fixed point of *NF* then, due to the contractance property (definition 4.1-1, property P1), *NF(B)* must be a proper subset of *B* (if equality holds then *B* would be a fixed point):

$NF(B) \subset B$

If $<I_1,...,I_n>$ denotes the *F*-box obtained by *NF(B)* then:

$$\langle I_1,\ldots,I_n \rangle \subset \langle I_1' \uplus I_1'',\ldots,I_n' \uplus I_n'' \rangle$$

and so there must be an i, between 1 and n, for which:

$$I_i \subset I_i' \uplus I_i''$$

Let $I_i=[a..b]$, $I_i'=[a'..b']$ and $I_i''=[a''..b'']$ (these are all closed intervals since B, B' and B'' are F-boxes) then, by definition 2.2.2-1 of union hull:

$$[a..b] \subset [a'..b'] \uplus [a''..b''] = [\min(\lfloor a' \rfloor, \lfloor a'' \rfloor)..\max(\lceil b' \rceil, \lceil b'' \rceil)] = [\min(a',a'')..\max(b', b'')]$$

Implying that one of the following inequalities must be true:

$$\min(a',a'') < a \qquad\qquad\qquad\qquad\qquad (3)$$
$$\max(b',b'') > b \qquad\qquad\qquad\qquad\qquad (4)$$

From (3) or (4) it follows that $I_i' \not\subseteq I_i$ or $I_i'' \not\subseteq I_i$ because:

if (3) is true and $a' \leq a''$ then $I_i' \not\subseteq I_i$

if (3) is true and $a'' \leq a'$ then $I_i'' \not\subseteq I_i$

if (4) is true and $b' \leq b''$ then $I_i'' \not\subseteq I_i$

if (4) is true and $b'' \leq b'$ then $I_i' \not\subseteq I_i$

But if $I_i' \not\subseteq I_i$ then $B_1 \not\subseteq NF(B)$ contradicting (1) and if $I_i'' \not\subseteq I_i$ then $B_2 \not\subseteq NF(B)$ contradicting (2). In any case a contradiction is derived which proves that B must be a fixed point of NF. ∎

Lemma B-5 If the elements of the domain of a monotonic narrowing function NF are those representable by a single n-ary F-box then the union of all its fixed points within an element B of its domain is its greatest fixed point:

$$\cup\text{Fixed-Points}_{NF}(B) \in \text{Fixed-Points}_{NF}(B)$$
$$\forall_{B_i \,\in\, \text{Fixed-Points}_{NF}(B)} \; B_i \subseteq \cup\text{Fixed-Points}_{NF}(B) \qquad\qquad ❑$$

Proof:
Consider the set of all fixed points of NF within the F-box B:

$$\text{Fixed-Points}_{NF}(B) = \{\, B_i \in \text{Domain}_{NF} \mid B_i \subseteq B \,\wedge\, NF(B_i) = B_i \,\}$$

This set must be a finite set because the number of F-boxes B_i which are subsets of an F-box B must be finite (see subsection 2.2.5). Thus if the number of its elements is m the set may be represented as:

$$\text{Fixed-Points}_{NF}(B) = \{\, B_i \in \text{Domain}_{NF} \mid B_i \subseteq B \,\wedge\, NF(B_i) = B_i \,\} = \{\, B_1,\ldots, B_m \,\}$$

Consider the union of all the elements of the above set:

$$\cup\text{Fixed-Points}_{NF}(B) = B_1 \cup \ldots \cup B_m$$

Let B' be the smallest F-box enclosing both B_{m-1} and B_m. B' must be within F-box B because both F-boxes B_{m-1} and B_m are within B. Moreover, since B_{m-1} and B_m are fixed points of NF, from Lemma B4, the F-box B' is also a fixed point of NF. Consequently it must be a member of the set of all fixed points of NF within B:

$$B' \in \{\, B_1,\ldots, B_m \,\}$$

So if we add the F-box B' to the union of all fixed points of NF, it will have no effect:

$$\cup\text{Fixed-Points}_{NF}(B) = B_1 \cup \ldots \cup B_m = B_1 \cup \ldots \cup B_m \cup B'$$

But if B' encloses both B_{m-1} and B_m then $B_{m-1} \cup B_m \subseteq B'$ and so the F-boxes B_{m-1} and B_m may be removed from the union of all fixed points of NF without changing its result:

$$\cup\text{Fixed-Points}_{NF}(B) = B_1 \cup \ldots \cup B_m \cup B' = B_1 \cup \ldots \cup B_{m-2} \cup B'$$

This way, the union of all fixed points of NF, which was represented by a union of m F-boxes (fixed points of NF), is now equivalently represented by a union of m-1 F-boxes (also all of them fixed points of NF).

Repeating the above procedure $m-1$ times, the union of all fixed points of NF will be represented by a single F-box. This F-box must be a fixed point of NF (all the F-boxes in the union are fixed points of NF) and must be its greatest fixed point since it includes all the other fixed points (it results from the union of all of them). $\triangledown$

Theorem 4.1-1 (Union of Fixed-Points). Let $P=(X,D,C)$ be a CCSP. Let NF be a monotonic narrowing function associated with a constraint of C. Let A be an element of $Domain_{NF}$. The union of all fixed-points of NF within A, denoted $\cup$Fixed-Points$_{NF}(A)$, is the greatest fixed-point of NF within A:

$\cup$Fixed-Points$_{NF}(A) \in$ Fixed-Points$_{NF}(A)$

$\forall_{A_i} \in$ Fixed-Points$_{NF}(A)$ $A_i \subseteq \cup$Fixed-Points$_{NF}(A)$ $\square$

Proof:

Let X be the n-ary tuple of variables $<x_1,...x_n>$.

If Assumption B1 is considered then, the elements of the domain of the monotonic narrowing function NF are those representable by unions of n-ary F-boxes and so, from Lemma B3, the union of all its fixed points within an element of its domain is its greatest fixed point.

If Assumption B2 is considered then, the elements of the domain of a monotonic narrowing function NF are those representable by a single n-ary F-box, from Lemma B5, the union of all its fixed points within an element of its domain is its greatest fixed point.

In either case, it is proved that the union of all the fixed points of NF within an element of its domain is its greatest fixed point.   ∎

Theorem 4.1-2 (Contraction Applying a Narrowing Function). Let $P=(X,D,C)$ be a CCSP. Let NF be a monotonic narrowing function associated with a constraint of C and A an element of $Domain_{NF}$. The greatest fixed-point of NF within A is included in the element obtained by applying NF to A:

$\cup$Fixed-Points$_{NF}(A) \subseteq NF(A)$

In particular, if NF is also idempotent then:

$\cup$Fixed-Points$_{NF}(A) = NF(A)$ $\square$

Proof:

Let A' be the union of all the fixed points of NF within element A:

$A' = \cup$Fixed-Points$_{NF}(A)$

From theorem 4.1-1 A' is the greatest fixed point of NF within A, so it is a fixed point of NF:

$$NF(A') = A' \qquad (1)$$

But A' is within element A:

$A' \subseteq A$

Therefore, from the monotonicity property of a narrowing function (definition 4.1-2, property P3) it follows:

$NF(A') \subseteq NF(A)$

And from (1) $NF(A')$ is the same as A' which proves that:

$$A' \subseteq NF(A) \qquad (2)$$

In particular, if NF is also idempotent then from definition 4.1-2, property P4:

$NF(NF(A)) = NF(A)$

Implying that $NF(A)$ is a fixed point of the narrowing function NF. But from (2), if A' is smaller or equal than $NF(A)$, it follows that the equality must hold: $A' = NF(A)$ otherwise A' would not be the greatest fixed point of NF within A.   ∎

Theorem 4.1-3 (Properties of the Propagation Algorithm). Let $P=(X,D,C)$ be a CCSP. Let set S_0 be a set of narrowing functions (obtained from the set of constraints C). Let A_0 be an element of Domain$_{NF}$ (where $NF \in S_0$) and d an element of D ($d \in D$). The propagation algorithm *prune*(S_0, A_0) (defined in figure 4.1) terminates and is correct:

$$\forall_{d \in A_0} d \text{ is a solution of the CCSP} \Rightarrow d \in prune(S_0, A_0)$$

If S_0 is a set of monotonic narrowing functions then the propagation algorithm is confluent and computes the greatest common fixed-point included in A_0. ❑

Proof:

The propagation algorithm *prune*(S_0, A_0) (defined in figure 4.1) is a procedure that obtains smaller domain elements A_i (see assumption B1 and B2) from an original element A_0 by consecutively applying narrowing functions from a set S_0 (obtained from the constraints of the CCSP) until obtaining an element A_n which is a fixed point of every narrowing function within S_0.

Due to the contractance property of the narrowing functions (definition 4.1-1, property P1), the propagation algorithm terminates. The reason is that due to contractance, a smaller (or equal) representable element (accordingly to assumptions B1 or B2) is obtained from each application of a narrowing function ($A_{i+1} \subseteq A_i$). Moreover, because the set of representable elements is finite (see subsection 2.2.5) this procedure is guaranteed to stop.

Due to the correctness property of the narrowing functions (definition 4.1-1, property P2), the propagation algorithm is correct. No solution is lost because the value combinations discarded by the application of a narrowing function do not satisfy at least one constraint of the CCSP (the constraint associated with the narrowing function).

When the algorithm stops, which was proved above, the obtained element A_n is a common fixed point of every narrowing function within S_0 (otherwise a narrowing function for which the element is not a fixed point would be applied). Moreover, if all narrowing functions within S_0 are monotonic then A_n must be the greatest common fixed-point included in A_0. Otherwise, if there would have been a common fixed-point A' included in A_0 and greater than A_n ($A_n \subset A' \subseteq A_0$) then, somewhere in the narrowing sequence from A_0 to A_n there would have been a step from A_i to A_{i+1} such that:

$$A_{i+1} \subset A' \subseteq A_i$$

But in this case, A' could not be a fixed point of the narrowing function NF for which $NF(A_i)=A_{i+1}$ because theorem 4.1-2 guarantees that $NF(A_i)$ includes all the fixed points of NF within A_i. This fact contradicts the assumption that A' is a common fixed-point of every narrowing function within S_0, proving that A_n must be the greatest common fixed-point included in A_0.

Because the above result was derived independently of the order for the application of the monotonic narrowing functions within S_0, it is valid to any particular order and so the propagation algorithm is confluent and computes the greatest common fixed-point included in A_0. ■

Theorem 4.2.1-1 (Projection Function based on the Inverse Interval Expression). Let $P=(X,D,C)$ be a CCSP. Let $c=(s,\rho) \in C$ be an n-ary primitive constraint expressed in the form $e_c \diamond e_0$ where $e_c \equiv e_1$ or $e_c \equiv \Phi(e_1,\ldots,e_m)$ (with Φ an exact m-ary basic operator and e_i a variable from s or a real constant). Let Ψx_i be the inverse interval expression of c with respect to the variable x_i ($e_i \equiv x_i$). The projection function $\pi_{x_i}^{\rho}$ of the constraint c wrt variable x_i is the mapping:

$$\pi_{x_i}^{\rho}(B) = \Psi x_i(B) \cap B[x_i] \qquad \text{where } B \text{ is an } n\text{-ary real box}$$

  ❑

Proof:

The projection function $\pi_{x_i}^{\rho}$ of the constraint c wrt variable x_i is, accordingly to definition 4.2-1:

$$\pi_{x_i}^{\rho}(B) = (\rho \cap B)[x_i] \qquad (1)$$

We will prove for both cases, where the primitive constraint is either expressed in the form $e_1 \diamond e_0$ or in the form $\Phi(e_1,\ldots,e_m) \diamond e_0$, that:

$$\pi_{x_i}^{\rho}(B) = \Psi x_i(B) \cap B[x_i] \qquad (2)$$

①   Proof that $\pi_{x_i}^{\rho}(B) = \Psi x_i(B) \cap B[x_i]$ with c expressed as $e_1 \diamond e_0$:

Accordingly to Assumption B3, the relation represented by the constraint is (with $n \leq 2$):

$$\rho = \{<r_1,\ldots,r_n> \mid r_1 \in D[x_1] \wedge \ldots \wedge r_n \in D[x_n] \wedge v_{e_1} = v_{e_0} + k \wedge k \in K\}$$

and so its intersection with the F-box B is:

$$\rho \cap B = \{<r_1,\ldots,r_n> \mid r_1 \in D[x_1] \cap B[x_1] \wedge \ldots \wedge r_n \in D[x_n] \cap B[x_n] \wedge v_{e_1} = v_{e_0} + k \wedge k \in K\}$$

Since the F-box B must be a subset of the original domains of the respective variables of the CCSP:

$$B \subseteq D[s]$$

Implying that for each i (with $1 \leq i \leq n$) $B[x_i] \subseteq D[x_i]$, and so $D[x_i] \cap B[x_i] = B[x_i]$, consequently:

$$\rho \cap B = \{<r_1,\ldots,r_n> \mid r_1 \in B[x_1] \wedge \ldots \wedge r_n \in B[x_n] \wedge v_{e_1} = v_{e_0} + k \wedge k \in K\}$$

The projection of the above set with respect to the variable x_i depends if this variable is the expression e_0 or is the expression e_1.

If $x_i \equiv e_0$ then $v_{e_0} = r_i$ (see Assumption B3) and so:

$$(\rho \cap B)[x_i] = \{ r_i \mid r_1 \in B[x_1] \wedge \ldots \wedge r_n \in B[x_n] \wedge v_{e_1} = r_i + k \wedge k \in K\}$$

which is equivalent to:

$$(\rho \cap B)[x_i] = \{ r_i \mid r_1 \in B[x_1] \wedge \ldots \wedge r_n \in B[x_n] \wedge r_i = v_{e_1} - k \wedge k \in K\}$$

and may be rewritten as:

$$(\rho \cap B)[x_i] = \{ v_{e_1} - k \mid r_1 \in B[x_1] \wedge \ldots \wedge r_{i-1} \in B[x_{i-1}] \wedge r_{i+1} \in B[x_{i+1}] \ldots \wedge r_n \in B[x_n] \wedge k \in K\} \cap B[x_i]$$

Considering $I_{e_1} = B[x_j]$ if $e_1 \equiv x_j$ or $I_{e_1} = [k_{e_1}]$ if $e_i \equiv k_{e_1}$, from the definition 3.1-1 of the basic interval arithmetic operator (-) (see also Appendix A, Assumption A1), it follows:

$$(\rho \cap B)[x_i] = (I_{e_1} - K) \cap B[x_i]$$

But $I_{e_1} - K$ is the Natural interval extension of the real expression $e_1 - k$ and so, accordingly to definition 4.2.1-2, it corresponds to the interval arithmetic evaluation of the inverse interval expression of c with respect to $x_i \equiv e_0$, denoted by Ψx_i. Therefore we have proven that, for this case:

$$(\rho \cap B)[x_i] = \Psi x_i(B) \cap B[x_i]$$

If $x_i \equiv e_1$ then $v_{e_1} = r_i$ (see Assumption B3) and so:

$$(\rho \cap B)[x_i] = \{ r_i \mid r_1 \in B[x_1] \wedge \ldots \wedge r_n \in B[x_n] \wedge r_i = v_{e_0} + k \wedge k \in K\}$$

which may be rewritten as:

$$(\rho \cap B)[x_i] = \{ v_{e_0} + k \mid r_1 \in B[x_1] \wedge \ldots \wedge r_{i-1} \in B[x_{i-1}] \wedge r_{i+1} \in B[x_{i+1}] \ldots \wedge r_n \in B[x_n] \wedge k \in K\} \cap B[x_i]$$

Considering $I_{e_0} = B[x_j]$ if $e_0 \equiv x_j$ or $I_{e_0} = [k_{e_0}]$ if $e_0 \equiv k_{e_0}$, from the definition 3.1-1 of the basic interval arithmetic operator (+) (see also Appendix A, Assumption A1), it follows:

$$(\rho \cap B)[x_i] = (I_{e_0} + K) \cap B[x_i]$$

But $I_{e_0}+K$ is the Natural interval extension of the real expression e_0+k and so, accordingly to definition 4.2.1-2, it corresponds to the interval arithmetic evaluation of the inverse interval expression of c with respect to $x_i \equiv e_1$, denoted by Ψx_i. Therefore we have proven that, for this case also:

$$(\rho \cap B)[x_i] = \Psi x_i(B) \cap B[x_i]$$

Consequently, if the above is true for any possible case then, from (1), we have proven that if c expressed as $e_1 \diamond e_0$:

$$\pi_{x_i}^{\rho}(B) = \Psi x_i(B) \cap B[x_i] \qquad \qquad \mathbf{❶}$$

②   Proof that $\pi_{x_i}^{\rho}(B) = \Psi x_i(B) \cap B[x_i]$ with c expressed as $\Phi(e_1,\ldots,e_m) \diamond e_0$:

Accordingly to Assumption B3, the relation represented by the constraint is (with $n \leq m+1$):

$$\rho = \{<r_1,\ldots,r_n> \mid r_1 \in D[x_1] \wedge \ldots \wedge r_n \in D[x_n] \wedge \Phi(v_{e_1},\ldots,v_{e_m})=v_{e_0}+k \wedge k \in K\}$$

and so its intersection with the F-box B is:

$$\rho \cap B = \{<r_1,\ldots,r_n> \mid r_1 \in D[x_1] \cap B[x_1] \wedge \ldots \wedge r_n \in D[x_n] \cap B[x_n] \wedge \Phi(v_{e_1},\ldots,v_{e_m})=v_{e_0}+k \wedge k \in K\}$$

Since the F-box B must be a subset of the original domains of the respective variables of the CCSP:

$$B \subseteq D[s]$$

Implying that for each i (with $1 \leq i \leq n$) $B[x_i] \subseteq D[x_i]$, and so $D[x_i] \cap B[x_i]=B[x_i]$, consequently:

$$\rho \cap B = \{<r_1,\ldots,r_n> \mid r_1 \in B[x_1] \wedge \ldots \wedge r_n \in B[x_n] \wedge \Phi(v_{e_1},\ldots,v_{e_m})=v_{e_0}+k \wedge k \in K\}$$

The projection of the above set with respect to the variable x_i depends if this variable is the expression e_0 or is some expression e_j with j between 1 and n.

If $x_i \equiv e_0$ then $v_{e_0}= r_i$ (see Assumption B3) and so:

$$(\rho \cap B)[x_i] = \{ r_i \mid r_1 \in B[x_1] \wedge \ldots \wedge r_n \in B[x_n] \wedge \Phi(v_{e_1},\ldots,v_{e_m})=r_i +k \wedge k \in K\}$$

which is equivalent to:

$$(\rho \cap B)[x_i] = \{ r_i \mid r_1 \in B[x_1] \wedge \ldots \wedge r_n \in B[x_n] \wedge r_i = \Phi(v_{e_1},\ldots,v_{e_m})-k \wedge k \in K\}$$

and may be rewritten as:

$$(\rho \cap B)[x_i] = \{\Phi(v_{e_1},\ldots,v_{e_m})-k \mid r_1 \in B[x_1] \wedge \ldots \wedge r_{i-1} \in B[x_{i-1}] \wedge r_{i+1} \in B[x_{i+1}] \ldots \wedge r_n \in B[x_n] \wedge k \in K\} \cap B[x_i]$$

Considering $I_{e_i}= B[x_j]$ if $e_i \equiv x_j$ or $I_{e_i}=[k_{e_i}]$ if $e_i \equiv k_{e_i}$, from the definition of any exact basic interval arithmetic operator (Φ) (see Appendix A, Assumption A1), it follows:

$$(\rho \cap B)[x_i] = (\Phi(I_{e_1},\ldots, I_{e_m})-K) \cap B[x_i]$$

$\Phi(I_{e_1},\ldots,I_{e_m})-K$ is the Natural interval extension of the real expression $\Phi(e_1,\ldots,e_m)-k$ and so, accordingly to definition 4.2.1-2, it corresponds to the interval arithmetic evaluation of the inverse interval expression of c with respect to $x_i \equiv e_0$, denoted Ψx_i. Therefore we have proven that, in this case:

$$(\rho \cap B)[x_i] = \Psi x_i(B) \cap B[x_i]$$

If $x_i \equiv e_j$, with j between 1 and n, then $v_{e_j}= r_i$ and from Assumption B3 there is a basic operator Φ_{e_j} such that:

$$(\rho \cap B)[x_i] = \{r_i \mid r_1 \in B[x_1] \wedge \ldots \wedge r_n \in B[x_n] \wedge r_i = \Phi_{e_j}(v_{e_0}+k,v_{e_1},\ldots,v_{e_{j-1}},v_{e_{j+1}},\ldots,v_{e_m}) \wedge k \in K\}$$

which may be rewritten as:

$$(\rho \cap B)[x_i] = \{\Phi_{e_j}(v_{e_0}+k,v_{e_1},\ldots,v_{e_{j-1}},v_{e_{j+1}},\ldots,v_{e_m}) \mid$$
$$r_1 \in B[x_1] \wedge \ldots \wedge r_{i-1} \in B[x_{i-1}] \wedge r_{i+1} \in B[x_{i+1}] \ldots \wedge r_n \in B[x_n] \wedge k \in K\} \cap B[x_i]$$

Considering $I_{e_i}= B[x_l]$ if $e_i\equiv x_l$ or $I_{e_i}=[k_{e_i}]$ if $e_i\equiv k_{e_i}$, from the definition of any exact basic interval arithmetic operator (Φ) (see Appendix A, Assumption A1), it follows:

$$(\rho\cap B)[x_i]= \Phi_{e_j}(I_{e_0}+K,I_{e_1},\ldots,I_{e_{j-1}},I_{e_{j+1}},\ldots,I_{e_m}) \cap B[x_i]$$

But $\Phi_{e_j}(I_{e_0}+K,I_{e_1},\ldots,I_{e_{j-1}},I_{e_{j+1}},\ldots,I_{e_m})$ is the Natural interval extension of the real expression $\Phi_{e_j}(e_0+k,e_1,\ldots, e_{j-1},e_{j+1},\ldots,e_m)$ and so, accordingly to definition 4.2.1-2, it corresponds to the interval arithmetic evaluation of the inverse interval expression of c with respect to $x_i\equiv e_j$, denoted by Ψx_i. Therefore we have proven that, for this case also:

$$(\rho\cap B)[x_i] = \Psi x_i(B) \cap B[x_i]$$

Consequently, if the above is true for any possible case then, from (1), we have proven that if c expressed as $\Phi(e_1,\ldots,e_m)\Diamond e_0$:

$$\pi_{x_i}^{\rho}(B) = \Psi x_i(B) \cap B[x_i] \qquad\qquad ❷$$

$$\blacksquare$$

Theorem 4.2.2-1 (Properties of the Interval Projection). Let $P=(X,D,C)$ be a CCSP. Let $c=(s,\rho)\in C$ be an n-ary constraint and B an n-ary F-box. Let $\prod_{x_i}^{\rho B}$ be the interval projection of c wrt variable $x_i\in s$ and B. The following properties are necessarily satisfied:

   (i)   if $\Diamond \equiv$ "$=$" then $\forall_{r\in B[x_i]}\, r\in\pi_{x_i}^{\rho}(B) \Rightarrow 0\in\prod_{x_i}^{\rho B}([r])$

   (ii)   if $\Diamond \equiv$ "$\leq$" then $\forall_{r\in B[x_i]}\, r\in\pi_{x_i}^{\rho}(B) \Rightarrow left(\prod_{x_i}^{\rho B}([r])) \leq 0$

   (iii)   if $\Diamond \equiv$ "$\geq$" then $\forall_{r\in B[x_i]}\, r\in\pi_{x_i}^{\rho}(B) \Rightarrow right(\prod_{x_i}^{\rho B}([r])) \geq 0$

We will say that a real value r satisfies the interval projection condition if the right side of the respective implication (i), (ii) or (iii) is satisfied.   ❑

Proof:

Consider that the n-ary constraint c is expressed in the form $f_E\Diamond 0$. Let F_n be the Natural interval extension of f with respect to f_E (see definition 3.2.1-2). Let B be the n-ary F-box. $\langle I_1,\ldots,I_n\rangle$. From definition 4.2.2-1, the interval projection of c wrt $x_i\in s$ and B is:

$$\prod_{x_i}^{\rho B}(I) = F_n(\langle I_1,\ldots,I_{i-1},I,I_{i+1},\ldots,I_n\rangle) \quad \text{(for every } I\subseteq I_i)$$

Moreover, if F_n is an interval extension of f then from definition 3.2.1-1, it follows:

$$\forall_{\langle r_1,\ldots,r_{i-1},r,r_{i+1},\ldots,r_n\rangle\in B}\, f(\langle r_1,\ldots,r_{i-1},r,r_{i+1},\ldots,r_n\rangle)\in F_n(\langle [r_1],\ldots,[r_{i-1}],[r],[r_{i+1}],\ldots,[r_n]\rangle) \quad (1)$$

And due to the monotonicity property of the interval function F_n (Lemma A5 from Appendix A):

$$F_n(\langle[r_1],\ldots,[r_{i-1}],[r],[r_{i+1}],\ldots,[r_n]\rangle)\subseteq F_n(\langle I_1,\ldots,I_{i-1},[r],I_{i+1},\ldots,I_n\rangle)=\prod_{x_i}^{\rho B}([r]) \qquad (2)$$

From (1) and (2) it follows:

$$\forall_{\langle r_1,\ldots,r_{i-1},r,r_{i+1},\ldots,r_n\rangle\in B}\, f(\langle r_1,\ldots,r_{i-1},r,r_{i+1},\ldots,r_n\rangle)\in \prod_{x_i}^{\rho B}([r]) \qquad (3)$$

On the other hand, from definition 4.2-1, the projection function wrt c and a variable $x_i\in s$ is:

$$\pi_{x_i}^{\rho}(B) = (\rho \cap B)[x_i]$$

and so, for every real value r within $\pi_{x_i}^{\rho}(B)$ there must be an tuple from B with $x_i=r$ satisfying c:

$$\forall_{r\in B[x_i]}\, r\in\pi_{x_i}^{\rho}(B) \Rightarrow \exists_{\langle r_1,\ldots,r_{i-1},r,r_{i+1},\ldots,r_n\rangle\in B^{\langle r_1,\ldots,r_{i-1},r,r_{i+1},\ldots,r_n\rangle\in\rho}}$$

which is equivalent to:

$$\forall_{r\in B[x_i]}\, r\in\pi_{x_i}^{\rho}(B) \Rightarrow \exists_{\langle r_1,\ldots,r_{i-1},r,r_{i+1},\ldots,r_n\rangle\in B}\, f(\langle r_1,\ldots,r_{i-1},r,r_{i+1},\ldots,r_n\rangle)\Diamond 0 \qquad (4)$$

Case $\Diamond \equiv$ "=" then (4) is:

$$\forall_{r \in B[x_i]} \; r \in \pi_{x_i}^{\rho}(B) \Rightarrow \exists_{<r_1,\ldots,r_{i-1},r,r_{i+1},\ldots,r_n> \in B}. f(<r_1,\ldots,r_{i-1},r,r_{i+1},\ldots,r_n>) = 0$$

If $f(<r_1,\ldots,r_{i-1},r,r_{i+1},\ldots,r_n>)$ denotes the real value zero then, any interval including it includes zero. Consequently, from (3), it follows the interval projection condition (i):

$$\forall_{r \in B[x_i]} \; r \in \pi_{x_i}^{\rho}(B) \Rightarrow 0 \in \prod_{x_i}^{\rho B}([r])$$

Case $\Diamond \equiv$ "$\leq$" then (4) is:

$$\forall_{r \in B[x_i]} \; r \in \pi_{x_i}^{\rho}(B) \Rightarrow \exists_{<r_1,\ldots,r_{i-1},r,r_{i+1},\ldots,r_n> \in B}. f(<r_1,\ldots,r_{i-1},r,r_{i+1},\ldots,r_n>) \leq 0$$

If $f(<r_1,\ldots,r_{i-1},r,r_{i+1},\ldots,r_n>)$ denotes a real value less or equal than zero then, the left bound of any interval including it must also be less or equal than zero. Consequently, from (3), it follows the interval projection condition (ii):

$$\forall_{r \in B[x_i]} \; r \in \pi_{x_i}^{\rho}(B) \Rightarrow left(\prod_{x_i}^{\rho B}([r])) \leq 0$$

Case $\Diamond \equiv$ "$\geq$" then (4) is:

$$\forall_{r \in B[x_i]} \; r \in \pi_{x_i}^{\rho}(B) \Rightarrow \exists_{<r_1,\ldots,r_{i-1},r,r_{i+1},\ldots,r_n> \in B}. f(<r_1,\ldots,r_{i-1},r,r_{i+1},\ldots,r_n>) \geq 0$$

If $f(<r_1,\ldots,r_{i-1},r,r_{i+1},\ldots,r_n>)$ denotes a real value greater or equal than zero then, the right bound of any interval including it must also be greater or equal than zero. Consequently, from (3), it follows the interval projection condition (iii):

$$\forall_{r \in B[x_i]} \; r \in \pi_{x_i}^{\rho}(B) \Rightarrow right(\prod_{x_i}^{\rho B}([r])) \geq 0 \qquad \blacksquare$$

Theorem 4.2.2-2 (Projection Function Enclosure based on the Interval Projection). Let $P=(X,D,C)$ be a CCSP. Let $c=(s,\rho) \in C$ be an n-ary constraint, B an n-ary F-box and x_i an element of s. Let a and b be respectively the leftmost and the rightmost elements of $B[x_i]$ satisfying the interval projection condition. The following property necessarily holds:

$$\pi_{x_i}^{\rho}(B) \subseteq [a..b] \qquad \square$$

Proof:

If a and b are respectively the leftmost and the rightmost elements of $B[x_i]$ satisfying the interval projection condition then outside this interval there are no elements of $B[x_i]$ satisfying this condition. However, from theorem 4.2.2-1, any element of $B[x_i]$ within the projection function $\pi_{x_i}^{\rho}(B)$ must satisfy the respective interval projection condition. Consequently outside the interval $[a..b]$ there are no elements of $B[x_i]$ within the projection function $\pi_{x_i}^{\rho}(B)$.

Since from definition 4.2-1 $\pi_{x_i}^{\rho}(B) = (\rho \cap B)[x_i] \subseteq B[x_i]$ it follows:

$$\pi_{x_i}^{\rho}(B) \subseteq [a..b] \qquad \blacksquare$$